The Material Point Method for Geotechnical Engineering

A Practical Guide

T0239601

The Material Point Method for Geotechnical Engineering

A Practical Guide

Edited by

James Fern
Alexander Rohe
Kenichi Soga
Eduardo Alonso

CRC Press is an imprint of the
Taylor & Francis Group, an **informa** business

CRC Press
Taylor & Francis Group
6000 Broken Sound Parkway NW, Suite 300
Boca Raton, FL 33487-2742

First issued in paperback 2020

© 2019 by Taylor & Francis Group, LLC
CRC Press is an imprint of Taylor & Francis Group, an Informa business

No claim to original U.S. Government works

Version Date: 20181203

ISBN 13: 978-0-367-73194-6 (pbk)
ISBN 13: 978-1-138-32331-5 (hbk)

This book contains information obtained from authentic and highly regarded sources. Reasonable efforts have been made to publish reliable data and information, but the author and publisher cannot assume responsibility for the validity of all materials or the consequences of their use. The authors and publishers have attempted to trace the copyright holders of all material reproduced in this publication and apologize to copyright holders if permission to publish in this form has not been obtained. If any copyright material has not been acknowledged please write and let us know so we may rectify in any future reprint.

Except as permitted under U.S. Copyright Law, no part of this book may be reprinted, reproduced, transmitted, or utilized in any form by any electronic, mechanical, or other means, now known or hereafter invented, including photocopying, microfilming, and recording, or in any information storage or retrieval system, without written permission from the publishers.

For permission to photocopy or use material electronically from this work, please access www.copyright.com (http://www.copyright.com/) or contact the Copyright Clearance Center, Inc. (CCC), 222 Rosewood Drive, Danvers, MA 01923, 978-750-8400. CCC is a not-for-profit organization that provides licenses and registration for a variety of users. For organizations that have been granted a photocopy license by the CCC, a separate system of payment has been arranged.

Trademark Notice: Product or corporate names may be trademarks or registered trademarks, and are used only for identification and explanation without intent to infringe.

Library of Congress Cataloging-in-Publication Data

Names: Fern, James, editor. | Rohe, Alexander, editor. | Soga, Kenichi, editor. | Alonso, Eduardo E., 1947- editor.
Title: The material point method for geotechnical engineering : a practical guide / editors, James Fern, Alexander Rohe, Kenichi Soga and Eduardo Alonso.
Description: Boca Raton : CRC Press, Taylor & Francis Group, [2019] | Includes bibliographical references and index.
Identifiers: LCCN 2018043536| ISBN 9781138323315 (hardback) | ISBN 9780429028090 (ebook)
Subjects: LCSH: Material point method--Simulation methods. | Geotechnical engineering--Simulation methods.
Classification: LCC TS161 .M365 2019 | DDC 620.001/13--dc23
LC record available at https://lccn.loc.gov/2018043536

Visit the Taylor & Francis Web site at
http://www.taylorandfrancis.com

and the CRC Press Web site at
http://www.crcpress.com

Contents

Foreword

Geotechnical engineering differs from other forms of engineering by the composition of ground, which is composed of a wide range of solids with discontinuities, pores fluids and structures. These different elements composing the ground interact with one another increasing the complexity of any prediction. In traditional geotechnical analysis, limit equilibrium methods (LEM) are useful techniques for estimating the onset of failure. However, these simple methods cannot provide any information on the deformation and are unable to simulate the real soil behaviour. Other well-established methods, such as the Finite Element Method (FEM), can model complex geometries and soil behaviours. However, FEM is limited to small deformation analyses and cannot model post-failure behaviours. New numerical methods are being developed in order to provide an engineering tool capable of simulating the entire instability process from the static stability analysis at small deformation to the dynamic post-failure run-out behaviour. This is the case of the material point method (MPM). MPM has been applied to a number of geotechnical problems and it has been extended to solve coupled flow-deformation problems in saturated conditions as it can simultaneously model fluid-like and solid-like behaviours. Fig. 1 depicts different applications of MPM in geotechnical engineering.

Sound geotechnical engineering practice is about predicting the most reliable behaviours of infrastructure during construction, operational use, maintenance, and deconstruction with adequate risk management. As pointed out by Dr Suzanne Lacasse in her Rankine Lecture, adding a probabilistic analysis and risk assessment enables assessing the inherent uncertainties in soil and material geometries and properties, probable loading conditions and numerical assumptions. Geotechnical engineering practices of simulation-based design then apply sufficiently large safety factors to bring risks to an acceptable level. This results in reliable and up-to-standard design of buildings, constructions and infrastructures with a very low probability of failure. However, excessive deformation, and even failure, of physical assets due to extreme events rarely occur. The often-unanticipated extreme events may be caused by exceptional ground or water conditions, extreme loading situations outside the design values, and external impacts of falling or penetrating objects as well as human errors during design and construction. Asset owners and managers increasingly demand quantifying in advance these improbable large deformation or failure events and associated risks in order to design appropriate counter measures to prevent them or to reduce their effects. Dr Evert Hoek pointed out in his 2014 lecture series: "There is no point in calculating the probability of failure if we

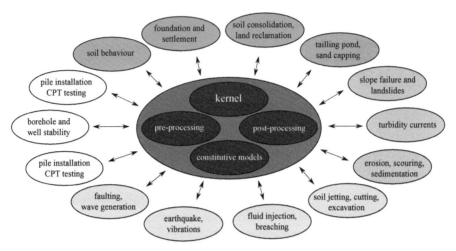

FIGURE 1: Range of applications of MPM in geotechnical engineering.

do not know what the consequence of that failure will be". MPM provides means of investigating the low-probability but high risk extreme events.

Making tools available will allow extending current geotechnical risk analysis to practice. It will firstly result in better contingency planning and measures and it will help to inform stakeholders and citizens on the geotechnical risks, potential damage and available mitigation measures. Secondly, business opportunities are currently missed to design more efficient infrastructure, decrease safety factors and reduce costs. For various structures, large deformations during their life cycle are acceptable when safety is guaranteed, damage is not disrupting construction and repair is affordable. These business opportunities become reality when a proper modelling approach and tools are available.

Ipo L. Ritsema and Simone van Schijndel
Directors at Deltares

Acknowledgement

The editors would like to thank the authors for their contributions to this book, and are grateful for the financial and technical support from Deltares, the Swiss National Science Foundation (SNSF) under grant agreement P1SKP2-158621 and from the Giovanni Lombardi Foundation.

The authors would like to acknowledge the financial support for the following chapters.

- Chapters 1, 7, 8, 9, and 13 received funding from Swiss National Science Foundation under grant agreements P1SKP2-158621 and P2SKP2-171774.

- Chapters 3 and 19 received funding from DFG for the project under grant agreement GR1024/29-1 and Menard/DYNIV for the site measurements in Stralsund.

- Chapter 4 received funding from the Dutch Organisation for Scientific Research (NWO), Deltares, Royal Boskalis Westminster N.V., Van Oord Dredging and Marine Contractors, Rijkswaterstaat, Stichting Flood Control IJkdijk, and Delft University of Technology.

- Chapter 14 received financial support from the Spanish government through the research project "PARTING" (BIA2013-48133-R) and fellowships (IJCI-2015-26342, BES-2014-068284), and to CIMNE by the CERCA Programme/Generalitat de Catalunya.

- Chapter 15 received financial support from the Spanish government through the research project "PARTING" (BIA2013-48133-R) and the fellowship IJCI-2015-26342 and to CIMNE by the CERCA Programme/Generalitat de Catalunya.

- Chapter 16 is part of the *Perspectief* research programme All-Risk with project number P15-21 4, which is (partly) financed by NWO Domain Applied and Engineering Sciences.

Contributors

E.E. Alonso
Polytechnic University of Catalonia
Barcelona, Spain

M. Alvarado
CIMNE - Centre de Metodes
    Numerics en Enginyeria
Barcelona, Spain

M. Calvello
University of Salerno
Salerno, Italy

F. Ceccato
University of Padova
Padova, Italy

A. Chmelnizkij
Technical University of
    Hamburg-Harburg
Hamburg, Germany

S. Cuomo
University of Salerno
Salerno, Italy

G. Di Carluccio
CIMNE - Centre de Metodes
    Numerics en Enginyeria
Barcelona, Spain

A. Elkadi
Deltares
Delft, the Netherlands

E.J. Fern
University of California
Berkeley, California USA

V. Galavi
Deltares
Delft, the Netherlands

P. Ghasemi
University of Salerno
Salerno, Italy

J. Grabe
Technical University of
    Hamburg-Harburg
Hamburg, Germany

M.A. Hicks
Delft University of Technology
Delft, the Netherlands

D. Liang
University of Cambridge
Cambridge, United Kingdom

D. Luger
Deltares
Delft, the Netherlands

M. Martinelli
Deltares
Delft, the Netherlands

M. Möller
Delft University of Technology
Delft, the Netherlands

J. Nuttall
Deltares
Delft, the Netherlands

N. Pinyol
CIMNE - Centre de Metodes
    Numerics en Enginyeria
Barcelona, Spain

G. Remmerswaal
Delft University of Technology
Delft, the Netherlands

I.L. Ritsema
Deltares
Delft, the Netherlands

A. Rohe
Deltares
Delft, the Netherlands

P. Simonini
University of Padova
Padova, Italy

K. Soga
University of California
Berkeley, California USA

F.S. Tehrani
Deltares
Delft, the Netherlands

R. Tielen
Delft University of Technology
Delft, the Netherlands

S. van Schijndel
Deltares
Delft, the Netherlands

P.J. Vardon
Delft University of Technology
Delft, the Netherlands

C. Vuik
Delft University of Technology
Delft, the Netherlands

E. Wobbes
Delft University of Technology
Delft, the Netherlands

A. Yerro
Virginia Polytechnic Institute and
    State University (Virginia Tech)
Blacksburg, Virginia USA

X. Zhao
University of Cambridge
Cambridge, United Kingdom

B. Zuada Coelho
Deltares
Delft, the Netherlands

List of Common Symbols

Symbol Description

Acronyms

BSMPM	B-spline material point method
BC	boundary condition
CC	Cam-Clay
CG	conjugate gradient
CPDI	convected particle domain interpolation
CPT	cone penetration test
CPTu	piezocone penetration test
CS	critical state
CSL	critical state line
CSSM	critical state soil mechanics
DDMP	dual domain material point method
DOF	degree of freedom
FEM	finite element method
GIMP	generalised interpolation material point
HP	hypoplastic model
IBM	implicit boundary method
IgA	isogeometric analysis
LAM	limit analysis method
LAS	local average subdivision
LEM	limit equilibrium method
LMP	liquid material point
MP	material point
MPM	material point method
RMPM	random material point method
MC	Mohr-Coulomb
MCSS	Mohr-Coulomb strain softening
MCC	modified Cam-Clay
NCL	virgin compression line
NMD	nodal mixed discretisation
MUSL	modified updated stress last
NS	Nor-Sand
OCC	original Cam-Clay
OCR	overconsolidation ratio
PIC	particle in-cell method
SQPSO	species-based quantum particle swarm optimisation
RBD	reliability based design
RMS	root mean square
SWE	shallow water equation
SWRC	soil water retention curve
SPD	symmetric positive definite
SMP	solid material point
THM	thermo-hydro-mechanical
TLS	Taylor least square
ULFEM	updated Lagrangian finite element method
VOF	volume of fluid

Roman symbols

a	aspect ratio
$\vec{a}$	acceleration
A^{e}	shear modulus constant
A_{ref}	reference area
$\vec{b}$	body force
c	cohesion
c_{h}	horizontal consolidation coefficient
c_{L}	specific heat of liquid
c_{NB}	Courant number

c_s	specific heat of solid	K_0	earth pressure coefficient at rest
c_u	undrained strength		
c_v	vertical consolidation coefficient	m	mass
		$\vec{m}$	diagonal of lumped mass matrix
D	dilatancy rate		
D_{crit}	critical depth for excavations	M	critical state stress ratio
		$\boldsymbol{M}$	(lumped) mass matrix
$\boldsymbol{D}$	stiffness matrix	$\boldsymbol{M}^C$	consistent mass matrix
e	void ratio	n	porosity
E	Young modulus	n^e	shear modulus exponent
E_u	undrained Young modulus	N	model parameter,
E_a^p	accumulated plastic strain		dilatancy coefficient,
$\vec{f}$	force		or NCL intercept
$\tilde{f}$	least-square approximation of function	N_g	total number of Gauss points
$\overline{f}$	volume average of function	N_n	total number of DOFs
f_l^{B-M}	liquid flow rate between band and matrix	p'	Cambridge mean effective stress
f_θ^{B-M}	heat flow rate between band and matrix	$\vec{p}$	momentum vector
		p_c	preconsolidation pressure
f_s	CPT sleeve friction	p_i	image pressure
F	yield function	p_g	pore gas pressure
g	gravity	p_l	pore liquid pressure
G	shear modulus	p_w	pore water pressure
G_s	specific gravity	δp_w	excess pore water pressure
h	height	P	potential function
h^*	normalised height	q	Cambridge deviatoric stress
H	hardening modulus		
$\dot{H}$	heat rate	q_c	CPT cone resistance
i	hydraulic gradient	q_h	heat flow conduction
I_C	relative pressure index	$\vec{q_l}$	Darcy flow vector
I_D	relative density index	Q	crushing pressure
I_L	liquidity index	$\vec{r}$	residual vector
I_L^{eq}	equivalent liquidity index	r^*	normalised run-out distance
I_p	plasticity index		
I_r	rigidity index	R	fitting parameter
I_R	relative dilatancy index	s	matrix suction
$\boldsymbol{I}$	identity matrix	S_r	degree of saturation
$\vec{I}$	identity vector	t	time
J	invariant of deviatoric stress tensor	T	temperature
		T^*	modified normalised time
$\boldsymbol{J}$	Jacobian matrix	u	displacement
k	hydraulic conductivity	u_1	pore water pressure at cone face
K	bulk modulus		

u_2	pore water pressure at cone shoulder	θ	Lode angle
u_3	pore water pressure at cone friction sleeve	θ_h	horizontal spatial variability
UCS	uniaxial compressive strength	θ_v	vertical spatial variability
v	specific volume	Θ	temperature
$\vec{v}$	velocity vector	κ	swelling modulus
v_0	maximal initial velocity	λ	CSL slope
v_{mp}	material point velocity	λ_{max}	maximum eigenvalue
V	volume	λ_{min}	minimum eigenvalue
V	normalised penetration velocity	Λ	plastic multiplier
w	water content	μ	friction coefficient
W	work	μ_l	liquid viscosity
$\dot{W}$	work rate	μ_w	water viscosity
$\mathbf{W_p}$	history variable	ν	Poisson ratio
x	global coordinate	ρ	unit mass
y	global coordinate	σ	stress
z	global coordinate	σ^r	radial stress
		σ^{tot}	total stress
		σ'	effective stress

Greek symbols

α	damping factor, or dilatancy coefficient	τ	shear stress
		τ_f	shear strength
α_l	liquid compressibility	φ	friction angle
β	volumetric thermal expansion coefficient, or mass scaling factor	$\vec{\phi}$	basis function vector
		χ	dilatancy coefficient
		$\vec{\phi}$	basis function vector
γ	unit weight, shear strain	ψ	dilatancy angle
		ψ_i	Taylor basis function
$\dot{\gamma}$	shear strain rate	ψ_l	liquid transfer coefficient
Γ	CSL intercept, or thermal conductivity coefficient	ψ_θ	heat transfer coefficient
		Ψ	state parameter
		Ξ	knot vector
δ	increment	ξ_i	knot
$\dot{\delta}$	displacement rate	ω_g	Gauss weight
$\delta_{i,j}$	Dirac delta function	Ω	domain
ε	strain		

Superscripts and subscripts

ε_{dev}	deviatoric strain	B	shear band
ε_r	radial strain	c	consolidation
ε_{vol}	volumetric strain	cs	critical state
ε_θ	circumferential strain	dev	deviatoric
ζ	softening shape factor	e	elastic
η'	effective stress ratio	ext	external
		eq	equivalent
η_w	water surface elevation	f	failure

g or G	gas	sat	saturated
i	image	t	time
int	internal	tc	triaxial compression
l or L	liquid	te	triaxial extension
m	mixture	T	transpose
mp	material point	u	undrained
max	maximum	v or V	void
min	minimum	vol	volumetric
M	matrix	w	water
N	normal	x	x-direction
p	plastic	y	y-direction
q	deviatoric stress	z	z-direction
res	residual	′	effective
s or S	solid		

Part I

Theory

1

Computational Geomechanics

E.J. Fern, K. Soga and E.E. Alonso

1.1 Introduction

The art of computational geomechanics refers to the process of computing the most probable outcomes of a given engineering problem. The science of computational geomechanics refers to the computation itself and it ensures that the predictions are exempt of computational errors. Whilst the science of computational geomechanics is well documented in the literature (i.e. [231, 232, 348]), the art of computational geomechanics is not so well documented; the document by Potts et al. [229] is one of the most comprehensive guideline documents available.

In order to carry out a prediction, a series of choices and idealisation have to be made and these choices are based on engineering judgement. Computational tools are powerful engineering aids, which open up new engineering opportunities, but they are not replacements for blind judgement. Although there has been great progress in computational geomechanics in the past years, some major issues still remain. These include data and knowledge deficiencies and limited computer capacity [134]. Computational geomechanics differs from other forms of computational mechanics by the complex behaviour of geomaterials and its spatial variability. A thorough understanding of geomechanics is necessary to carry out any numerical predictions in geotechnical engineering [263].

Potts [230] raised the following question in his Rankine Lecture: "Is numerical analysis a virtual dream or a practical reality?" This question was motivated by the lack of popularity of numerical predictions in practice at that time. The lack of knowledge, know-how and guidance in computational geomechanics can hamper the use of numerical methods in practice. Although many engineers have been exposed to numerical methods during their education or career, the complexity of an analysis and its practical potential are not always fully appreciated [230].

The increasing computational capacity of personal and high-performance computers allows the development of new numerical methods, which permits modelling not only small deformation problems but also large deformation ones. Alonso demonstrated in his Rankine Lecture that these new methods could be used to explain the triggering mechanism and motion of landslides. This book aims to provide guidance on the use of the material point method (MPM) for geotechnical engineering. This chapter outlines the process of carrying out numerical predictions. Although this book is dedicated to MPM, some elements presented in this chapter are applicable to other numerical methods.

1.2   General context

Numerical analyses are time consuming and expensive but they permit exploring problems with complicated geometries, complex material behaviours, variable loading sequences, and interactions between existing and new infrastructure for which conventional methods are incapable of solving. They permit carrying out predictions of a probable outcome considering the interaction between different mechanics and structures (i.e. hydro-mechanical coupling, thermo-hydro-mechanical coupling, soil-structure interactions). Numerical analyses are used (1) to gain additional insight in a given problem and (2) to explore different probable outcomes.

Numerical predictions in geotechnical engineering are important due to the complex mechanical behaviour of geomaterials but they are hampered by the difficulty in making accurate predictions due to inherent variability and uncertainty of the ground conditions [161]. The inaccuracy is not related to the science of computational geomechanics but to the uncertainty of the input parameters. The absence of information on the site conditions is a common and major issue for any prediction. This is due to difficulties in subsurface exploration and its associated cost. However, it does not make computational geomechanics worthless but it points out some deficiency in obtaining the required information.

In order to avoid confusion with the different terms used in this book, the following definitions are adopted.

A model is the geometrical and mechanical idealisation of a given problem. It includes, although not exclusively, the idealised geometry and stratigraphy, the boundary conditions, the contact conditions and the mesh. The model should be simple enough to permit computational-efficient simulations but it must also contain sufficient details to capture the correct mechanics.

A simulation is the actual computation of a specific probable outcome using a given model, a type of analysis, specific constitutive models and various contact conditions. Therefore, different simulations can be generated from a given model. It includes the computation choices such as defining the critical time step (Courant number), damping conditions and the duration of the simulation. These numerical features are explained in Chapter 6.

An analysis is the whole process of mathematical idealisation of a real problem, including the computation of a prediction and its interpretation.

1.3 Anatomy and classification of predictions

In an attempt to provide guidance, Lambe [161] described the process of carrying out numerical simulations and called it *the anatomy of prediction* (Fig. 1.1). Each step of the process is a seldom discrete isolated component but an interrelationship between steps does exist as the choices made in one step impact the following ones. Therefore, the numerical prediction obtained at the end of the process is dependent on the choices and assumptions made in every step, and on the state of knowledge prior to the computational idealisation. Numerical simulations are not best guesses but approximated solutions of a predefined set of governing equations solved in a specific sequence. The anatomy of prediction serves as the layout of this chapter and each step is discussed in a separate section.

Any numerical prediction relies on the field situation at the time of the prediction. For this reason, Lambe [161] introduced a classification system called *the classification of prediction* (Table 1.1). It is based on the moment the prediction is made and the amount of information available at the time of the prediction.

Class-A predictions are simulation carried out prior to the event and in absence of any field data. This is typically the case for feasibility studies.

Class-B predictions are carried out during the event in absence of any site data. If some site data is available, then the predictions become a class-B1 prediction. This is typically the case for tender designs.

Class-C predictions are carried out after the event of interest but in absence of site data and a class-C1 prediction in presence of site data. These are the most abundant type of simulations in the academic literature as they permit the validation of a given method and assumptions.

Potts and Zdravković [348] give an example of a class-A prediction and a class-C prediction for the stabilisation of the Pisa Tower. The class-A predictions aimed to estimate the number of counter-weights required for the stabilisation and to check undesirable and unexpected responses of the tower. The class-C prediction permitted a better understanding of the ground response.

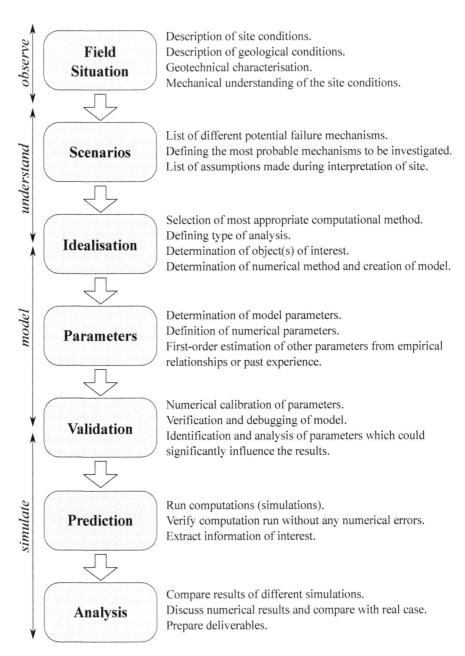

FIGURE 1.1: Anatomy of prediction.

TABLE 1.1

Classification of prediction.

Prediction type	Prediction made	Field results
Class A	before event	-
Class B	during event	unknown
Class B1	during event	known
Class C	after event	unknown
Class C1	after event	known

1.4   Field situation

The first step in any numerical analysis is to describe the field situation, which is composed of the object of interest and its surrounding environment. It may seem at first a trivial step but it is often complex and evolves with time. This is especially the case for design work, which goes through different design stages, and for landslides for which the site investigations can spread over a few years. The content and extent of the field situation is case specific. It can be short and straightforward for simple cases such as a reduced-scale laboratory experiment, for which examples are given in Chapters 9, 10 and 21, but it can also be complex as is the case for the Vajont Landslide (Chapter 14). The field situation typically includes the following aspects, albeit not exclusively.

The geometry of the object of interest and its surrounding environment have to be described. It is necessary to decide which details of the geometry are relevant for the analysis and which can be neglected. For instance, the description of a tunnel should include the inner diameter, the excavated diameter and the dimensions and types of the linings. It will then be decided which diameter is the most relevant for the analysis, how to model the different linings and if contact conditions or hinges are necessary to model the mechanical behaviour of the lining.

The geological and environmental conditions are used to define the stratigraphy and geotechnical model. The geological model is an idealisation of the location of the different soil and rock units with respect to their history (deposition epoch, constituents, etc.). The geology provides hints on the state of the ground. The geotechnical model is an idealisation of the location of the different soil and rock units in terms of mechanical characteristics. For instance, a given clay can have different stiffness moduli and undrained strengths at different depth. This gives one geological unit but two geotechnical units.

The environmental conditions also need to be described. For example, the presence of hydrocarbons in the ground can decrease the strength of the soil or rock [174]. Environmental conditions also include variations of the groundwater table associated with seasonal rainfall and groundwater flows.

The geotechnical characterisation describes soil and rock units in terms of geotechnical characteristics. These can be grain-size distribution, texture, Atterberg limits and/or minimum and maximum void ratios. It also describes the mechanical behaviour of different soil and rock units obtained from site and laboratory tests. Based on the geotechnical characterisation, constitutive models are selected and calibrated. Bishop [35] pointed out that errors can be introduced during soil sampling and testing, which do not reproduce sufficiently accurately the soil conditions and stress state of the natural ground. Unless equal attention is paid to each factor, an elaborate mathematical treatment can lead to a fictitious impression of accuracy [35].

The geotechnical characterisation also needs to establish the in-situ stress state, pore pressure profiles, and consolidation states. This includes determining the orientation and magnitude of the principal stresses. This is especially important in tectonically active areas where both horizontal stresses can be larger than the vertical stress giving earth pressure coefficients greater than 1.0 ($\sigma'_{h1} > \sigma'_{h2} > \sigma'_v \rightarrow K_{0,1} > K_{0,2} > 1.0$).

The natural variability of the geotechnical characteristics has encouraged some researchers to include this variability with random fields and carry out statistical analysis for probable outcomes. An example of such an analysis is presented in Chapter 16.

The loading sequence of the object of interest has to be known and clearly described in order to idealise the different stages and loading schemes of a problem. The plastic behaviour of soil and rock is stress-path dependent.

The accuracy of a prediction depends on the quality of the computational method as well as on the data used to make the prediction [161]. The input data is often the weakness of many numerical analyses [134]. Therefore, the quality of the computed prediction depends on the engineer's understanding of the site making this first step an important step in any numerical analysis.

1.5   Scenarios

Once the field conditions are established, it is necessary to build a mechanical understanding of the problem and to define the objectives of the numerical analysis. It is rarely possible to answer all the questions with a single analysis and a list of potential scenarios must be generated. This permits defining which mechanism should be simulated and establish the limitation of the analysis. In some cases, several mechanisms might be of interest in which case several series of simulations might be required. It is also important to have a certain idea of the expected outcome of the numerical simulation as this facilitates the validation of the model and the debugging process.

It is necessary at this stage to define how the simulations will be presented and which elements and variables are of interest for the analysis so that the modelling efforts go into obtaining these deliverables. This could be the internal forces of a tunnel lining, in which case the lining can be modelled with beam elements rather than a continuum material. It could also be the development of the shear strains in a slope in order to identify the surface of rupture. In some cases, elements or material points (MPs) may need to be monitored during the computations and the desired information recorded in a file, as is demonstrated in Chapter 12.

1.6 Idealisation

It is rarely possible to perform an analysis in which the full knowledge and detail of the object of interest permits a complete, detailed and accurate prediction. Not only is it rarely possible, it is rarely desirable as the understanding is enhanced by intelligent simplifications and idealisation with which the engineer can gain insight [207]. Moreover, it is difficult to create a good model that enables a realistic analysis of the physical processes involved in a real project and that provides a realistic prediction of design quantities [45]. A certain level of idealisation and interpretation is required.

1.6.1 Computational methods

Once the object of interest and the prevailing mechanisms are defined, a computational method needs to be selected. This choice is driven by the technical capacity of the method to capture the correct mechanics and to produce the desired deliverables. The choice of the method is also dependent on the availability of the method (software) and the available computational resources as some methods are computationally expensive. The choice of the numerical methods influences the idealisation of the problem.

A plethora of different computational methods exists ranging from the simplest (e.g. empirical methods) to more advanced methods (e.g. FEM or MPM). Each method has its advantages and limitations. Soga et al. [278] reviewed the different methods for large deformation simulations and a few families of continuum-mechanics methods are summarised as follows.

The empirical methods (EMs) are based on past experiences and theoretical developments. They are often simple enough to be computed by hand or with a spreadsheet. However, EMs rarely consider interactions between different entities of the system and they are often not adequate for large deformation analyses. They rely on simple mechanical idealisation of the material. Nevertheless, they provide a simple tool for basic predictions and can be used as a verification tool for more advanced numerical predictions. EMs can produce conservative predictions in comparison with more advanced numerical methods [230].

The limit equilibrium methods (LEMs) usually decompose the problem into slices or blocks and check whether the ground is in a state of equilibrium. This is achieved by solving Newton's governing equations and imposing a failure criterion. The use of simple failure criterion implies that the failure is determined from the stress state only and neglects the stress path to failure, and that the criterion holds everywhere along the failure surface [49] but the resulting stresses do not necessarily satisfy equilibrium at every point [276]. The failure surface has to be defined by the user, which often requires many trials in order to determine the critical case. These methods also idealise the drainage conditions as drained or undrained, excluding partially drained conditions from the analysis [230]. LEM is computationally cheap and it can easily be carried out by hand or with a spreadsheet, even for a large number of cases. Fig. 1.2 (a) shows an example of the LEM applied to a strip footing.

The EMs and LEMs separate deformation (serviceability) from strength (failure) and struggle to include any form of interaction in the analysis. However, the long history of these methods resulted in widely available tools and extensive collective experience concerning their reliability [263, 276]. Chapter 11 discusses the use of LEMs and MPM.

The limit analysis methods (LAMs) are based on the plastic bounding theorems and normality (associated flow rule, see Chapter 8). The lower-bound theorem seeks for admissible stress fields such that equilibrium is satisfied [276] but not the compatibility conditions [233]. It is possible to have different stress fields that satisfy the same conditions. The upper-bound theorem seeks the kinematically admissible velocity field satisfying the velocity boundary conditions [276]. It reaches a solution in absence of equilibrium giving unsafe estimates [233]. Human estimate will always be an upper bound as soil will always find cleverer modes of failure [207]. Fig. 1.2 (b-c) shows an example of the LAM applied to a strip footing.

The limit theorems of plasticity provide more flexible and comprehensive solutions because the upper bound introduces a kinematic compatibility consistent with the hypothesis of associated failure criterion, which is absent in LEM. The combination of LAM with FEM (e.g. [276]) and optimisation algorithms has resulted in fast computer programmes for stability analysis.

(a) Limit equilibrium

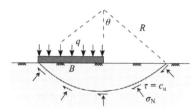

(b) Lower-bound theorem

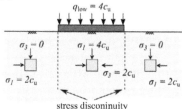

(c) Upper-bound theorem

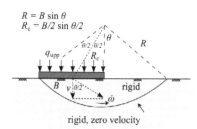

(d) Material point method

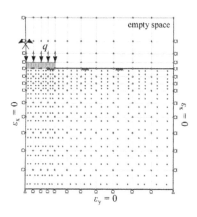

FIGURE 1.2: Schematic description of (a) the limit equilibrium method, (b-c) the limit analysis method and (d) the material point method of a strip footing (original example from [276]).

The finite element method (FEM) solves Newton's governing equations in order to obtain the stress-strain states with respect to loading, and discretises the continuum into non-overlapping components. It also allows the use of complex soil models. FEM is the most popular numerical method in geotechnical engineering and it is well documented (e.g. [162, 231, 232]). However, FEM struggles with large deformation simulations due to mesh distortion problems. However, an advanced algorithm such as the arbitrary Lagrangian-Eulerian FEM (ALE FEM) offers freedom in changing the computational mesh but it requires a significant computational effort in some cases.

The finite difference method (FDM) approximates derivatives in the partial differential equation by linear combinations of function values at the grid points. It is capable of simulating large deformation problems by updating the coordinates of the grid. It also uses prismatic elements which makes the modelling of complex geometries difficult.

The material point method (MPM) is an Eularian-Lagrangian method, which discretises the continuum medium in material mass points. It moves these points in a computational grid by solving Newton's governing equations. MPM does not suffer from mesh distortion and it has the advantage of being able to solve the governing equations for the different phases with distinct layers of MPs [21,334]. However, MPM is computationally expensive and suffers from numerical instabilities related to MPs crossing cells. Chapter 2 presents the fundamental theory of MPM, and Chapters 3 and 4 the different MPM formulations. Fig. 1.2 (d) shows an example of the MPM applied to a strip footing.

The smoothed particle hydrodynamics (SPH) method idealises the continuum material with spaced particles over which the particle properties are smoothed by a kernel function. It is suitable for large deformation simulations but it is computationally expensive, partly because the boundary conditions have to be modelled with a series of particles. Chapter 21 compares the use of SPH and MPM for hydraulic engineering.

1.6.2   Types of analysis

Different types of analysis can be carried out depending on the nature of the problem being solved and the desired deliverables. A few types of simulations are as follows.

Drained, undrained and partially drained analyses define how the excess pore water pressure is dissipated in the simulation, which plays a central role in the mechanical behaviour of soils and rocks.

- A *drained analysis* assumes the absence of excess pore water pressure because of the absence of water (dry soil) or because the deformation is slow enough that the excess pore pressure is rapidly dissipated. It thus depends on the hydraulic conductivity of the soil. Chapter 9 presents simulations of granular column collapses for dry sand and Chapter 15 of pile installations in saturated drained conditions.

- An *undrained analysis* assumes that the deformation is fast and excess pore pressure is accumulated in the ground. There is a high potential for error in undrained simulations as there is not a single constitutive model which can reliably predict the undrained effective stress path, although some models predict more realistic behaviours than others. For this reason, total stress analyses are sometimes favoured as the total stress paths are known and predictable. In undrained effective stress analyses, the effective parameters are used in the constitutive model and the bulk modulus of the water is added to the stiffness of the soil to make the ground incompressible. Excess pore pressure can be computed in effective stress analyses. It is possible to carry out an undrained total stress analysis for which the total stress paths are correct but for which it is not possible to obtain the excess pore pressure. In such simulations, the total stress parameters are used for the constitutive model and these differ from the effective stress parameters. Total stress analyses are usually carried out for situations in which the undrained response is of interest and a single undrained shear strength can be assigned to the soil element [207].

Despite the difficulty in carrying out undrained analyses, these are often the conditions in which failures take place. Many chapters of this book present undrained analysis using both the total stress and the effective stress approach. For instance, Chapter 11 presents undrained effective stress simulations with a simple constitutive model, and Chapters 10 and 18 with an advanced constitutive model. Chapters 13 and 15 show examples of total stress analyses.

- A *partially drained or consolidation analysis* considers the dissipation of excess pore pressure with time. It is thus an effective stress analysis. This permits transitioning from undrained conditions to drained conditions. This drainage condition can only be investigated by means of numerical analyses [230]. Chapters 18 and 20 show examples of partially drained analyses.

Strength-reduction analyses are the process of progressively reducing the strength of the soil until a failure occurs. The ratio between the strength at failure and the initial strength of the soil gives a factor of safety. This kind of analysis permits identifying the critical strength required for the soil to be stable. An example of strength-reduction analysis is given in Chapter 15.

Inverse analyses are used to back-calculate model parameters such that the simulation will match the field data. The danger with such analyses is that the model parameters become a set of random variables in which only one set of values is obtained without considering mechanical consistency and for which the model parameters may lose their physical meaning. Therefore, the back-calculation of model parameter values must be carried out with caution. An example of inverse analysis is presented in Chapter 10.

Phased analyses are conducted by running a sequence of different types of analysis in an attempt to reproduce field conditions. Nearly all simulations are phased in one way or another as the initial stresses are first generated and then the model is loaded. However, it is possible in some cases to also perform consolidation analysis prior to or after loading, or to carry out a strength-reduction analysis at the end of the simulation in order to obtain a factor of safety.

1.6.3   Model

Once the objectives of the simulations and the type of analysis are defined, the model itself can be generated.

The geometry of the model has to be defined. The components of the model
need to be simplified sufficiently for computational reasons and to facilitate the understanding of the problem. However, it also needs to capture
enough details in order to predict the correct mechanics. For instance, the
model stratigraphy of the ground has to be defined such that it reflects the
real stratigraphy. The dimensions of the model have to be large enough
to include items of interest and to prevent any influence of the boundary
conditions.

The configuration of the model has to be determined and it is a function of
the symmetry of the problem. It can be (1) plane strain, (2) axisymmetric
or (3) three-dimensional (3D).

- *Plane strain models* are two-dimensional idealisations of the object
  of interest. They assume a nil deformation in the out-of-plane direction ($\varepsilon_z = 0$) and, hence, the shear stresses can only be non-nil in
  the plane of analysis. This assumption is representative of long objects such as tunnels, dykes, strip foundations and some excavation.
  It is not rare to simplify the models according the symmetry of the
  initial configuration. However, this simplification is only possible if
  the final configuration is also symmetric. This condition is important
  for large deformation simulation in which the final configuration can
  differ substantially from the initial configuration. Fig. 1.3 shows a 2D
  plane strain idealisation of a dam. Whilst the initial configuration of
  the dam is symmetric, the pressure of the reservoir on the dam is
  not. Hence, the model cannot be further simplified with the symmetry of the dam. Additionally, the final configuration of dam is not
  symmetric.

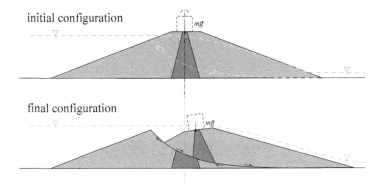

FIGURE 1.3: Initial and final configurations of a numerical model.

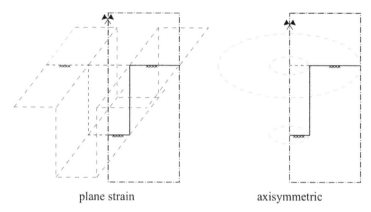

plane strain   axisymmetric

FIGURE 1.4: Two-dimensional idealisation of numerical models.

- *Axisymmetric models* are 2D models, which idealise the object of interest according to a symmetry axis and they assume that the radial strains are nil ($\varepsilon_\theta = 0$). Hence, the shear stresses can only be non-nil in the plane of analysis. In order to carry out an axisymmetric analysis, the initial and final configurations and all loads must be axisymmetric. Fig. 1.4 shows the same idealised excavation in plane strain and axisymmetric conditions but represents two different problems. The plane strain is an idealisation of a trench and the axisymmetric of a shaft. Chapter 5 gives more details on the axisymmetric formulation, whereas Chapter 15 shows examples of plane strain models of excavations.

- *3D models* do not make any assumptions on the out-of-plane direction. However, these simulations require significantly more computational resources as well as being more difficult to analyse. Therefore, 3D simulations are only carried out when 2D models cannot be used. Chapter 19 gives an example of a 3D simulation for which a 2D model could not be used.

The mesh is the discretisation of the space into elements. The mesh is sometimes referred to as the background or computational grid as it is only used for computation reasons and it is not dependent on the geometry of the objects. The elements can also be referred to as cells. However, the mesh must follow the shape of the boundary conditions in MPM. The continuum medium is discretised with MPs and this can be done independently from the mesh, although some codes (i.e. Anura3D [14]) use the background grid to discretise the continuum bodies into MPs. Thus, the size of the mesh defines the amount of mass that a given MP holds. There is no ubiquitous set of rules on how a mesh should be defined as this depends on the problem being solved. However, a good mesh should be as regular as possible with smooth transition between different element sizes. Elongated elements should be avoided as much as possible as it influences the accuracy of the computation. The boundaries of the mesh should be far enough to avoid any influence on the area of interest.

Whilst the mesh deforms with the material in FE simulations, the background computational grid in MPM is generally fixed. It is possible to make this grid move in order to reduce numerical instabilities related to MPs crossing into another cell. Examples of the moving mesh are given in Chapters 17 and 18.

Different element types can be used to model the different components of the model. The ground is typically modelled with continuum elements. Some thin layers or discontinuities can also be modelled as interface elements in FEM or as contact conditions in MPM (see Chapter 6). Structural elements are typically modelled as beam, plate, struts or anchor elements. However, it may be interesting to model them as a continuum in other cases. This depends on the desired output type as well as their thickness.

The boundary conditions have to be defined such that they mimic the field conditions. Different types of boundary conditions exist. The most common are displacement boundary conditions, which constrain displacements in a given direction by imposing a nil velocity field. It is also possible to assign pressure conditions, which are necessary when modelling unsaturated soil (see Chapter 12). Conditions can also be imposed within the model and for part of the simulation. This is typically the case when initialising the stresses. Chapter 9 shows an example of such a technique to model the granular column collapse. Inflow and outflow conditions can be imposed in order to model liquid flows. These conditions permit the creation and destruction of MPs during the simulation and an example is given in Chapter 21. It is also possible to remove MPs in a specific area in order to model excavations (see Chapter 15). The removal of MPs can be spontaneous replicating an excavation and, after that, neighbouring MPs can move into the excavated zone. The removal of MPs can also be applied continuously, removing any MP moving into the predefined area. More information on the boundary conditions is given in Chapter 6.

Constitutive models are mathematical idealisations of the material behaviour, which characterise the stress-strain relationship and define the stiffness of the material. However, there is not a single constitutive model which can adequately predict all the mechanical behaviours of a real soil. A few examples of differences between real and remoulded soil are related to soil structure, homogeneity, anisotropy, beddings and discontinuities. Burland [48] pointed out that, although the constitutive models are largely developed for remoulded soils, they still offer tremendous insight in the mechanical behaviour of real soil.

Constitutive models have evolved significantly throughout the history of geomechanics and Potts et al. [229] classifies them as follows.

- The *first generation* of models are simple elastic-plastic models (Fig. 1.5 a). The behaviour is assumed to be elastic until a failure occurs, after which plastic deformation takes place. The elastic phase can be linear (e.g. [139]) or non-linear (e.g. [90]). The failure threshold is defined by a failure criterion and it is solely predicted by the stress state irrespective of the stress path. Different failure criteria exist such as Mohr-Coulomb [296], Matsuoka-Nakai [197], Drucker-Prager [86] or Hoek-Brown [133, 135]. These models are simple enough for hand calculations but they are also popular in numerical predictions because of their simplicity in understanding and computational efficiency.

- The *second generation* of models are incremental stress-strain relationships (Fig. 1.5 b). These models consider the stress-path dependency to failure and permit the inclusion of plasticity in the hardening phase. Moreover, they are based on mechanical theories which impose a mechanical consistency between stresses and strains. This is not the case for first generation models. Such models are the Original and Modified Cam-Clay models [248, 249], which are presented in Chapter 8.

- The *third generation* of models are the latest generation of models (Fig. 1.5 c). They include extended hardening concepts, bounding surface or multi-surface plasticity, and bubble models. Many of these models are extensions of the Cam-Clay models such as Nor-Sand [148] (see Chapter 8) or Cam-Clay with sub-loading surface [124].

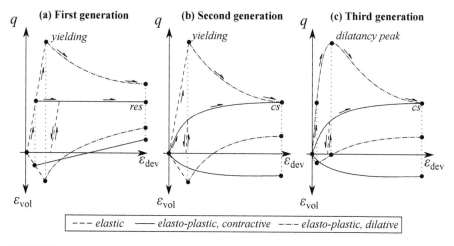

FIGURE 1.5: Schematic description of first, second and third generation of constitutive models for drained triaxial compression test.

A constitutive model is chosen such that it captures at best the mechanics necessary for the prediction. However, it must be compatible with the computational method [232] and the loading sequence. For instance, the ability to reproduce the cyclic loading behaviours is necessary for earthquake engineering, whereas the ability to model unloading paths is necessary for excavation problems. The choice of constitutive model also depends on the available data from which the model parameters are calculated.

The accuracy of a numerical prediction depends heavily on the constitutive model. This does not imply that complex models need to be used. For instance, the bedrock is often modelled as linear elastic when the object of interest is located well above it. However, the selection of a constitutive model requires a thorough understanding of the mechanics of interests, and extensive knowledge of computational geomechanics. Some of the most popular constitutive models can give non-sensible predictions, which is a serious pitfall for practice [230].

Loads have to be defined in the model and this can be done in different ways. The compatibility conditions of MPM impose that the loads are applied at the nodes in contact with a continuum material or a structural element. However, the large deformation nature of MPM allows MPs to move from one element to another and hence MPs may no longer be subjected to this load. It is thus possible to transfer this load from a given node to another one as MPs move (Fig. 1.6). It is also possible to introduce loads in the model as a pressure condition as is the case for gas and liquid pressures. An alternative loading scheme is to model a fictitious continuum body with its weight acting as the load. The advantage of such an approach is that this load can move within the model as the soil deforms (Fig. 1.3).

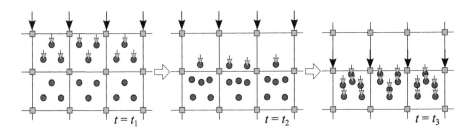

FIGURE 1.6: Schematic description of loads applied on material points.

1.7 Parameters

The quantification of parameters is a difficult step in the analysis as it strongly depends on the type of analysis, constitutive model and model specificities, which include numerical damping of the system. There are considerable discussions on which parameter values to use in a simulation. This is partly due to the lack of physical meaning of some parameters or an inability to quantify the model parameters from experimental data. Two families of parameters can be defined – mechanical and numerical parameters.

Mechanical parameters are parameters used to predict the mechanical behaviour, typically for the constitutive model, but it can also be for contact conditions. The quantification of these parameters depends on the choice of the model and the available data. Mechanical parameters can sometimes be estimated from site and laboratory tests providing that these parameters have physical meanings and are quantifiable. However, one should make sure that the laboratory and site tests are representative of the field conditions. It is well known in rock mechanics that there is a significant difference between the mechanical behaviour of intact rock and rock mass. Soils can also contain some large scale fissures which alter the mechanical behaviour of the soil mass [49]. Sampling techniques can disturb the specimen and, thus, the laboratory test no longer reflects the field conditions (e.g. [69, 164]). Therefore, the determination of the model parameters requires an in-depth knowledge of geomechanics, constitutive modelling and laboratory testing. Potts and Zdravković [232], Lees [162] and Jefferies and Been [150] provide additional information on how parameters can be derived from laboratory tests.

In absence of data, it is sometimes possible to derive model parameters from empirical or theoretical relationships. In some rare cases, catalogues of material characteristics are provided in which parameters are given. However, the use of catalogued material parameters can be misleading as the material is often site specific and state dependent.

Numerical parameters are used to solve the governing equations and they are as important as mechanical parameters. However, the absence of physical meaning makes them more difficult to quantify. A few numerical parameters are described below.

The *characteristic element length* defines the size of an element. However, it does not control the mesh quality. The size of elements plays an important role in the analysis as the solution is mesh-size dependent. Issues related to the size of the mesh are presented in Chapters 9 and 19.

The *damping coefficient* reduces the out-of-balance force in an attempt to reduce numerical oscillations. It is also used to reduce the amount of deformation taking place in the ground. This is often done during the gravity loading phase. Chapter 6 gives more information on different types of damping and typical values of the damping coefficient.

The *Courant number* sizes the critical time step and, hence, influences the computational time and accuracy (see Chapter 3). Small Courant numbers give small critical time steps but increase the computational time. Small time steps generally improve the accuracy of the predictions but very small time steps can also decrease the accuracy for thermo-hydro-mechanical coupled problems.

The *number of load steps* and the *load step duration* divide the simulations into a series of steps (see Chapter 2). Loading conditions, material and numerical parameters and type of analysis can change at the beginning of each time step and the results are saved at the end of each steps. The step duration can change throughout the simulation in order to reduce the size of the output files and focus on the moment of interest. This could be the onset of failure and post-failure behaviour of a dyke failure.

Tolerances and substepping are used as criteria for convergence and these are used throughout the codes. Small values of tolerance generally increase the accuracy of the prediction but they can result in a substantial increase in the computational time. For instance, most constitutive models use substepping techniques in order to guarantee the accuracy of the prediction and for which a tolerance is given and an error is calculated (see [275] for more information). However, errors can be immense when stresses are very low ($\sigma' \to 0$). The error is calculated by dividing a very small number by another very small number, which can give a big error. This causes unnecessary substepping and computational cost. This is the case for the free surface of a granular flow.

A more in-depth discussion on the numerical parameters is given in Chapters 3 and 6.

1.8   Verification and validation

Verification and validation are two important steps in the analysis [45].

The verification of a prediction consists of making sure that all conditions and parameters of the model are correctly input in the computational code, that the code is performing as expected, and that the sequence of events (i.e. phased analysis) is correctly set. It is making sure that the computer model reflects the mathematical idealisation.

The validation of the prediction consists of making sure that the numerical predictions are representative of the real problem. This is often done comparing the numerical predictions with experimental data, design charts, simplified models, or past experiences

Constitutive models are often validated by simulating laboratory tests with a single stress-point simulation, which provides a certain level of confidence. However, the ability to replicate a given test with a single stress-point simulation does not guarantee that the correct model parameters are used and that the observed behaviour in the single stress-point simulation will be the same in a larger model.

Model boundaries are introduced to limit the extent of the model and reduce the computational time. However, these model boundaries can influence the results as they impose a given stress or strain condition. They can also cause undesired reflections of waves as is demonstrated in Chapter 20. The verification of the boundary influence is done by carrying out a series of simplified model simulations with different dimensions or by running the same simulations with different mesh sizes and comparing the results.

Initial conditions have to be verified after the stress initialisation. This is important as soil is stress-path and pressures dependent. In undrained analyses, it is essential to recreate the pore water pressure distribution, which might require a separate groundwater flow calculation to be performed.

It is important to be aware of the limitations and assumptions of a given simulations as well as some standard pitfalls such as the effect of strain softening or dilative soil response in undrained conditions, which can lead to high effective stresses and soil strength [229]. Locking effects, which are mismatches between the kinematic deformation possibilities and flow rules, can also lead to unrealistic overshoots of stresses.

1.9 Predictions and analysis

Once the modelling process is completed and the results are obtained, the results can be interpreted in order to give the numerical prediction an engineering value. The translation from computer model results to engineering meanings can lead to some discrepancies and misinterpretations in addition to the uncertainties and assumptions of the model. Brinkgreve and Engin [45] point out that the most common misinterpretations are related to the interpretation of factors of safety, the behaviour of structural elements, overlooking of essential details (especially for 3D models), and the general lack of knowledge and understanding of the modelling assumptions.

The translation from numerical predictions to engineering meanings can be straightforward but it can also be more challenging. For instance, the computations of factor of safety is difficult and the criterion of failure has to be clearly defined (see Chapter 11).

Portraying a prediction of the most probable outcome is often the interpretation of a series of simulations in which different scenarios are investigated. Results are often presented in terms of contour plots of a given variable. Whilst this type of deliverable permits building an understanding, it can be difficult to extract important information and the values for engineering. The preliminary determination of a variable or object of interest, when designing the numerical model, facilitates predictions and discussions.

1.10   Closure

This chapter explains the process of carrying out numerical predictions. It describes each step of the process as well as introducing different types of simulations. Whilst the chapter does not provide the scientific details of numerical analyses (which is the matter of the following chapters), it offers a broad picture of what numerical predictions are and the different levels of idealisation which have to be made. It is shown that the initial state of knowledge of a given problem prior to the analysis influences the numerical model and, hence, the prediction.

As a rule of thumb, numerical models and analyses should go from the simplest ones to more complicated ones as numerous levels of idealisation are required. This approach allows the validation in a step-wise process in order to finally obtain a robust model and prediction. It is important to keep in mind that dummy parameter values might permit the computation to run but they often lead to incorrect predictions creating an illusion of reality. It is important to challenge the model by running a series of tests that verify the model is correctly setup. Finally, it is important to keep in mind that numerical predictions are by definition an idealised interpretation of reality.

2

Fundamentals of the Material Point Method

A. Yerro and A. Rohe

2.1 Basic concept

The material point method (MPM) is one amongst a vast number of computational methods being developed to simulate soil-water-structure interaction problems in history-dependent materials involving large deformations. MPM combines the advantages of mesh-based and point-based approaches: mesh distortion can be eliminated and history is stored in material points (MPs). MPM is an advanced formulation of the finite element method (FEM) where the continuum body is represented by a set of Lagrangian points, called MPs. The MPs are moving through an Eulerian computational mesh. The MPs carry all physical properties of the continuum such as velocities, stresses, strains, density, momentum, material parameters, and other state parameters, whereas the computational mesh is used to solve the balance equations without storing any permanent information.

In MPM, the computational domain is spatially discretised in two frames as shown in Fig. 2.1. First, the continuum body is divided into a set of MPs. Each MP represents an initially defined part of the domain and the mass of such subdomain. One of the basic and most important features of MPM is that the mass of each MP remains strictly constant which implies that mass conservation is automatically satisfied, whilst the volume of the MP can change enabling material compression or extension. In standard MPM, mass is considered to be concentrated at the corresponding MP. Other quantities, such as velocities, strains and stresses, are also initialised and carried by MPs. Each MP moves attached to the deformations of the body and this provides the Lagrangian description of the continuum. The second frame is the computational mesh which is equivalent to a conventional finite element (FE) mesh. It is constructed to cover the full domain of the problem, including empty spaces into which MPs are expected to move during computation. The discretised momentum balance equations are typically solved at the nodes of this computational mesh, whereas mass conservation and constitutive equations

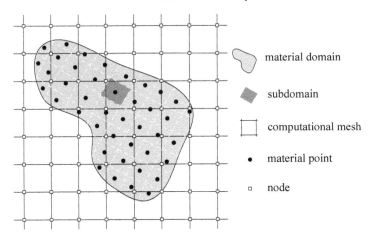

FIGURE 2.1: Spatial discretisation of a continuum body with nodes of the computational mesh and material points (modified from [332]).

are solved at the MPs. The information required to solve the balance equations on the computational mesh at any instance of the analysis is transferred from the MPs to the nodes of the mesh by using mapping functions, i.e. the typical shape functions as used also in FEM. After solving the balance equations by using an incremental time integration scheme the quantities carried by the MPs are updated by interpolation of the mesh results, using the same mapping functions. The information associated with the mesh is not required for the next step of the analysis. Therefore, it can be discarded avoiding any mesh distortion.

The application of engineering boundary conditions is easily facilitated. For instance, stresses and displacements, or their rates, can be applied on any mesh node of the boundary or directly on the MPs. An additional benefit of MPM is the continuum description of the material for which well-known constitutive models, which describe the stress-strain relationship of materials, can be applied. Shortcomings of MPM are for example its mesh dependency, which is inherent to any FE formulation, its computational costs, and stability issues caused by MPs crossing element boundaries. An overview of the historical development and applications of MPM can be found in later sections of this chapter.

The MPM algorithm for a single calculation step of a time increment is illustrated in Fig. 2.2. At the beginning of each step, the components of the momentum balance equations are defined by mapping information from the MPs to the nodes of the computational mesh by means of interpolation or shape functions (Fig. 2.2 a). The equations of motion are solved for the primary unknown variables, i.e. the nodal accelerations (Fig. 2.2 b). These nodal values are used to update acceleration, velocity and position of MPs as well as to compute strains and stresses at the MPs (Fig. 2.2 c). At that point,

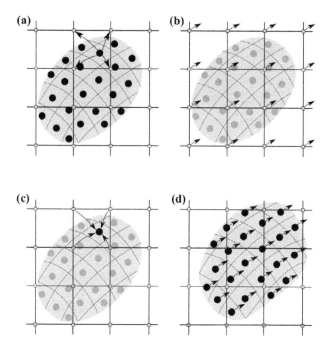

FIGURE 2.2: MPM algorithm for a single calculation step of a time increment: (a) map information from MPs to nodes, (b) solve balance equations, (c) map velocity field to MPs, and (d) update position of MPs (modified from [101]).

no permanent information is stored any more at the mesh. Thus, it can be freely redefined at the end of each time step, but commonly it is kept fixed. The MPs assignment of MPs to elements is updated after mesh adjustment (Fig. 2.2 d).

Soil is a mixture of three constituents (solid, liquid and gas) that interact with each other determining the mechanical and hydraulic response of the material. However, taking rigorously into account these interactions may be in many cases unnecessarily complicated, computationally expensive, and even not feasible for engineering applications. Fig. 2.3 shows the concept of multiple constituents used in MPM. The columns represent the number of phases of the continuum and the rows the number of MP sets describing each phase.

Dry soil, pure liquid and saturated soil in fully drained or undrained conditions can be modelled with the one-phase single-point formulation. In this case only, one set of MPs is needed to carry all required information of the material. Typical applications are undrained slope failure, collapse of dry granular materials (Chapter 9), silo discharge problems, water reservoirs and flows (Chapter 21), shallow foundations in undrained conditions, undrained cone penetration tests (CPTs) (Chapter 5) and submerged slope failure in highly

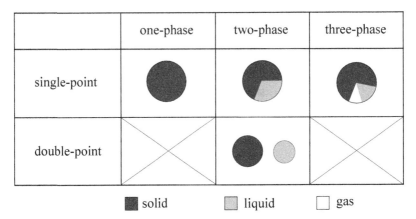

FIGURE 2.3: Overview of multi-phase MPM approaches.

permeable sandy soils. The full formulation of the one-phase single-point approach is presented in Section 2.2.

In the two-phase single-point approach, the saturated porous media are discretised by a single set of MPs which moves according to the solid velocity field. Each MP describes a representative volume element of fully saturated soil, carrying the information of both phases, solid and liquid together. While the MPs are attached to the solid skeleton giving a Lagrangian description of its movement, an Eulerian approach with respect to the solid represents the motion of the liquid phase. Typical applications are consolidation, pile installation (jacking, driving, vibrating) with generation and dissipation of excess pore pressures, CPTs in partially drained conditions (Chapter 18), and failure of saturated slopes due to infiltration or loading (Chapter 10). The full formulation of the two-phase single-point approach can be found in Section 2.3.

In the three-phase single-point approach, the soil is understood as a material composed of three distinct constituents, i.e. solid, liquid and gas. The solid phase constitutes the soil skeleton of the media while liquid and gas phases fill the voids. All phases are combined in a single MP and the balance and momentum equations are formulated and numerically solved as aforementioned. Typical applications are rainfall infiltration and drought effects in slope failure and collapse analyses for unsaturated soils, collapse behaviour of low-density soils, and unrestrained swelling of expansive clays. Chapter 12 shows examples of the three-phase single-point formulation. The full formulation of the three-phase single-point approach can be found in Section 2.4.

In the two-phase double-point approach, the solid-liquid mixture is modelled using two distinct sets of MPs, one for each constituent respectively. Fig. 2.4 shows an example of a slope which is characterised by two materials – soil (dry and saturated parts) and free surface water. The soil is a porous material composed of two constituents, the solid skeleton and groundwater respectively, whereas the free surface water is a pure liquid. In general, the

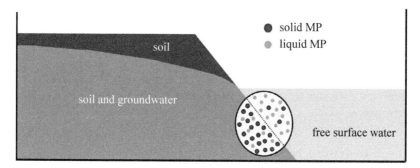

FIGURE 2.4: Schematic description of an example of the double-point ide-alisation with a partly saturated slope in contact and interaction with free surface water.

water can flow out of the soil body into the water reservoir or vice versa, and interacts with the solid skeleton through drag forces. The soil can behave as a solid porous material with liquid in its pores (solid-like response) or as a liquefied material in which soil grains float in the liquid (liquid-like response). The material can also change between these states. Application examples are transient seepage flow through a porous material (Chapter 20) or the free-fall of a poro-elastic body under water. Advanced examples involving state transition are the free-fall of a coarse-grained soil under water, collapse of a submerged sand column (liquefaction and breaching), collapse of a slope due to seepage flow with subsequent sediment transport (Chapter 11), internal insta-bility (suffusion), piping, erosion problems, slurry and debris flows including separation. The full formulation of the two-phase double-point approach can be found in Section 2.5.

2.2   One-phase single-point formulation

Soil is a multi-phase porous material, which is characterised by a solid skele-ton filled with liquids or gases. In several cases, it can be regarded as an homogeneous one-phase material. Saturated soil can be in drained, partially drained and undrained conditions depending on the soil permeability and the loading rate. In drained conditions, a negligible excess pore pressure is gen-erated and dissipates rapidly. Therefore, it can be neglected. In undrained conditions, the loading rate is so fast that there is a significant generation of excess pore pressures, but negligible relative movement between solid and liquid phase. Therefore, pore pressure dissipation can be neglected. Both cases allow for a simplified modelling approach in which only the solid velocity field is considered, and the one-phase single-point formulation can be applied.

However, the excess pore pressure generation and dissipation are no longer negligible in partially drained conditions. Therefore, the two-phase formulation is necessary in order to take into account the fully-coupled behaviour of solid and liquid and is presented in Section 2.3.

The governing equations of the one-phase single-point formulation are general and not exclusive to geomechanical problems. Therefore, the formulation can be applied to any kind of one-phase solid or pure liquid material.

2.2.1   Governing equations

The continuum satisfies the fundamental conservation laws of thermodynamics. These are the conservation of mass, the conservation of momentum and the conservation of energy. They must also satisfy the corresponding initial and boundary conditions. Furthermore, constitutive equations, which characterise the stress-strain relationship of the continuum material, are required. The formulation of MPM is given in the Lagrangian description of motion. The matrix notation is used throughout this chapter.

Conservation of mass

When sources and sinks are neglected, i.e. no mass is entering or leaving the domain Ω covered by the material, the change of mass in time is zero. The conservation of mass can then be described with the differential equation (Eq. 2.1).

$$\frac{\mathrm{d}\rho}{\mathrm{d}t} + \rho \operatorname{div}(\vec{v}) = 0 \qquad (2.1)$$

where $\mathrm{d}/\mathrm{d}t$ is the total or material time derivative, ρ the mass density of the material, and $\vec{v}$ the velocity vector. One of the basic assumptions of MPM is that the mass of each MP remains constant during deformation. This implies that the mass conservation is automatically satisfied and Eq. 2.1 can be used to update density.

Conservation of momentum

The conservation of momentum implies the conservation of both linear and angular momentum. The conservation of linear momentum represents the equation of motion of a continuum, i.e. Newton's second law of motion. It relates the motion or the kinematics of a continuum to the internal and external forces acting upon it. The conservation of linear momentum of the continuum can be described with the differential equation given in Eq. 2.2.

$$\rho \frac{\mathrm{d}\vec{v}}{\mathrm{d}t} = \operatorname{div}\left(\boldsymbol{\sigma}^{\mathrm{T}}\right) + \rho \vec{g} \qquad (2.2)$$

where $\mathrm{d}\vec{v}/\mathrm{d}t$ is the acceleration vector, $\boldsymbol{\sigma}$ the stress matrix and $\vec{g}$ the gravity vector. A dynamic formulation is considered, which means that the acceleration term is taken into account.

The conservation of angular momentum implies that the stress matrix is symmetric ($\boldsymbol{\sigma} = \boldsymbol{\sigma}^T$).

Conservation of energy

Heat effects and any source of thermal energy are neglected and mechanical work is considered as the only source of energy. Therefore, the conservation of energy gives Eq. 2.3.

$$\rho \frac{dE}{dt} = \dot{\boldsymbol{\varepsilon}}^T \boldsymbol{\sigma} \tag{2.3}$$

where E is the internal energy per unit mass and $\dot{\vec{\varepsilon}}$ is the deformation rate matrix. Chapter 14 and Appendix A discuss the inclusion of heat in the conservation of energy.

Initial conditions and boundary conditions

Let $\partial\Omega$ represent the boundary of the domain Ω covered by the material, which consists of two parts, namely $\partial\Omega_u$ and $\partial\Omega_\tau$. The boundary $\partial\Omega_u$ denotes the prescribed displacement boundary whereas $\partial\Omega_\tau$ denotes the prescribed stress boundary or prescribed traction boundary as indicated in Fig. 2.5. The displacement BCs (also essential BCs or Dirichlet BCs) can be written as given in Eq. 2.4.

$$\vec{u}(\vec{x}, t) = \vec{U}(t) \quad \text{on} \quad \partial\Omega_u(t) \tag{2.4}$$

where $\vec{U}(t)$ is the prescribed surface displacement vector.

The traction BCs, also called the natural BCs or the Neumann BCs, are defined by Eq. 2.5.

$$\boldsymbol{\sigma}(\vec{x}, t)\,\vec{n} = \vec{\tau}(t) \quad \text{on} \quad \partial\Omega_\tau(t) \tag{2.5}$$

where $\vec{n}$ is the outward unit normal vector of the boundary surface $\partial\Omega_\tau$, and $\vec{\tau}(t)$ is the prescribed surface traction vector.

The initial conditions are given in Eq. 2.6.

$$\vec{u}(\vec{x}, t_0) = \vec{U}_0 \quad , \qquad \vec{v}(\vec{x}, t_0) = \vec{V}_0 \quad , \qquad \boldsymbol{\sigma}(\vec{x}, t_0) = \boldsymbol{\sigma}_0 \tag{2.6}$$

Constitutive relation and stress variables

The mechanical constitutive equation of a material specifies the dependency of the stress in a body on kinematic variables such as the strain or the velocity gradient $\vec{L} = \nabla\vec{v}$, also called the rate-of-deformation.

The Cauchy stress $\boldsymbol{\sigma}$ is used for the expression of stresses. In case of large deformations (or rotations), the variance of the Cauchy stress can lead to incompatibility with the used strain measure. The use of the Cauchy stress is therefore restricted to small strain rates. Unlike the Cauchy stress, its rate $\dot{\boldsymbol{\sigma}}$ is not objective. Alternatively, the objective Jaumann stress rate $\dot{\boldsymbol{\sigma}}^J$ can be used (Eq. 2.7).

$$\dot{\boldsymbol{\sigma}}^J = \dot{\boldsymbol{\sigma}} + \boldsymbol{\sigma} \cdot \boldsymbol{\omega} - \boldsymbol{\omega} \cdot \boldsymbol{\sigma} \tag{2.7}$$

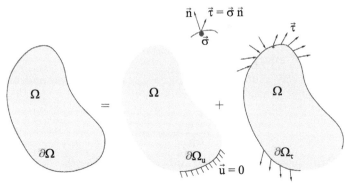

domain boundary   displacement boundary   traction boundary

FIGURE 2.5: Displacement and traction boundary conditions (modified from [5]).

where $\boldsymbol{\omega}$ denotes the spin or rate-of-rotation matrix, which is the skew-symmetric part of the velocity gradient matrix $\boldsymbol{L}$. MPM is formulated in Lagrangian description which enables the use of the Jaumann stress rate for large deformations and rotations.

Taking advantage of the symmetry of the stress tensor ($\boldsymbol{\sigma} = \boldsymbol{\sigma}^{\mathrm{T}}$), the stress and strain tensors can be represented in a reduced vector form (Eq. 2.8).

$$\vec{\sigma}(\vec{x}, t) = [\sigma_{11} \ \sigma_{22} \ \sigma_{33} \ \sigma_{12} \ \sigma_{23} \ \sigma_{31}]^{\mathrm{T}}$$

$$\vec{\varepsilon}(\vec{x}, t) = [\varepsilon_{11} \ \varepsilon_{22} \ \varepsilon_{33} \ \gamma_{12} \ \gamma_{23} \ \gamma_{31}]^{\mathrm{T}}$$

(2.8)

where the first three vector components (σ_{ij} and ε_{ij} with $i = j$) are the normal components in x_i coordinate direction, and the last three components for $i \neq j$ represent the shear components in x_i–x_j plane. Eq. 2.8 is also known as the Voigt notation, which has storage and computational advantages in the numerical implementation. The last three terms of the strain vector are represented by $\gamma_{ij} = 2\varepsilon_{ij}$, to ensure that energy is preserved and that different expressions using matrices or vectors are equal (Eq. 2.9).

$$\varepsilon^T \boldsymbol{\sigma} = \vec{\varepsilon} \, \vec{\sigma}$$

(2.9)

In general, the stress rate $\dot{\vec{\sigma}}$ can be a function of the deformation rate $\dot{\vec{\varepsilon}}$, stress state $\vec{\sigma}$, the temperature T, and a vector of internal variables $\vec{\chi}$ (Eq. 2.10).

$$\dot{\vec{\sigma}} = f\left(\dot{\vec{\varepsilon}}, \vec{\sigma}, T, \vec{\chi}\right)$$

(2.10)

An incremental stress-strain relation can be expressed as Eq. 2.11.

$$\dot{\vec{\sigma}} = \boldsymbol{D} \, \dot{\vec{\varepsilon}}$$

(2.11)

where $\boldsymbol{D}$ represents the constitutive matrix or tangent stiffness matrix. The constitutive behaviour of soils is further discussed in Chapter 8.

2.2.2 Numerical implementation

Weak form

Before discretising the equation for the conservation of linear momentum, its strong form (Eq. 2.2) has to be transformed into the weak form (or virtual work equation). The momentum equation is therefore multiplied by a test function which is the virtual velocity vector $\delta\vec{v}$, and is integrated over the current domain Ω covered by the continuum (Eq. 2.12).

$$\int_\Omega \delta\vec{v}\,\rho\frac{d\vec{v}}{dt}d\Omega = \int_\Omega \delta\vec{v}\,\mathrm{div}(\boldsymbol{\sigma})d\Omega + \int_\Omega \delta\vec{v}\,\rho\vec{g}\,d\Omega \tag{2.12}$$

where $\delta\vec{v} = 0$ on $\partial\Omega_u$.

The first term on the right hand side of Eq. 2.12 is given by Eq. 2.13.

$$\int_\Omega \delta\vec{v}\,\mathrm{div}(\boldsymbol{\sigma})d\Omega = \int_\Omega \mathrm{div}(\delta\vec{v}\,\boldsymbol{\sigma})d\Omega - \int_\Omega \mathrm{div}(\delta\vec{v})\boldsymbol{\sigma}\,d\Omega \tag{2.13}$$

Using the Gauss (or divergence) theorem, the first term on the right hand side of Eq. 2.13 gives Eq. 2.14.

$$\int_\Omega \mathrm{div}(\delta\vec{v}\,\boldsymbol{\sigma})d\Omega = \int_{\partial\Omega} \delta\vec{v}(\boldsymbol{\sigma}\vec{n})dS = \int_{\partial\Omega_\tau} \delta\vec{v}\,\vec{\tau}\,dS \tag{2.14}$$

where dS denotes the surface integral which is only non-zero at the traction boundary $\partial\Omega_\tau$.

After substituting all previous terms into Eq. (2.12), the weak form of the linear momentum equation gives Eq. 2.15.

$$\int_\Omega \delta\vec{v}\,\rho\frac{d\vec{v}}{dt}d\Omega = \int_{\partial\Omega_\tau} \delta\vec{v}\,\vec{\tau}\,dS - \int_\Omega \mathrm{div}(\delta\vec{v})\boldsymbol{\sigma}\,d\Omega + \int_\Omega \delta\vec{v}\,\rho\vec{g}\,d\Omega \tag{2.15}$$

Eq. 2.15 is used in the formulation of the discrete equations in Section 2.2.2. The weak form for the two- and three-phase formulations, presented later in the chapter, can be derived accordingly but the full development falls out of the scope of this chapter.

Space discretisation

The domain Ω is decomposed into finite subdomains Ω_{el}, which are the finite elements (Fig. 2.6). The collection of these subdomains forms the calculation mesh, and n_{el} denotes the total number of elements in the mesh. Each element is connected with its surrounding elements by its nodes, and n_n denotes the total number of nodes in the mesh. The state variables have pre-defined interpolation or shape functions within the element and the solution is obtained at the nodes. Hence, the equilibrium is satisfied at the nodes.

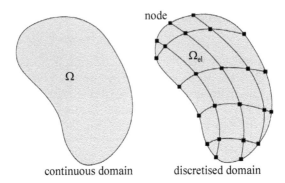

continuous domain   discretised domain

FIGURE 2.6: Element discretisation of the domain (modified from [5]).

The discretised form in MPM is obtained by approximating the displacement $\vec{u}$, velocity $\vec{v}$ and acceleration $\vec{a}$ (Eq. 2.16).

$$\vec{u}(\vec{x},t) \approx \boldsymbol{N}(\vec{x})\underline{\vec{u}}(t) \quad , \qquad \vec{v}(\vec{x},t) \approx \boldsymbol{N}(\vec{x})\underline{\vec{v}}(t) \quad , \qquad \vec{a}(\vec{x},t) \approx \boldsymbol{N}(\vec{x})\underline{\vec{a}}(t) \quad (2.16)$$

The vectors $\underline{\vec{u}}(t)$, $\underline{\vec{v}}(t)$ and $\underline{\vec{a}}(t)$ contain the nodal values of displacement, velocity and acceleration, respectively. The corresponding virtual quantities are approximated accordingly, i.e. $\delta\vec{u} \approx \boldsymbol{N}\delta\underline{\vec{u}}$. The interpolation function, or shape function matrix, $\boldsymbol{N}$ has the following form.

$$\boldsymbol{N}(\vec{x}) = [\boldsymbol{N}_1(\vec{x}) \quad \boldsymbol{N}_2(\vec{x}) \quad \dots \quad \boldsymbol{N}_{n_{\mathrm{n}}}(\vec{x})] \quad (2.17)$$

with

$$\boldsymbol{N}_i(\vec{x}) = \begin{bmatrix} N_i(\vec{x}) & 0 & 0 \\ 0 & N_i(\vec{x}) & 0 \\ 0 & 0 & N_i(\vec{x}) \end{bmatrix} \quad \text{for each node} \quad i = 1,\dots,n_{\mathrm{n}} \quad (2.18)$$

The strain-displacement matrix $\boldsymbol{B}$, which contains the gradients of the shape functions, can be written as Eq. 2.19.

$$\boldsymbol{B}(\vec{x}) = [\boldsymbol{B}_1(\vec{x}) \quad \boldsymbol{B}_2(\vec{x}) \quad \dots \quad \boldsymbol{B}_{n_{\mathrm{n}}}(\vec{x})] \quad (2.19)$$

with

$$\boldsymbol{B}_i(\vec{x}) = \begin{bmatrix} \dfrac{\partial N_i(\vec{x})}{\partial x_1} & 0 & 0 \\[2ex] 0 & \dfrac{\partial N_i(\vec{x})}{\partial x_2} & 0 \\[2ex] 0 & 0 & \dfrac{\partial N_i(\vec{x})}{\partial x_3} \\[2ex] \dfrac{\partial N_i(\vec{x})}{\partial x_2} & \dfrac{\partial N_i(\vec{x})}{\partial x_1} & 0 \\[2ex] 0 & \dfrac{\partial N_i(\vec{x})}{\partial x_3} & \dfrac{\partial N_i(\vec{x})}{\partial x_2} \\[2ex] \dfrac{\partial N_i(\vec{x})}{\partial x_3} & 0 & \dfrac{\partial N_i(\vec{x})}{\partial x_1} \end{bmatrix} \quad \text{for each node} \quad i = 1,\dots,n_{\mathrm{n}} \quad (2.20)$$

The shape functions and B-matrices are given for the three-dimensional case with the global coordinate system (x_1, x_2, x_3). Then, the virtual work equation (Eq. 2.15) can be written (Eq. 2.21).

$$\delta \vec{\underline{v}}^{\mathrm{T}} \int_{\Omega} \boldsymbol{N}^{\mathrm{T}} \rho \boldsymbol{N} \, \underline{\vec{a}} \, \mathrm{d}\Omega = \delta \vec{\underline{v}}^{\mathrm{T}} \int_{\partial \Omega_\tau} \boldsymbol{N}^{\mathrm{T}} \vec{\tau} \, \mathrm{d}S - \delta \vec{\underline{v}}^{\mathrm{T}} \int_{\Omega} \boldsymbol{B}^{\mathrm{T}} \boldsymbol{\sigma} \, \mathrm{d}\Omega + \delta \vec{\underline{v}}^{\mathrm{T}} \int_{\Omega} \boldsymbol{N}^{\mathrm{T}} \rho \vec{g} \, \mathrm{d}\Omega \quad (2.21)$$

where $\vec{\tau}$ and $\vec{g}$ are vectors containing the components of tractions and gravitational acceleration, respectively. Vector $\delta \vec{\underline{v}}$ contains the virtual nodal velocities that are arbitrary except on boundaries where a velocity is prescribed. Hence, Eq. 2.21 can be written as Eq. 2.22.

$$\int_{\Omega} \boldsymbol{N}^{\mathrm{T}} \rho \boldsymbol{N} \, \underline{\vec{a}} \, \mathrm{d}\Omega = \int_{\partial \Omega_\tau} \boldsymbol{N}^{\mathrm{T}} \vec{\tau} \, \mathrm{d}S - \int_{\Omega} \boldsymbol{B}^{\mathrm{T}} \boldsymbol{\sigma} \, \mathrm{d}\Omega + \int_{\Omega} \boldsymbol{N}^{\mathrm{T}} \rho \vec{g} \, \mathrm{d}\Omega \quad (2.22)$$

In the numerical implementation, the integrals are evaluated for each element, while looping over all elements of the mesh. The global matrices are formed by assembling the element matrices. The numerical integration (Eq. 2.22) gives Eq. 2.23.

$$\boldsymbol{M} \vec{a} = \vec{f}^{\mathrm{ext}} - \vec{f}^{\mathrm{int}} \quad (2.23)$$

with

$$\boldsymbol{M} = \int_{\Omega} \boldsymbol{N}^{\mathrm{T}} \rho \boldsymbol{N} \mathrm{d}\Omega \quad (2.24)$$

$$\vec{f}^{\mathrm{ext}} = \vec{f}^{\mathrm{trac}} + \vec{f}^{\mathrm{grav}} = \int_{\partial \Omega_\tau} \boldsymbol{N}^{\mathrm{T}} \vec{\tau} \mathrm{d}S + \int_{\Omega} \boldsymbol{N}^{\mathrm{T}} \rho \vec{g} \, \mathrm{d}\Omega \quad (2.25)$$

$$\vec{f}^{\mathrm{int}} = \int_{\Omega} \boldsymbol{B}^{\mathrm{T}} \boldsymbol{\sigma} \, \mathrm{d}\Omega \quad (2.26)$$

where $\vec{f}^{\mathrm{ext}}$, $\vec{f}^{\mathrm{int}}$, $\vec{f}^{\mathrm{trac}}$ and $\vec{f}^{\mathrm{grav}}$ are the external, internal, traction and gravity forces, respectively. $\boldsymbol{M}$ is the mass matrix.

Material point discretisation

Until the previous step, both the MPM and FEM solution algorithms are identical. The main difference is that in MPM the continuum is also divided into a finite number of subdomains called material points (Fig. 2.1) with n_{MP} denoting the number of MPs in each element. Each MP moves attached to the solid skeleton of the continuum that it represents, providing a Lagrangian description of motion and carrying all information (e.g. displacement, velocity, stresses) and having a constant mass. From a numerical point of view, MPs can be compared to Gaussian integration points in FEM. However, MPs can move throughout the computational mesh and are not fixed within the elements. Elements filled with MPs are called active elements and their nodes contribute to the momentum balance equations. Conversely, empty elements, which do not have any MPs, are ignored reducing the computational cost.

Each MP is initially placed at a predefined local position inside the element and, hence, the local position vector $\vec{\xi}_{\mathrm{MP}}$ of material point MP is initialised. The global position vector $\vec{x}_{\mathrm{MP}}$ is then obtained (Eq. 2.27).

$$\vec{x}_{\mathrm{MP}}(\vec{\xi}_{\mathrm{MP}}) \approx \sum_{i=1}^{n_{\mathrm{n,el}}} N_{\mathrm{i}}(\vec{\xi}_{\mathrm{MP}})\vec{x}_{\mathrm{i}} \tag{2.27}$$

where $n_{\mathrm{n,el}}$ is the number of nodes per element, $N_{\mathrm{i}}(\vec{\xi}_{\mathrm{MP}})$ the shape function of node i evaluated at the local position of material point MP, and $\vec{x}_{\mathrm{i}}$ the nodal coordinates.

The volumes associated with MPs are calculated such that all MPs inside an element initially contain the same amount of the element volume (Eq. 2.28).

$$V_{\mathrm{MP}} = \frac{1}{n_{\mathrm{MP}}} \int_{\Omega_{\mathrm{el}}} \mathrm{d}\Omega_{\mathrm{el}} \approx \frac{1}{n_{\mathrm{MP}}} \sum_{\mathrm{q}=1}^{n_{\mathrm{q,el}}} w_{\mathrm{q}} \left| \vec{J}(\vec{\xi}_q) \right| \tag{2.28}$$

where V_{MP} is the volume associated with material point MP, n_{MP} is the number of MPs in the element, $n_{q,el}$ the number of Gauss points in the element, w_{q} the local integration weight associated with Gauss point q, and $\vec{J}$ the Jacobian matrix. The mass of each MP is then calculated (Eq. 2.29).

$$m_{\mathrm{MP}} = \rho_{\mathrm{MP}} V_{\mathrm{MP}} \tag{2.29}$$

where ρ_{MP} is the mass density of the material to which material point MP is associated.

To simplify computations, a lumped mass matrix $\boldsymbol{M}_{\mathrm{lumped}}$ can be used instead of the consistent mass matrix $\boldsymbol{M}$. The lumped mass matrix is a diagonal matrix in which each entry m_{i} is obtained by summing over the corresponding row of the consistent mass matrix, giving Eq. 2.30.

$$\boldsymbol{M}_{\mathrm{lumped}} = \sum_{\mathrm{MP}=1}^{n_{\mathrm{MP}}} m_{\mathrm{MP}} \boldsymbol{N}(\vec{\xi}_{\mathrm{MP}}) \tag{2.30}$$

The matrix inversion becomes trivial, although the consequence of using a lumped mass matrix induces some dissipation of kinetic energy [42].

In MPM, stresses are integrated at MPs, which means that the quadrature points coincide with the MPs and the integration weight associated with the MP is its volume V_{MP}. Then, the stress integration can be approximated (Eq. 2.31).

$$\int_{\Omega_{\mathrm{el}}} \boldsymbol{B}^{\mathrm{T}}\boldsymbol{\sigma} \, \mathrm{d}\Omega_{\mathrm{el}} \approx \sum_{\mathrm{MP}=1}^{n_{\mathrm{MP}}} V_{\mathrm{MP}} \boldsymbol{B}^{\mathrm{T}}(\vec{\xi}_{\mathrm{MP}}) \, \boldsymbol{\sigma}_{\mathrm{MP}} \tag{2.31}$$

External forces are mapped to the MPs located next to the traction boundary elements. These so-called boundary MPs carry the surface traction throughout the computation. The boundary traction vector $\vec{\tau}$ is interpolated from the nodes of the boundary elements to the boundary MPs. Hence, the traction at a boundary material point *MP* becomes Eq. 2.32.

$$\vec{\tau}(\vec{x}_{bMP}) \approx \sum_{i=1}^{n_{n,bel}} N_i(\vec{\xi}_{bMP})\vec{\tau}(\vec{x}_i) \tag{2.32}$$

where $n_{n,bel}$ is the number of nodes of the boundary element, N_i the shape function of node i of the boundary element and $\vec{\xi}_{bMP}$ are the coordinates of the boundary material point bMP. These coordinates represent the projection of the MP on the boundary element. S_{el} is the traction surface of the boundary element. Then, the integrals of Eq. 2.24 can be approximated for each element (Eq. 2.33).

$$\boldsymbol{M} \approx \sum_{MP=1}^{n_{MP}} m_{MP} \boldsymbol{N}(\vec{\xi}_{MP}) \tag{2.33}$$

$$\vec{f}^{\,trac} \approx \sum_{bMP=1}^{n_{bMP}} \frac{S_{el}}{n_{bMP}} \boldsymbol{N}^T (\vec{\xi}_{MP}) \left(\sum_{i=1}^{n_{n,bel}} N_i(\vec{\xi}_{bMP})\vec{\tau}(\vec{x}_i) \right) \tag{2.34}$$

$$\vec{f}^{\,grav} \approx \sum_{MP=1}^{n_{MP}} m_{MP} \boldsymbol{N}^T(\vec{\xi}_{MP})\vec{g} \tag{2.35}$$

$$\vec{f}^{\,int} \approx \sum_{MP=1}^{n_{MP}} V_{MP} \boldsymbol{B}^T(\vec{\xi}_{MP})\,\boldsymbol{\sigma}_{MP} \tag{2.36}$$

Time discretisation

Time is discretised into k time instants Δt. Then, the time discretisation can be expressed as Eq. 2.37.

$$t^{k+1} = t^k + \Delta t \tag{2.37}$$

The momentum balance equation (Eq. 2.23) is posed at time t^k and can therefore be written as Eq. 2.38.

$$\boldsymbol{M}^k \vec{a}^k = \vec{f}^{\,ext,k} - \vec{f}^{\,int,k} \tag{2.38}$$

where the acceleration $\vec{a}^k$ is the unknown.

An explicit Euler time integration scheme is used to update the velocity. This is a first-order numerical procedure for solving ordinary differential equations with a given initial value. Being $\vec{v}^k$ the velocity at time t^k, the velocity at the next time step t^{k+1} is calculated using the acceleration at time t^k as Eq. 2.39.

$$\vec{v}^{k+1} = \vec{v}^k + \Delta t\, \vec{a}^k \tag{2.39}$$

The displacements $\vec{u}$ at time t^{k+1} are calculated using the updated nodal velocity $\vec{v}^{k+1}$ as Eq. 2.40.

$$\vec{u}^{k+1} = \vec{u}^k + \Delta t \ \vec{v}^{k+1} \tag{2.40}$$

This algorithm is based on the work presented by Sulsky et al. [285] and is identical to Lagrangian FEM. However, this approach can cause ill-conditioned mass matrices in MPM as nodal masses can reach zero in case a MP has just entered a previously empty element and is still close to the element boundary. As a remedy, Sulsky et al. [287] introduced a slightly modified algorithm described below. The important aspect is to use the momentum instead of velocity where possible, thus, avoiding dividing by (almost zero) nodal masses.

2.2.3 Computational cycle

The MPM computational cycle of each time step can be summarised as follows.

1. Calculate the nodal mass using the shape functions to form the lumped mass matrix M_i^k at time t^k (Eq. 2.33).

2. Evaluate the internal and external forces $\vec{f}_i^{\text{ext},k}$ and $\vec{f}_i^{\text{int},k}$ at the nodes (Eqs. 2.34 to 2.36).

3. Solve the momentum balance equation and determine the nodal accelerations $\vec{a}_i^k$ following from Eq. 2.38 as given in Eq. 2.41.

$$\vec{a}_i^k = \left[M_i^k\right]^{-1}\left(\vec{f}_i^{\text{ext},k} - \vec{f}_i^{\text{int},k}\right) \tag{2.41}$$

4. Update the velocity at the MPs considering Eq. 2.39 and using the nodal accelerations and shape functions as given in 2.42.

$$\vec{v}_{\text{MP}}^{k+1} = \vec{v}_{\text{MP}}^k + \Delta t \sum_{i=1}^{n_{n,el}} N_i(\vec{\xi}_{\text{MP}}^k) \ \vec{a}_i^k \tag{2.42}$$

5. Update the nodal momentum (Eq. 2.43).

$$\vec{P}_i^{k+1} = \sum_{el=1}^{n_{el,i}} \sum_{\text{MP}=1}^{n_{\text{MP}}} m_{\text{MP}} N_i(\vec{\xi}_{\text{MP}}^k) \ \vec{v}_{\text{MP}}^{k+1} \tag{2.43}$$

6. Update the nodal velocities (Eq. 2.44).

$$\vec{v}_i^{k+1} = \frac{\vec{P}_i^{k+1}}{M_i^k} \tag{2.44}$$

7. Compute the incremental nodal displacement (Eq. 2.45).

$$\Delta \vec{u}_i^{k+1} = \Delta t \ \vec{v}_i^{k+1} \tag{2.45}$$

8. Compute the strain increment (Eq. 2.46).

$$\Delta \vec{\varepsilon}_{MP}^{k+1} = \boldsymbol{B}(\vec{\xi}_{MP}^{k}) \Delta \vec{u}_i^{k+1} \qquad (2.46)$$

9. Update the stresses using the material constitutive model (Eq. 2.11).

10. Update the volume and mass density of MPs (Eq. 2.47).

$$V_{MP}^{k+1} = (1 + \Delta \varepsilon_{vol,MP}^{k+1}) V_{MP}^{k} \qquad (2.47a)$$

$$\rho_{MP}^{k+1} = \frac{\rho_{MP}^{k}}{\left(1 + \Delta \varepsilon_{vol,MP}^{k+1}\right)} \qquad (2.47b)$$

11. Considering Eq. 2.40, update MP displacements and positions (Eq. 2.48).

$$\vec{u}_{MP}^{k+1} = \vec{u}_{MP}^{k} + \sum_{i=1}^{n_{n,el}} N_i(\vec{\xi}_{MP}^{k}) \, \Delta \vec{u}_i^{k+1} \qquad (2.48a)$$

$$\vec{x}_{MP}^{k+1} = \vec{x}_{MP}^{k} + \sum_{i=1}^{n_{n,el}} N_i(\vec{\xi}_{MP}^{k}) \, \Delta \vec{u}_i^{k+1} \qquad (2.48b)$$

12. Discard nodal values as all updated information is carried by the material points. Initialise the computational mesh for the next step.

2.3 Two-phase single-point formulation

The study of saturated soils requires the implementation of hydro-mechanical formulations capable of modelling the interaction between the solid skeleton (i.e. solid phase) and the pore water (i.e. liquid phase). Several coupled two-phase formulations have been discussed since the pioneering work by Terzaghi and Biot [32,33,296]; Zienkiewicz et al. [348–350] provide a comprehensive discussion. Within the MPM framework, two distinct approaches are developed to study the interaction between solid and liquid phases – (1) the two-phase single-point approach and (2) the two-phase double-point approach (Fig. 2.3). The first is described in this section and the second in Section 2.5.

The two-phase single-point approach is consistent with a purely macro-scale continuum theory where the behaviour of the porous medium is described based on the kinematics of the solid skeleton [32]. It adopts a single set of MPs to represent the saturated porous medium. Each MP represents a volume of saturated soil and carries the information of both the solid and liquid phases

(Fig. 2.7). In this manner, the volume associated with a MP V_{MP} is the total volume of the mixture, which is equal to the sum of the partial volumes of solid $V_{\mathrm{S}}^{\mathrm{MP}}$ and liquid $V_{\mathrm{L}}^{\mathrm{MP}}$ (Eq. 2.49).

$$V_{\mathrm{MP}} = V_{\mathrm{S}}^{\mathrm{MP}} + V_{\mathrm{L}}^{\mathrm{MP}} \qquad (2.49)$$

where the subscripts S and L refer to the solid and liquid phase, respectively.

MPs move attached to the solid skeleton giving a Lagrangian description of the solid movement, while the motion of the liquid is described with respect to the solid motion. In this approach, the solid mass of each MP remains constant, but the total mass can change as a result of in- or outflow of liquid. Therefore, only the conservation of solid is automatically fulfilled throughout the calculation.

The two-phase single-point formulation has been successfully applied to a number of geotechnical problems such as the simulation of CPT in partially drained conditions [55, 109], and to model landslides and slope failures [102, 278, 335, 336].

2.3.1 Governing equations

The two-phase single-point approach is formulated within a continuum framework by solving a set of physical laws comprising the dynamic momentum balance of the liquid phase, the dynamic momentum balance of the mixture, and mass balances and constitutive relationships of both phases involved (i.e. solid and liquid). In this case, all dynamic terms are taken into account, and the acceleration of the solid skeleton $\vec{a}_{\mathrm{S}}$ and the acceleration of the pore liquid $\vec{a}_{\mathrm{l}}$ are the primary unknowns. This formulation proved to be able to capture the physical response of saturated soil under dynamic as well as static loading [309].

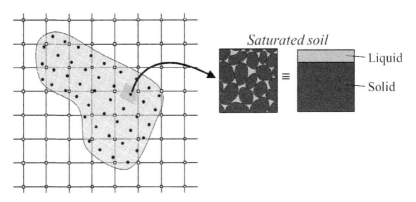

FIGURE 2.7: Two-phase single-point MPM approach (modified from [332]).

Conservation of momentum

The dynamic momentum conservation for the mixture can be written as Eq. 2.50.

$$n_S \rho_S \vec{a}_S + n_L \rho_w \vec{a}_L = \text{div}(\boldsymbol{\sigma}) + \rho_m \vec{g} \tag{2.50}$$

where ρ_S is the solid density, ρ_L the liquid density, n_S the volumetric concentration ratio of solid, n_L the volumetric concentration ratio of liquid, and ρ_m the density of the mixture ($\rho_m = n_S \rho_S + n_L \rho_L$). Note that for saturated soils, n_L is equivalent to porosity of the solid skeleton n and $n_L + n_S = 1$.

The momentum balance of the liquid phase per unit of liquid volume is given in Eq. 2.51.

$$\rho_L \vec{a}_L = \nabla p_L - \vec{f}_L^d + \rho_L \vec{g} \tag{2.51}$$

where $\vec{f}_L^d$ is the drag force, and p_L is the liquid pressure.

Conservation of mass

The mass conservation of the solid phase is expressed as given in Eq. 2.52.

$$\frac{d(n_S \rho_S)}{dt} + \text{div}(n_S \rho_S \vec{v}_S) = 0 \tag{2.52}$$

where $\vec{v}_S$ is the velocity vector of the solid phase.

Similarly, the conservation of liquid mass can be written as Eq. 2.53.

$$\frac{d(n_L \rho_L)}{dt} + \text{div}(n_L \rho_S \vec{v}_L) = 0 \tag{2.53}$$

where $\vec{v}_L$ is the (true) velocity of the liquid phase

When considering incompressible solid grains and disregarding the spatial variations in density and porosity, the expressions for the conservation of mass of the solid and liquid phases reduce to Eq. 2.54.

$$-\frac{D^S n_L}{Dt} + n_S \text{div}(\vec{v}_S) = 0 \tag{2.54}$$

$$\rho_L \frac{D^S n_L}{Dt} + n_L \frac{D^S \rho_L}{Dt} + n_L \rho_L \text{div}(\vec{v}_L) = 0 \tag{2.55}$$

The material derivative with respect to the solid has been used in the previous development (Eq. 2.56).

$$\frac{D^S(\bullet)}{Dt} = \frac{d(\bullet)}{dt} + \vec{v}_L \nabla(\bullet) \tag{2.56}$$

Substituting Eq. 2.54 into Eq. 2.55 eliminates the term $D^S n_L / Dt$ and gives Eq. 2.57.

$$n_S \text{div}(\vec{v}_S) + \frac{n_L}{\rho_L} \frac{D^S \rho_L}{Dt} + n_L \text{div}(\vec{v}_L) = 0 \tag{2.57}$$

The liquid is considered as weakly compressible material. Therefore, the volumetric strain $\varepsilon_{vol,L}$ of the liquid is defined by Eq. 2.58.

$$\frac{D^S \varepsilon_{vol,L}}{Dt} = -\frac{1}{\rho_L} \frac{D^S \rho_L}{Dt} \tag{2.58}$$

Substituting Eq. 2.58 into Eq. 2.57 and rearranging terms gives the mass balance of the mixture (i.e. saturated porous media) (Eq. 2.59).

$$\frac{D^S \varepsilon_{vol,L}}{Dt} = \frac{1}{n_L}[n_S \text{div}(\vec{v}_S) + n_L \text{div}(\vec{v}_L)] \tag{2.59}$$

Eq. 2.59 is also known as the storage equation and represents the volumetric strain rate of the pore liquid. By rearranging terms of Eq. 2.54, the mass balance of the solid phase becomes Eq. 2.60, which represents the variation of volumetric concentration ratio of the liquid phase (i.e. porosity).

$$\frac{D^S n_L}{Dt} = n_S \text{div}(\vec{v_S}) \tag{2.60}$$

Constitutive relation

The constitutive equations for both phases are required to fully describe the behaviour of saturated soils. Assuming the validity of Terzaghi's effective stress concept, the mechanical behaviour of the solid skeleton can be modelled in terms of effective stresses. The general form of a suitable stress-strain relationship is presented in Eq. 2.61. Eq. 2.62 is the constitutive relationship considered for the pore liquid, which relates volumetric strain of the liquid to liquid pressure, where K_L is the elastic bulk modulus of the liquid.

$$\frac{D^S \vec{\sigma}'}{Dt} = \boldsymbol{D} \frac{D^S \vec{\varepsilon}}{Dt} \tag{2.61}$$

$$\frac{D^S p_L}{Dt} = K_L \frac{D^S \varepsilon_{vol,L}}{Dt} \tag{2.62}$$

where $\boldsymbol{D}$ is the tangent stiffness matrix.

Boundary conditions

The proposed formulation requires that the boundary of the domain is the union of the following components.

$$\partial \Omega = \partial \Omega_u \cup \partial \Omega_\tau \cup \partial \Omega_{v_L} \cup \partial \Omega_p$$

where $\partial \Omega_{v_L}$ and $\partial \Omega_p$ are the prescribed velocity and prescribed pressure boundaries of the liquid phase, respectively, whereas $\partial \Omega_u$ is the prescribed displacement (velocity) boundary of the solid phase and $\partial \Omega_\tau$ is the prescribed total stress boundary.

The following conditions must be satisfied at the boundary.

$$\partial \Omega_u \cap \partial \Omega_\tau = 0 \quad \text{and} \quad \partial \Omega_{v_L} \cap \partial \Omega_p = 0$$

2.3.2 Hypotheses

The spatial variation of density and porosity is neglected assuming incompressible soil grains. In addition, the flow is considered laminar and stationary in the slow velocity regime. Hence, the interaction force between solid and liquid phases (i.e. drag force $\vec{f}_L^d$, term in Eq. 2.51 is governed by Darcy's law (Eq. 2.63).

$$\vec{f}_L^d = \frac{n_L \mu_L}{\kappa_L} (\vec{v}_L - \vec{v}_S) \tag{2.63}$$

This hypothesis can be controversial in high velocity flows where drag forces can become non-linear. In Eq. 2.63, μ_L is the dynamic viscosity of the liquid and κ_L is the liquid intrinsic permeability, which are assumed constant throughout the simulation. The isotropic intrinsic permeability of the liquid κ_L can also be expressed in terms of Darcy permeability k_L (Eq. 2.64).

$$\kappa_L = k_L \frac{\mu_L}{\rho_L g} \tag{2.64}$$

In the framework of the two-phase single-point approach, the solid mass conservation is automatically fulfilled because the solid mass remains constant in each MP. However, this condition is not naturally satisfied for the liquid as the liquid can move apart from the solid skeleton depending on solid volumetric strain changes (porosity changes). Consequently, the liquid mass in MPs can change and the conservation of the liquid mass is totally controlled by the accuracy in which the liquid mass balance is solved. Fluxes due to spatial variations of liquid mass are neglected in the two-phase single-point formulation ($\nabla n_L \rho_L \approx 0$). Hence, the total mass balance results in Eq. 2.62, and describes the volumetric strain rate of the liquid phase. This hypothesis is reasonable when gradients of porosity are relatively small, but can induce errors when two materials with very different porosity are in contact. In addition, to obtain Eqs. 2.62 and 2.64, the liquid is assumed to be weakly compressible, and shear stresses in the liquid phase are neglected.

2.3.3 Numerical implementation

Since each MP represents a portion of the saturated porous medium V_{MP} and the mass is concentrated at the corresponding MP, the density of liquid ρ_L, solid ρ_S and mixture ρ_m can be expressed as Eqs. 2.65 to 2.67.

$$\rho_L (\vec{x}, t) = \sum_{MP=1}^{n_{MP}} m_L^{MP} \delta (\vec{x} - \vec{x}_{MP}) \tag{2.65}$$

$$\rho_S (\vec{x}, t) = \sum_{MP=1}^{n_{MP}} m_S^{MP} \delta (\vec{x} - \vec{x}_{MP}) \tag{2.66}$$

$$\rho_m (\vec{x}, t) = \sum_{MP=1}^{n_{MP}} m_m^{MP} \delta (\vec{x} - \vec{x}_{MP}) \tag{2.67}$$

where $m_{\mathrm{L}}^{\mathrm{MP}} = n_{\mathrm{L}}\rho_{\mathrm{L}}V_{\mathrm{MP}}$, $m_{\mathrm{S}}^{\mathrm{MP}} = n_{\mathrm{S}}\rho_{\mathrm{S}}V_{\mathrm{MP}}$, and $m_{\mathrm{m}}^{\mathrm{MP}} = \rho_{\mathrm{m}}V_{\mathrm{MP}}$ are the liquid, solid and total mass of a MP. δ is the Dirac delta function and the superscript *MP* indicates that a variable is referring to a MP.

Similarly to the one-phase MPM approach, the momentum balance equations (Eqs. 2.51 and 2.50) are discretised in space by means of the Galerkin method considering standard nodal shape functions (Eq. 2.17). Assuming quadratures at the MPs by using Eqs. 2.65 to 2.67, the final system of equations can be written in a compact form (Eqs. 2.68 and 2.69).

$$\widetilde{\boldsymbol{M}}_{\mathrm{L}}\vec{a}_{\mathrm{L}} = \vec{f}_{\mathrm{L}}^{\,\mathrm{ext}} - \vec{f}_{\mathrm{L}}^{\,\mathrm{int}} - \boldsymbol{Q}_{\mathrm{L}}(\vec{v}_{\mathrm{L}} - \vec{v}_{\mathrm{S}}) \tag{2.68}$$

$$\boldsymbol{M}_S\vec{a}_{\mathrm{S}} + \boldsymbol{M}_{\mathrm{L}}\vec{a}_{\mathrm{L}} = \vec{f}^{\,\mathrm{ext}} - \vec{f}^{\,\mathrm{int}} \tag{2.69}$$

where $\vec{a}_{\mathrm{L}}$, $\vec{a}_{\mathrm{S}}$, $\vec{v}_{\mathrm{L}}$, and $\vec{v}_{\mathrm{S}}$ are the nodal acceleration and velocity vectors for both phases. $\boldsymbol{M}_{\mathrm{S}}$, $\boldsymbol{M}_{\mathrm{L}}$, and $\widetilde{\boldsymbol{M}}_{\mathrm{L}}$ are liquid and solid lumped mass matrices; $\vec{f}_{\mathrm{L}}^{\,\mathrm{ext}}, \vec{f}_{\mathrm{L}}^{\,\mathrm{int}}, \vec{f}^{\,\mathrm{ext}}, \vec{f}^{\,\mathrm{int}}$ are internal and external nodal force vectors of the liquid phase and the mixture, and $\boldsymbol{Q}_{\mathrm{L}}$ is the drag force matrix. The expression for all matrices and force vectors are given in Eqs. 2.70 to 2.77.

$$\widetilde{\boldsymbol{M}}_{\mathrm{L}} \approx \sum_{\mathrm{MP}=1}^{n_{\mathrm{MP}}} \widetilde{m}_{\mathrm{L}}^{\mathrm{MP}}\boldsymbol{N} \tag{2.70}$$

$$\boldsymbol{M}_{\mathrm{L}} \approx \sum_{\mathrm{MP}=1}^{n_{\mathrm{MP}}} m_{\mathrm{L}}^{\mathrm{MP}}\boldsymbol{N} \tag{2.71}$$

$$\boldsymbol{M}_{\mathrm{S}} \approx \sum_{\mathrm{MP}=1}^{n_{\mathrm{MP}}} m_{\mathrm{S}}^{\mathrm{MP}}\boldsymbol{N} \tag{2.72}$$

$$\boldsymbol{Q}_{\mathrm{L}} \approx \sum_{\mathrm{MP}=1}^{n_{\mathrm{MP}}} \boldsymbol{N}^{\mathrm{T}}\frac{n_{\mathrm{L}}^{\mathrm{MP}}\mu_{\mathrm{L}}^{\mathrm{MP}}}{\kappa_{\mathrm{L}}^{\mathrm{MP}}}\boldsymbol{N}V_{\mathrm{MP}} \tag{2.73}$$

$$\vec{f}_{\mathrm{L}}^{\,\mathrm{ext}} \approx \int_{\partial\Omega^{p_{\mathrm{L}}}} \boldsymbol{N}^{\mathrm{T}}\vec{\bar{p}}_{\mathrm{L}}\mathrm{d}\partial\Omega^{p_{\mathrm{L}}}\bigg|_{\mathrm{MP}} + \sum_{\mathrm{MP}=1}^{n_{\mathrm{NP}}} m_{\mathrm{L}}^{\mathrm{MP}}\boldsymbol{N}^{\mathrm{T}}\vec{g} \tag{2.74}$$

$$\vec{f}^{\,\mathrm{ext}} \approx \int_{\partial\Omega^{\tau}} \boldsymbol{N}^{\mathrm{T}}\vec{\tau}\mathrm{d}\partial\Omega^{\tau}\bigg|_{\mathrm{MP}} + \sum_{\mathrm{MP}=1}^{n_{\mathrm{MP}}} m_{\mathrm{m}}^{\mathrm{MP}}\boldsymbol{N}^{\mathrm{T}}\vec{g} \tag{2.75}$$

$$\vec{f}_{\mathrm{L}}^{\,\mathrm{int}} \approx \sum_{\mathrm{MP}=1}^{n_{\mathrm{MP}}} \boldsymbol{B}^{\mathrm{T}}p_{\mathrm{L}}^{\mathrm{MP}}\vec{m}V_{\mathrm{MP}} \tag{2.76}$$

$$\vec{f}^{\,\mathrm{int}} \approx \sum_{\mathrm{MP}=1}^{n_{\mathrm{MP}}} \boldsymbol{B}^{\mathrm{T}}\vec{\sigma}_{\mathrm{MP}}V_{\mathrm{MP}} \tag{2.77}$$

where $\widetilde{m}_{\mathrm{L}}^{\mathrm{MP}} = \rho_{\mathrm{L}}V_{\mathrm{MP}}$, $\vec{\tau}$ is the prescribed traction vector, $\vec{\bar{p}}_{\mathrm{L}}$ is the prescribed liquid pressure, $\boldsymbol{N}$ is the matrix of nodal shape functions evaluated at local

MP positions, $\boldsymbol{B}$ is the matrix that includes the gradients of the nodal shape functions, and $\vec{m} = (1\ 1\ 1\ 0\ 0\ 0)^{\mathrm{T}}$.

While the momentum balances (Eqs. 2.68 and 2.69) are discretised and solved at the nodes of the mesh, the rest of the governing equations such as mass balances (Eqs. 2.59 and 2.60) and constitutive equations (Eqs. 2.61 and 2.62) are posed locally at the MPs to update secondary variables.

2.3.4 Computational cycle

The MPM solution scheme for each time step can be summarised as follows.

1. Liquid nodal acceleration $\vec{a}_{\mathrm{L}}$ is calculated by solving momentum balance of the liquid (Eq. 2.68).

2. $\vec{a}_{\mathrm{L}}$ is used to obtain the nodal acceleration of the solid $\vec{a}_{\mathrm{S}}$ from the momentum balance of the mixture (Eq. 2.69).

3. Velocities and momentum of the MPs are updated from nodal accelerations of each phase.

4. Nodal velocities are then calculated from nodal momentum and used to compute the strain rate at the MP location.

5. Constitutive laws of liquid and soil (Eqs. 2.62 and 2.61) give the increment of excess pore pressure and effective stress respectively.

6. Displacement and position of each MP are updated according to the updated velocity of the solid phase.

7. Porosity of each MP is updated (Eq. 2.60) according to volumetric deformation of the solid skeleton.

8. Nodal values are discarded, the MPs carry all the updated information, and the computational grid is initialised for the next time step.

2.4 Three-phase single-point formulation

Unsaturated porous media consist of a combination of solid skeleton and pore fluids, i.e. liquid and gas. The three-phase single-point MPM approach assumes that all three phases can be represented with a unique continuum (Fig. 2.8). This formulation was presented by Yerro et al. [333, 334], and can be understood as the extension of the two-phase single-point formulation (Section 2.3). Within this framework, every MP represents a portion of the

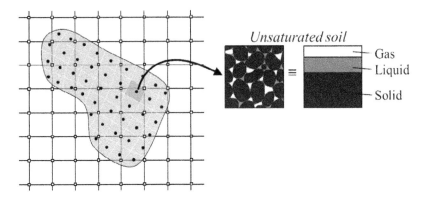

FIGURE 2.8: Scheme of three-phase single-point MPM approach (modified from [334]).

solid-liquid-gas mixture. Hence, the volume V_{MP} of each MP can be computed as the summation of solid, liquid and gas volumes (Eq. 2.78).

$$V_{\mathrm{MP}} = V_{\mathrm{S}}^{\mathrm{MP}} + V_{\mathrm{L}}^{\mathrm{MP}} + V_{\mathrm{G}}^{\mathrm{MP}} \tag{2.78}$$

where the subscript S denotes that a variable refers to the solid phase, L refers to the liquid phase and G refers to the gas phase.

MPs move attached to the solid skeleton carrying water and gas properties as local variables. Therefore, only the movement of the solid phase is described in a Lagrangian manner. The solid mass of each MP remains constant and is automatically conserved throughout the calculation, while the mass of liquid and gas can change as a result of in- or outflow of liquid and gas.

The three-phase single-point formulation has been applied to study wetting problems and rainfall effects on the stability of slopes [333,334] and a practical example is given in Chapter 12.

2.4.1   Governing equations

To simulate the dynamic behaviour of unsaturated porous media, momentum balances, mass balances, and constitutive equations for the three phases must be integrated in the formulation.

Conservation of momentum

The dynamic momentum balances of gas, liquid, and mixture are the main governing equations with the accelerations of each phase being primary unknowns, i.e. $\vec{a}_{\mathrm{S}}$, $\vec{a}_{\mathrm{L}}$ and $\vec{a}_{\mathrm{G}}$ (Eqs. 2.79 to 2.81).

$$\rho_{\mathrm{G}}\vec{a}_{\mathrm{G}} = \nabla p_{\mathrm{G}} - \vec{f}_{\mathrm{G}}^{\mathrm{d}} + \rho_{\mathrm{G}}\vec{g} \tag{2.79}$$

$$\rho_{\mathrm{L}}\vec{a}_{\mathrm{L}} = \nabla p_{\mathrm{L}} - \vec{f}_{\mathrm{L}}^{\mathrm{d}} + \rho_{\mathrm{L}}\vec{g} \tag{2.80}$$

$$n_S \rho_S \vec{a}_S + n_L \rho_L \vec{a}_L + n_G \rho_G \vec{a}_G = \text{div}(\boldsymbol{\sigma}) + \rho_m \vec{g} \tag{2.81}$$

where n_S, n_L and n_G are the volumetric concentration ratio of solid, liquid and gas; ρ_S, ρ_L, ρ_G are densities of all phases and ρ_m is density of the mixture ($\rho_m = n_S \rho_S + n_L \rho_L + n_G \rho_G$); $\vec{f}_G^d$ and $\vec{f}_L^d$ are gas and liquid drag forces; and p_G and p_L are gas and liquid pressures. Within this framework, the porosity of the solid skeleton (n) can be written as $n = n_L + n_G$, and the volumetric concentration ratio of liquid and gas can be expressed in terms of degree of saturation (S_r) and porosity as $n_L = S_r n$ and $n_G = (1 - S_r)n$.

Conservation of mass

Considering a reference volume of the unsaturated porous medium, the mass conservation of each phase (ph) is considered (Eq. 2.82).

$$\frac{d(n_{ph}\rho_{ph})}{dt} + \text{div}(n_{ph}\rho_{ph}\vec{v}_{ph}) = 0 \tag{2.82}$$

Following the same procedure explained in Section 2.3.1, Eq. 2.83 is derived from the solid mass conservation to describe porosity n variations ($n = n_L + n_G$) (Eq. 2.83).

$$\frac{D^S n}{Dt} = n_S \text{div}(\vec{v}_S) \tag{2.83}$$

Including Eq. 2.83 into liquid and gas mass balances gives Eqs. 2.84 and 2.85.

$$\frac{D^S (\rho_L n_L)}{Dt} = (\vec{v}_S - \vec{v}_L)\nabla(n_L\rho_L) - n_L\rho_L\text{div}(\vec{v}_L) \tag{2.84}$$

$$\frac{D^S (\rho_G n_G)}{Dt} = (\vec{v}_S - \vec{v}_G)\nabla(n_G\rho_G) - n_G\rho_G\text{div}(\vec{v}_G) \tag{2.85}$$

Eqs. 2.84 and 2.85 can be expressed as Eqs. 2.86 and 2.87 by taking into account the definitions of liquid and gas volumetric concentration ratios in terms of porosity and degree of saturation, including Eq. 2.83, and rearranging terms.

$$n\frac{D^S (\rho_L S_r)}{Dt} = \text{div}\left[\rho_L n S_r(\vec{v}_S - \vec{v}_L)\right] - \rho_L S_r\text{div}(\vec{v}_S) \tag{2.86}$$

$$n\frac{D^S (\rho_G(1 - S_r))}{Dt} = \text{div}\left[\rho_L n(1 - S_r)(\vec{v}_S - \vec{v}_G)\right] - \rho_L(1 - S_r)\text{div}(\vec{v}_S) \tag{2.87}$$

Finally, the material derivatives from previous expressions are commonly solved assuming gas and liquid pressures as state variables, which yields to the following system of equations, in which the rates of p_L and p_G are primary unknowns (Eqs. 2.88 and 2.89).

$$n\frac{d(\rho_L S_r)}{dp_L}\frac{D^S p_L}{Dt} + n\frac{d(\rho_L S_r)}{dp_G}\frac{D^S p_G}{Dt} = \\ \text{div}\left[\rho_L n S_r(\vec{v}_S - \vec{v}_L)\right] - \rho_L S_r\text{div}(\vec{v}_S) \tag{2.88}$$

$$n\frac{\mathrm{d}\left(\rho_{\mathrm{G}}(1-S_{\mathrm{r}})\right)}{\mathrm{d}p_{\mathrm{L}}}\frac{\mathrm{D}^{\mathrm{S}}p_{\mathrm{L}}}{\mathrm{D}t}+n\frac{\mathrm{d}\left(\rho_{\mathrm{G}}(1-S_{\mathrm{r}})\right)}{\mathrm{d}p_{\mathrm{G}}}\frac{\mathrm{D}^{\mathrm{S}}p_{\mathrm{G}}}{\mathrm{D}t}=$$
$$\mathrm{div}\left[\rho_{\mathrm{L}}n(1-S_{\mathrm{r}})(\vec{v}_{\mathrm{S}}-\vec{v}_{\mathrm{G}})\right]-\rho_{\mathrm{L}}(1-S_{\mathrm{r}})\mathrm{div}(\vec{v}_{\mathrm{S}}) \tag{2.89}$$

Constitutive relation

Mechanical constitutive relationships must be considered as well. In unsaturated soils, two stress variables can be used to capture the soil behaviour, e.g. net stress $\vec{\sigma}_{\mathrm{net}}$ and suction s . The incremental stress-strain equation becomes Eq. 2.90.

$$\frac{\mathrm{D}^{\mathrm{S}}\vec{\sigma}_{\mathrm{net}}}{\mathrm{D}t}=\boldsymbol{D}^{\mathrm{ep}}\frac{\mathrm{D}^{\mathrm{S}}\vec{\varepsilon}}{\mathrm{D}t}+\vec{h}'\frac{\mathrm{D}^{\mathrm{S}}s}{\mathrm{D}t} \tag{2.90}$$

where $\boldsymbol{D}^{\mathrm{ep}}$ is the tangent stiffness matrix, $\vec{h}'$ is a constitutive vector. Both are defined by the constitutive model.

Finally, hydraulic constitutive equations, such as the water retention curve, the variation of liquid permeability with respect to the degree of saturation, or the variation of fluid densities with respect to liquid and gas pressures are also taken into account to calculate derivatives from Eqs. 2.88 and 2.89 and update corresponding properties.

Boundary conditions

The boundary conditions required in this formulation are similar to those described in Section 2.3.1 for the two-phase approach, but adding the corresponding terms for the gas phase.

2.4.2   Hypotheses

This approach considers that gas and liquid flows are laminar and stationary in slow velocity regime. Hence, the liquid drag force $\vec{f}_{\mathrm{L}}^{\mathrm{d}}$ can be expressed by Eq. 2.63. Similarly, the gas drag force $\vec{f}_{\mathrm{G}}^{\mathrm{d}}$ can be expressed with Eq. 2.91.

$$\vec{f}_{\mathrm{G}}^{\mathrm{d}}=\frac{n_{\mathrm{G}}\mu_{\mathrm{G}}}{\kappa_{\mathrm{G}}}\left(\vec{v}_{\mathrm{G}}-\vec{v}_{\mathrm{S}}\right) \tag{2.91}$$

where μ_{G} is the dynamic viscosity of the gas and κ_{G} the gas intrinsic permeability.

In the three-phase MPM framework, like in the two-phase single-point formulation, the solid mass remains constant in each MP throughout the calculation, and as a result solid mass is automatically conserved. However, liquid and gas can leave the solid skeleton. Hence, the conservation of both fluid masses fully depends on liquid and gas momentum balances. In contrast to the two-phase single-point approach, here fluxes due to spatial variations of liquid and gas mass are not neglected, and all terms from mass balances are taken into account. In unsaturated soils, the gradients of liquid and gas are governed by the porosity of the solid skeleton and the degree of saturation S_{r}. While porosity changes can be assumed to be relatively small in many problems, gradients of degree of saturation can become extremely high (depending on the water retention curve). In order to solve the material derivatives in Eqs. 2.84 and 2.85, liquid and gas pressures are considered to be state variables, and constitutive relationships are taken into account to determine partial derivatives with respect to p_L and p_G. Liquid and gas are generally assumed to be weakly compressible, and shear stresses in both fluid phases are neglected.

2.4.3 Numerical implementation

Momentum balance equations (Eqs. 2.79 to 2.81) are discretised following the same procedure as in Section 2.2.2, which gives Eqs. 2.92 to 2.94, respectively.

$$\boldsymbol{M}_\mathrm{G}\vec{a}_\mathrm{G} = \vec{f}_\mathrm{G}^{\,\mathrm{ext}} - \vec{f}_\mathrm{G}^{\,\mathrm{int}} - \boldsymbol{Q}_\mathrm{G}(\vec{v}_\mathrm{G} - \vec{v}_\mathrm{S}) \tag{2.92}$$

$$\boldsymbol{M}_\mathrm{L}\vec{a}_\mathrm{L} = \vec{f}_\mathrm{L}^{\,\mathrm{ext}} - \vec{f}_\mathrm{L}^{\,\mathrm{int}} - \boldsymbol{Q}_\mathrm{L}(\vec{v}_\mathrm{L} - \vec{v}_\mathrm{S}) \tag{2.93}$$

$$\widetilde{\boldsymbol{M}}_\mathrm{S}\vec{a}_\mathrm{S} + \widetilde{\boldsymbol{M}}_\mathrm{L}\vec{a}_\mathrm{L} + \widetilde{\boldsymbol{M}}_\mathrm{G}\vec{a}_\mathrm{G} = \vec{f}^{\,\mathrm{ext}} - \vec{f}^{\,\mathrm{int}} \tag{2.94}$$

where $\vec{a}_\mathrm{G}$, $\vec{a}_\mathrm{L}$, $\vec{a}_\mathrm{S}$, $\vec{v}_\mathrm{G}$, $\vec{v}_\mathrm{L}$, and $\vec{v}_\mathrm{S}$ are the nodal acceleration and velocity vectors for each phase. $\boldsymbol{M}_\mathrm{G}$, $\boldsymbol{M}_\mathrm{L}$, $\widetilde{\boldsymbol{M}}_\mathrm{G}$, $\widetilde{\boldsymbol{M}}_\mathrm{L}$, and $\widetilde{\boldsymbol{M}}_\mathrm{S}$ are gas, liquid, and solid lumped mass matrices. $\vec{f}_\mathrm{G}^{\,\mathrm{ext}}$, $\vec{f}_\mathrm{G}^{\,\mathrm{int}}$, $\vec{f}_\mathrm{L}^{\,\mathrm{ext}}$, $\vec{f}_\mathrm{L}^{\,\mathrm{int}}$, $\vec{f}^{\,\mathrm{ext}}$, $\vec{f}^{\,\mathrm{int}}$ are internal and external nodal force vectors of the gas and liquid phase, and the mixture, and $\boldsymbol{Q}_\mathrm{G}$ and $\boldsymbol{Q}_\mathrm{L}$ are the drag force matrices. The expressions for all matrices and force vectors are given in Eqs. 2.95 to 2.107.

$$\boldsymbol{M}_\mathrm{G} \approx \sum_{\mathrm{MP}=1}^{n_\mathrm{MP}} \widetilde{m}_\mathrm{G}^{\mathrm{MP}} \boldsymbol{N} \tag{2.95}$$

$$\boldsymbol{M}_\mathrm{L} \approx \sum_{\mathrm{MP}=1}^{n_\mathrm{MP}} \widetilde{m}_\mathrm{L}^{\mathrm{MP}} \boldsymbol{N} \tag{2.96}$$

$$\widetilde{\boldsymbol{M}}_\mathrm{G} \approx \sum_{\mathrm{MP}=1}^{n_\mathrm{MP}} m_\mathrm{G}^{\mathrm{MP}} \boldsymbol{N} \tag{2.97}$$

$$\widetilde{\boldsymbol{M}}_\mathrm{L} \approx \sum_{\mathrm{MP}=1}^{n_\mathrm{MP}} m_\mathrm{L}^{\mathrm{MP}} \boldsymbol{N} \tag{2.98}$$

$$\widetilde{\boldsymbol{M}}_\mathrm{S} \approx \sum_{\mathrm{MP}=1}^{n_\mathrm{MP}} m_\mathrm{S}^{\mathrm{MP}} \boldsymbol{N} \tag{2.99}$$

$$\boldsymbol{Q}_\mathrm{G} \approx \sum_{\mathrm{MP}=1}^{n_\mathrm{MP}} \boldsymbol{N}^\mathrm{T} \frac{n_\mathrm{G}^{\mathrm{MP}} \mu_\mathrm{G}^{\mathrm{MP}}}{\kappa_\mathrm{G}^{\mathrm{MP}}} \boldsymbol{N} V_\mathrm{MP} \tag{2.100}$$

$$\boldsymbol{Q}_\mathrm{L} \approx \sum_{\mathrm{MP}=1}^{n_\mathrm{MP}} \boldsymbol{N}^\mathrm{T} \frac{n_\mathrm{L}^{\mathrm{MP}} \mu_\mathrm{L}^{\mathrm{MP}}}{\kappa_\mathrm{L}^{\mathrm{MP}}} \boldsymbol{N} V_\mathrm{MP} \tag{2.101}$$

$$\vec{f}_{\mathrm{G}}^{\,\mathrm{ext}} \approx \int_{\partial\Omega^{p_{\mathrm{G}}}} \boldsymbol{N}^{\mathrm{T}}\vec{\bar{p}}_{\mathrm{G}}\mathrm{d}\partial\Omega^{p_{\mathrm{G}}}\bigg|_{\mathrm{MP}} + \sum_{\mathrm{MP}=1}^{n_{\mathrm{MP}}} \boldsymbol{N}^{\mathrm{T}}\widetilde{m}_{\mathrm{G}}^{\mathrm{MP}}\vec{g} \tag{2.102}$$

$$\vec{f}_{\mathrm{L}}^{\,\mathrm{ext}} \approx \int_{\partial\Omega^{p_{\mathrm{L}}}} \boldsymbol{N}^{\mathrm{T}}\vec{\bar{p}}_{\mathrm{L}}\mathrm{d}\partial\Omega^{p_{\mathrm{L}}}\bigg|_{\mathrm{MP}} + \sum_{\mathrm{MP}=1}^{n_{\mathrm{MP}}} \boldsymbol{N}^{\mathrm{T}}\widetilde{m}_{\mathrm{L}}^{\mathrm{MP}}\vec{g} \tag{2.103}$$

$$\vec{f}^{\,\mathrm{ext}} \approx \int_{\partial\Omega^{\vec{\tau}}} \boldsymbol{N}^{\mathrm{T}}\vec{\tau}\mathrm{d}\partial\Omega^{\vec{\tau}}\bigg|_{\mathrm{MP}} + \sum_{\mathrm{MP}=1}^{n_{\mathrm{MP}}} \boldsymbol{N}^{\mathrm{T}}m_{\mathrm{MP}}\vec{g} \tag{2.104}$$

$$\vec{f}_{\mathrm{G}}^{\,\mathrm{int}} \approx \sum_{\mathrm{MP}=1}^{n_{\mathrm{MP}}} \boldsymbol{B}^{\mathrm{T}}p_{\mathrm{G}}^{\mathrm{MP}}\vec{m}V_{\mathrm{MP}} \tag{2.105}$$

$$\vec{f}_{\mathrm{L}}^{\,\mathrm{int}} \approx \sum_{\mathrm{MP}=1}^{n_{\mathrm{MP}}} \boldsymbol{B}^{\mathrm{T}}p_{\mathrm{L}}^{\mathrm{MP}}\vec{m}V_{\mathrm{MP}} \tag{2.106}$$

$$\vec{f}^{\,\mathrm{int}} \approx \sum_{\mathrm{MP}=1}^{n_{\mathrm{MP}}} \boldsymbol{B}^{\mathrm{T}}\boldsymbol{\sigma}_{\mathrm{MP}}V_{\mathrm{MP}} \tag{2.107}$$

It is important to highlight that the gradients of liquid and gas masses ($\nabla(n_{\mathrm{L}}\rho_{\mathrm{L}})$ and $\nabla(n_{\mathrm{G}}\rho_{\mathrm{G}})$) require solving the mass balance equations at the MPs (Eqs. 2.84 and 2.85). In this approach, interpolated liquid and gas masses at the nodes of the computational mesh are used to compute such quantities at the MPs.

2.4.4   Computational cycle

The solution is integrated explicitly in time. The computational cycle for each time step and can be summarised as follows.

1. Gas and liquid nodal accelerations $\vec{a}_{\mathrm{G}}$ and $\vec{a}_{\mathrm{L}}$ are calculated by solving Eqs. 2.92 and 2.93.

2. $\vec{a}_{\mathrm{G}}$ and $\vec{a}_{\mathrm{L}}$ are used to obtain the nodal acceleration of the solid $\vec{a}_{\mathrm{S}}$ from Eq. 2.94.

3. Velocities and momentum of the MPs are updated from nodal accelerations of each phase.

4. Nodal velocities are then calculated from nodal momentums and used to compute the strain rate at the MP location.

5. Liquid and gas pore pressure rates at the MPs are calculated by solving liquid and gas mass balances (Eqs. 2.88 and 2.89).

6. The increment of constitutive stresses is calculated applying the constitutive law of the soil (Eq. 2.90).

7. Displacement and position of each MP are updated according to the velocity of the solid phase.

8. Additional properties (e.g. porosity, degree of saturation, permeability) are updated at the MPs according to hydraulic constitutive relationships.

9. Nodal values are discarded, the MPs carry all the updated information, and the computational grid is initialised for the next time step.

2.5 Two-phase double-point formulation

The basis of the two-phase double-point formulation follows from mixture theory extended by the concept of volume fractions [304] to model the interaction between solid and liquid phases. It postulates that each point in space of a body is simultaneously occupied by a finite number of particles, one for each component of the mixture. In this way, the saturated body can be represented as a superposition of two continua, i.e. solid skeleton and pore fluid, following their own movement with the restriction imposed by the interaction between phases. The two-phase double-point formulation was first proposed within the MPM framework [3, 20, 192, 328].

In this approach, the solid skeleton and the liquid phase are represented separately by two sets of Lagrangian MPs (Fig. 2.9) – solid material points (SMPs) and liquid material points (LMPs). While SMPs move attached to the solid skeleton, LMPs follow the liquid motion, both carrying properties from their respective phase. As a result, the required number of MPs to discretise a saturated porous domain increases substantially compared to the two-phase single-point formulation.

An important advantage of this approach compared to the two-phase single-point formulation is that the mass of all MPs (i.e. LMPs and SMPs) remains constant. Therefore, the conservation of both solid and liquid mass is fulfilled throughout the calculation. Another important feature of the double-point formulation is that LMPs embody either liquid within the pores or free surface liquid. In this framework three possible subdomains can be identified (Fig. 2.9) as follows.

- Porous medium in saturated conditions, when SMPs and LMPs share the same grid element.

- Porous medium in dry conditions, when only SMPs are located in the grid element.

- Free surface liquid, when only LMPs are located in the grid element.

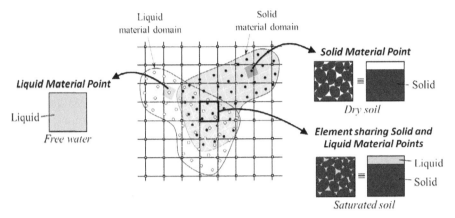

FIGURE 2.9: Scheme of two-phase double-point MPM approach.

The two-phase double-point formulation has been applied to simulate the submerged column collapse [194], the fluidisation of a vertical column test [39], the interaction between water jet and soil bed [345], the simulation of crater development around a damaged pipeline [193], and a dyke failure [195]. In addition, a comparison between the two-phase single-point and double-point formulations is presented in Ceccato et al. [62].

2.5.1 Governing equations

The governing equations of the two-phase double-point approach comprise the momentum and mass balances of both solid and liquid phases with their respective constitutive equations. Note that solid momentum balance is considered for convenience instead of momentum balance of the mixture. Similarly to the two-phase single-point approach, all dynamic terms are considered and acceleration of both phases ($\vec{a}_S$ and $\vec{a}_L$) are the primary unknowns.

Conservation of momentum

The liquid and solid momentum balances per unit of saturated porous media are given in Eqs. 2.108 and 2.109.

$$n_L \rho_L \vec{a}_L = \text{div}(\bar{\boldsymbol{\sigma}}_L) - \vec{f}_L^{\,d} + n_L \rho_L \vec{g} \qquad (2.108)$$

$$n_S \rho_S \vec{a}_S = \text{div}(\bar{\boldsymbol{\sigma}}_S) + \vec{f}_L^{\,d} + n_S \rho_S \vec{g} \qquad (2.109)$$

where $\bar{\boldsymbol{\sigma}}_S = \boldsymbol{\sigma}' - n_S \boldsymbol{\sigma}_L$ and $\bar{\boldsymbol{\sigma}}_L = n_L \boldsymbol{\sigma}_L$ correspond to the partial stresses for solid and liquid phases respectively, $\boldsymbol{\sigma}'$ is the effective stress tensor, and $\boldsymbol{\sigma}_L$ is the stress tensor of the liquid phase (equivalent to pore pressure p_L in saturated porous media).

Conservation of mass

The solid and liquid mass balances are also taken into account (Eqs. 2.52 and 2.53 respectively). Following the same procedure explained in Section 2.3.1, Eq. 2.110 is derived from the solid mass conservation to describe n_L variations. The storage equation is also derived leading to Eq. 2.111.

$$\frac{\mathrm{D}^S n_L}{\mathrm{D}t} = n_S \mathrm{div}(\vec{v}_S) \tag{2.110}$$

$$\frac{\mathrm{D}^L \varepsilon_{\mathrm{vol},L}}{\mathrm{D}t} = \frac{1}{n_L} \left[n_S \mathrm{div}(\vec{v}_S) + n_L \mathrm{div}(\vec{v}_L) + (\vec{v}_L - \vec{v}_S)\nabla n_L \right] \tag{2.111}$$

The volumetric strain rate depends on three terms. The first two are consistent with Eq. 2.59, but the third one which is a function of the relative velocity and the gradient of solid concentration ratio is usually neglected in the two-phase single-point formulation for small variation of the porosity throughout the domain. However, the third term in the two-phase double-point formulation becomes crucial because the variations of solid concentration ratio between free surface water and the porous medium becomes significantly large. $\mathrm{D}^L(\bullet)/\mathrm{D}t$ denotes the material time derivative with respect to the liquid phase.

Constitutive relation

Constitutive equations are required to update stresses in SMPs and LMPs. Eq. 2.61 is considered to model the behaviour of solid skeleton (in SMPs), and Eq. 2.112 to update volumetric stresses of the liquid (equivalent to p_L) (in LMPs). In addition, when LMPs represent pure liquid or fluidised mixture, the deviatoric part of the stress tensor $\vec{\sigma}_{\mathrm{dev},L}$ is also accounted for (Eq. 2.113).

$$\frac{\mathrm{D}^L \sigma_{\mathrm{vol},L}}{\mathrm{D}t} = K_L \frac{\mathrm{D}^L \varepsilon_{\mathrm{vol},L}}{\mathrm{D}t} \tag{2.112}$$

$$\sigma_{\mathrm{dev},L} = 2\mu_L \frac{\mathrm{D}^L \varepsilon_{\mathrm{vol},L}}{\mathrm{D}t} \tag{2.113}$$

2.5.2 Hypotheses

The two-phase double-point formulation comprises two essential differences compared to the single-point approach. Both are related to the fact that the behaviour of the continuum described in the double-point framework can vary from a dry porous medium to pure liquid. This leads to extreme changes in flow regime and huge gradients of volumetric concentration ratios in transition zones.

The first difference is that the drag force $\vec{f}_L^{\mathrm{d}}$ is generalised and Eq. 2.63 is extended to account for laminar and steady flow in high velocity regime and gradients of porosity (n_L), leading to Eq. 2.114 where β is the non-Darcy flow coefficient [95] and can be computed with Eq. 2.115.

$$\vec{f}_L^d = \frac{n_L^2 \mu_L}{\kappa_L} (\vec{v}_L - \vec{v}_S) + \beta n_L^3 \rho_L |\vec{v}_L - \vec{v}_S| (\vec{v}_L - \vec{v}_S) + \boldsymbol{\sigma}_L \nabla n_L \qquad (2.114)$$

$$\beta = \frac{B}{\sqrt{\kappa_L A n_L^3}} \qquad (2.115)$$

Moreover, the intrinsic permeability κ_L is computed and updated as a function of the effective porosity n_L with the Kozeny-Carman formula [27] as given in Eq. 2.116.

$$\kappa_L = \frac{\frac{D^2}{A} n_L^3}{(1 - n_L)^2} \qquad (2.116)$$

The second difference is given by the storage equation (Eq. 2.111). Now, all convective terms are accounted for, including the spatial variations of liquid volumetric concentration ratio $((\vec{v}_L - \vec{v}_S) \nabla n_L)$. Therefore, the liquid fluxes due to changes in porosity are also taken into account. The liquid is considered as slightly compressible.

2.5.3 Numerical implementation

In the double-point formulation, each set of MPs (i.e. LMPs and SMPs) represents a portion of either liquid or solid phase. The corresponding mass is concentrated at the position of the MP. Therefore, the liquid and solid densities can be written in terms of mass (Eqs. 2.65 and 2.66).

The momentum balance equations (Eqs. 2.109 and 2.108) are discretised at the nodes of the computational mesh and integrated at the MPs following the procedure explained in Section 2.2.2, which leads to the following system of equations.

$$\boldsymbol{M}_L \vec{a}_L = \vec{f}_L^{ext} - \vec{f}_L^{int} - \vec{f}_L^{n} - \boldsymbol{Q}_L(\vec{v}_L - \vec{v}_S) \qquad (2.117)$$

$$\boldsymbol{M}_S \vec{a}_S = \vec{f}_S^{ext} - \vec{f}_S^{int} - \vec{f}_S^{n} + \boldsymbol{Q}_L(\vec{v}_L - \vec{v}_S) \qquad (2.118)$$

where $\vec{a}_S$ and $\vec{a}_L$ are the solid and liquid nodal acceleration, $\vec{M}_S$ and $\vec{M}_L$ the solid and liquid lumped mass matrices. $\vec{f}_L^{ext}$, $\vec{f}_L^{int}$, $\vec{f}_S^{ext}$ and $\vec{f}_S^{int}$ are the internal and external nodal force vectors of the liquid and solid phase, respectively. $\vec{f}_L^{n}$ and $\vec{f}_S^{n}$ are the vectors of solid-liquid interaction forces related to the gradient of the concentration ratio, and $\vec{Q}_L$ is the nodal drag force matrix. The expressions for all matrices and force vectors are given in Eqs. 2.119 to 2.127.

$$\boldsymbol{M}_L \approx \sum_{MP=1}^{n_{LMP}} m_L^{MP} \boldsymbol{N} \qquad (2.119)$$

$$\boldsymbol{M}_S \approx \sum_{MP=1}^{n_{MP}} m_S^{MP} \boldsymbol{N} \qquad (2.120)$$

$$\vec{f}_{L}^{ext} \approx \int_{\partial\Omega^{pL}} \boldsymbol{N}^{T}\vec{\bar{p}}_{L}d\partial\Omega^{pL}\bigg|_{MP} + \sum_{MP=1}^{n_{MP}} m_{L}^{MP}\boldsymbol{N}^{T}\vec{g} \tag{2.121}$$

$$\vec{f}_{S}^{ext} \approx \int_{\partial\Omega^{\tau}} \boldsymbol{N}^{T}\left(\vec{\tau} + n_{S}^{MP}\vec{\bar{p}}_{L}\right)d\partial\Omega^{\tau}\bigg|_{MP} + \sum_{MP=1}^{n_{MP}} m_{S}^{MP}\boldsymbol{N}^{T}\vec{g} \tag{2.122}$$

$$\vec{f}_{L}^{int} \approx \sum_{MP=1}^{n_{MP}} \boldsymbol{B}^{T}\bar{\sigma}_{L}^{MP}V_{MP} \tag{2.123}$$

$$\vec{f}_{S}^{int} \approx \sum_{MP=1}^{n_{MP}} \boldsymbol{B}^{T}\bar{\sigma}_{S}^{MP}V_{MP} \tag{2.124}$$

$$\vec{f}_{L}^{n} \approx \sum_{MP=1}^{n_{MP}} \boldsymbol{N}^{T}\nabla n_{L}^{MP}\sigma_{L}^{MP}V_{MP} \tag{2.125}$$

$$\vec{f}_{S}^{n} \approx \sum_{MP=1}^{n_{MP}} \boldsymbol{N}^{T}\nabla n_{S}^{MP}\sigma_{L}^{MP}V_{MP} \tag{2.126}$$

$$\boldsymbol{Q}_{L} \approx \sum_{MP=1}^{n_{MP}} \boldsymbol{N}^{T}\left(\frac{(n_{L}^{MP})^{2}\mu_{L}^{MP}}{\kappa_{L}^{MP}} + \beta(n_{L}^{MP})^{3}\rho_{L}^{MP}\left|\vec{v}_{L} - \vec{v}_{S}\right|\right)\boldsymbol{N}V_{MP} \tag{2.127}$$

Note that $\vec{\tau}$ is the prescribed traction vector, $\vec{\bar{p}}_{L}$ is the prescribed liquid pressure, $\boldsymbol{N}$ is the matrix of nodal shape functions, $\boldsymbol{B}$ is the matrix that includes the gradients of the nodal shape functions.

One numerical difficulty of this formulation is to calculate the gradient of concentration ratio (∇n_{L}). Note that this gradient is required at the nodes (in the drag force $\vec{f}^{d}$, Eq. 2.114), but also at the MPs (in the mass balance of the mixture, Eq. 2.111). The effective liquid concentration ratio field n_{L} is a discontinuous function at the transition between free surface water and porous medium and is characterised by two constant values: 1 in the free surface water and the value of the porosity n in the porous medium. Martinelli [192] proposed the concept of a transition zone in which an interpolated liquid concentration ratio is introduced which gives a smooth transition of n_{L} between free surface water and groundwater.

2.5.4 Fluidisation, sedimentation and free surface liquid

This formulation can distinguish between mixtures characterised by low and high porosities (Fig. 2.10). Fig. 2.10 a shows a low-porosity mixture, where the grains of the solid skeleton are in contact and the behaviour can be described by constitutive models developed for granular materials (solid-like response). Conversely, as shown in Fig. 2.10 b, in a high-porosity mixture the grains are not in contact and float together with the liquid phase. In this case, the

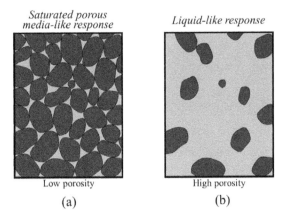

FIGURE 2.10: Behaviour of solid-liquid mixtures: (a) solid-like response (low porosity) as a saturated porous medium and (b) liquid-like response (high porosity) as a free surface liquid.

effective stresses are equal to zero and the response of the mixture is described by the Navier-Stokes equation (liquid-like response) .

In the current formulation, the two previously mentioned states are distinguished through the maximum porosity n_{max} of the SMPs, which is the maximum value of the porosity for a given soil in its loosest state. During the fluidisation process of a fully saturated soil, when the porosity is lower than the maximum porosity ($n_L = 1 - n_S < n_{max}$), the decrease in the mean effective stress results in an increase of the porosity. When the contact forces between the grains vanish, the mean effective stress becomes nil. However, the fluidisation occurs only if the grains are significantly separated, so that the porosity of the SMPs is larger than n_{max}. In the reverse process, i.e. the sedimentation of a fluidised mixture, the porosity decreases due to the fact that the solid grains get closer to each other. However, the effective stresses recur only if the porosity is smaller than n_{max}, i.e. the grains are close enough to be in contact.

Eq. 2.108 is used to describe the behaviour of liquid in the solid-liquid mixture. In case of liquid-like response, the deviatoric part of the stress tensor of the liquid is computed using the liquid strain rate tensor (Eq. 2.113) and a liquid viscosity which takes into account the solid concentration ratio of the mixture. In case of solid-like behaviour the deviatoric stress tensor for the porous liquid is set to zero. In this formulation, all LMPs belonging to the liquid free surface are detected and the liquid stress of such MPs is set to zero.

2.5.5   Computational cycle

The solution is integrated explicitly in time and the computational cycle for each time step can be summarised as follows.

1. Nodal acceleration of the liquid $\vec{a}_L$ is calculated by solving the discretised form of Eq. 2.117.

2. $\vec{a}_L$ is used to obtain the nodal acceleration of the solid $\vec{a}_S$ from the discretised form of Eq. 2.118.

3. Velocities and momentum of the MPs are updated from nodal accelerations of each phase.

4. Nodal velocities are then calculated from nodal momentum and used to compute the strain rate at the MP location.

5. Liquid and soil constitutive laws give the increment of liquid stress and effective stress respectively in LMP and SMP (Eqs. 2.112, 2.113, and 2.61).

6. All LMPs that belong to the liquid free surface are detected and stresses are set to zero.

7. Displacement and position of each MP is updated according to the corresponding velocity field.

8. Nodal values are discarded, the MPs carry all the updated information, and the computational grid is initialised for the next time step.

2.6 Closure

MPM can be used to numerically simulate soil-water-structure interaction problems in history-dependent materials involving large deformations. Herein, soil is considered as a multi-phase porous material. Depending on the field of application either a one-, two- or three-phase MPM formulation is required to properly model the material's physical behaviour. For cases where phase transition (fluidisation and solidification, erosion and deposition) or in- and outflow of groundwater and free surface water play a role the MPM formulation has to be extended by using two sets of MPs (double-point formulation).

3

Different Formulations and Integration Schemes

A. Chmelnizkij and J. Grabe

3.1 Standard material point method

Since the original, or standard, formulation of the material point method (MPM) [285] presented in Chapter 2, there have been many improvements and modifications of the method in an attempt to overcome some numerical drawbacks. These relate to convergence issues and non-physical noise caused by the interpolation between nodes and material points (MPs), which is inherent to the method itself. These are related to the choice of the trail functions and MPs crossing cells in the grid.

Fig. 3.1 shows a schematic description of the linear shape functions at the node i, which have an influence over the interval $[i - 1, i + 1]$. The standard MPM uses linear shape functions for the spatial discretisation. Dhakal et al. [83] found numerical issues for the calculation of a shock propagation in terms of two different constant stress levels to the left σ_L and σ_R to the right of node i. They observed that a low number of MPs per element leads to a non-physical flip of the initial stress distribution and an unrealistic stress accumulation, but a high number of MPs produces oscillations. The reason is that the information of a MP is lumped in one point. Therefore, the information is not distributed in the way of a proper continuum.

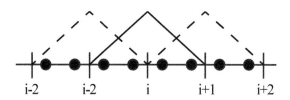

FIGURE 3.1: Discretisation by linear shape functions with 2 MPs/cell.

The oscillations occur due to the computation of the internal forces when using linear shape functions and the gradient of the shape functions. The spatial derivatives of the shape functions (Fig. 3.1) are piecewise constant with a discontinuity at the corresponding central node (Fig. 3.2). The calculation of internal forces contains this gradient of the shape functions and, therefore, a discontinuous function needs to be integrated. The accuracy of such integrations highly depends on the quadrature scheme.

The internal force at the node i can be calculated using the standard MP integration for the internal forces. Considering that the gradient of the shape function has opposite signs in element e and e + 1 as shown in Eq. 3.2 the internal force results in a difference of the stresses of element e and e + 1.

$$f_i^{\text{int}} \approx \sum_{p=1}^{N_p^e} \sigma_p \Omega_p - \sum_{p=1}^{N_p^{e+1}} \sigma_p \Omega_p \qquad (3.1)$$

where N_p^e represents the number of MPs in the eth element, σ_p the stress and Ω_p the volume of the pth MP. Assuming that the system is approximately in equilibrium but the MPs are still slightly moving ($\sigma_p \approx \sigma \ \forall p$) and the volumes of the MPs are similar ($\Omega_p \approx \Omega \ \forall p$), Eq. 3.1 can be rewritten as Eq. 3.2.

$$f_i^{\text{int}} \approx \sigma \Omega (N_p^e - N_p^{e+1}) \qquad (3.2)$$

The internal force in the node i (Eq. 3.2) only depends on the number of MPs in the elements e and e + 1. The force f_i^{int} is zero at equilibrium. If one MP in element e, which is close to node i, crosses the border and enters element e + 1, a sudden change in the internal force of $2\sigma\Omega$ occurs. In general, if the volumes of MPs are kept constant every unequal distribution of MPs in adjacent elements will lead to such problems. The reason is that the volumes of MPs in one element do not correspond to the volume of the element and therefore a wrong weighting in the quadrature formula is used. This is a well-known numerical issue with the standard MPM and it is called grid crossing. There are many approaches to circumvent or mitigate grid crossing issues. The use of higher order nodal shape functions is one approach, which leads to smoother gradients of the shape functions at the nodes. A different approach is to introduce a characteristic function for the MPs, which is non-zero inside a certain interval and represents the volume occupied by the MP, and zero

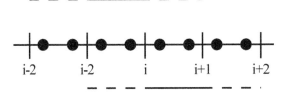

FIGURE 3.2: Piecewise constant gradients of linear shape functions.

elsewhere. This approach allows the MP to cross the grid partially and not at once. Another approach is to modify only the gradients of the shape functions and not the shape functions themselves. These different methods will be discussed in the following sections.

3.2   Generalised interpolation material point method

The generalised interpolation material point method (GIMP) [24] uses a characteristic function $\chi_p(x)$, which is non-zero over a specific interval and zero elsewhere. The interval L_p represents the partition of the total domain Ω occupied by a MP (Fig. 3.3). Therefore, the MP is no longer associated with a point, but rather with an interval as shown in Fig. 3.3.

The characteristic functions should be a partition of unity (i.e. Eq. 3.3) as this property is important to ensure the conservation of physical quantities. For the piece-wise constant characteristic function shown in Fig. 3.3, L_p can be chosen in such a way that it will be a partition of unity for the initial configuration.

$$\sum_p \chi_p(x) = 1 \tag{3.3}$$

Using Eq. 3.3, it is easy to prove that the total mass of the system is conserved. In Eq. 3.4 additionally, the interchangeability of summation and integration is used.

$$M_{\text{tot}} = \sum_{p=1}^{N_p^{\text{tot}}} \int_\Omega \chi_p(x)\rho d\Omega = \int_\Omega \sum_{p=1}^{N_p^{\text{tot}}} \chi_p(x)\rho d\Omega = \int_\Omega \rho d\Omega \tag{3.4}$$

where N_p^{tot} is the total number of MPs and ρ the density of the material.

Other properties of the MP can also be treated as a distributed value in the interval L_p of the characteristic function, e.g. stresses or momentum. This combination of characteristic functions for MP properties and nodal shape functions for the grid interpolation is similar to the distortion of nodal shape

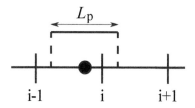

FIGURE 3.3: Characteristic function $\chi_p(x)$ for one MP.

functions in the Petrov-Galerkin method. Fig. 3.4 considers the same configuration as Fig. 3.3, marking the contribution of the MP to the elements. The
MP already contributes its properties to both elements and, therefore, grid
crossing is taking place. In contrast to the standard MPM, the MP is not
associated with just one element.

The contribution of a MP property to an element can be illustrated as
the area under the characteristic function. The dark grey area in Fig. 3.4
represents the contribution of the MP to element e while the light grey area is
associated with the contribution to element $e+1$. The interval L_p represents
the shape of the MP in one dimension. If L_p is kept constant, gaps and overlaps
of characteristic functions can occur due to different movements of MPs. This
might lead to the violation of property Eq. 3.3 and numerical noise at the
nodes as in the standard MPM can occur. Tracking the deformation of MPs is
cumbersome in multi-dimensional problems and contains problems similar to
the element distortion in the classical Lagrangian FEM. The standard MPM
can be derived from GIMP by setting the characteristic function equal to the
Dirac delta function (Eq. 3.5).

$$\chi_p = \delta(x - x_p)V_p \tag{3.5}$$

where V_p represents the volume of the MP, e.g. $V_p = \int_\Omega \chi_p d\Omega$.

In general, an arbitrary MP property f_p can be approximated at any point
by the characteristic functions (Eq. 3.6).

$$f(x) = \sum_{p=1}^{N_p^{tot}} f_p \chi_p(x) \tag{3.6}$$

This representation can be used to replace nodal and element values in the governing equations. Considering the weak formulation of the momentum equation in FEM after spatial discretisation (Eq. 3.7), the stress σ, density ρ and
change of momentum $\dot{\vec{p}}$ can be replaced by Eq. 3.6. The test functions of the
weak formulation are denoted by $\delta \vec{v}$.

$$\int_\Omega \rho \dot{\vec{p}} \, \delta\vec{v} \, d\Omega + \int_\Omega \sigma \, \nabla\delta\vec{v} d\Omega = \int_\Omega \rho \, \vec{b} \, \delta\vec{v} \, d\Omega + \int_{\partial\Omega} \vec{\tau} \, \delta\vec{v} \, d(\partial\Omega) \tag{3.7}$$

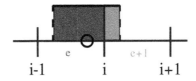

FIGURE 3.4: Treating the grid crossing problem in GIMP.

Eq. 3.7 results in a new conservation equation in terms of material MP values (Eq. 3.8).

$$\sum_p \int_\Omega \frac{\dot{\vec{p}}_p \chi_p}{V_p} \vec{\delta v} \, d\Omega + \sum_p \int_\Omega \sigma_p \chi_p \nabla \vec{\delta v} d\Omega =$$
$$\sum_p \int_\Omega \frac{m_p \chi_p}{V_p} \vec{b} \, \vec{\delta v} d\Omega + \int_{\partial\Omega} \vec{\tau} \, \vec{\delta v} d(\partial\Omega) \tag{3.8}$$

The integrals in Eq. 3.8 are zero outside the support of χ_p. The test functions can be replaced for example by piece-wise linear nodal shape functions as shown in Fig. 3.1 denoted by $N = (N_1, N_2, ..., N_{n_{tot}})$.

3.2.1 Alternative formulation

It is assumed in Section 3.2 that the characteristic function is 1 in a compact support L_p for one-dimensional or Ω_p for multi-dimensional cases, and zero elsewhere (Eq. 3.9).

$$\chi_p(x) = \begin{cases} 1 & \text{if } x \in \Omega_p \\ 0 & \text{else} \end{cases} \tag{3.9}$$

The characteristic function (Eq. 3.9) is a simple example for a function which can be defined as a partition of unity for the initial configuration. This GIMP approach is referred to as contiguous particle GIMP or cpGIMP. The weight function $S_{i,p}$ and its gradient $\nabla S_{i,p}$ are given in Eq. 3.10.

$$S_{i,p} = \frac{1}{V_p} \int_{\Omega_p} \chi_p(x) N_i(x) d\Omega \tag{3.10a}$$

$$\nabla S_{i,p} = \frac{1}{V_p} \int_{\Omega_p} \chi_p(x) \nabla N_i(x) d\Omega \tag{3.10b}$$

Eq. 3.8 can be rewritten as Eq. 3.11 by applying Eq. 3.10 to the disctretised momentum equation and assuming that $\Omega_p \subset \Omega$.

$$\sum_p S_{i,p} \dot{\vec{p}} = -\sum_p \sigma_p \nabla S_{i,p} V_p + \sum_p m_p \vec{b} S_{i,p} + \int_{\partial\Omega} \vec{\tau} \, \vec{\delta v} d(\partial\Omega) \tag{3.11}$$

Normally, the domain Ω_p is time-dependent and it is necessary to update the compact support of $\chi_p(x)$ in each time step. This task can be computationally expensive, especially for multi-dimensional problems. In one-dimensional problems, the current length of the interval L_p can be calculated using the deformation gradient $\boldsymbol{F}$ (Eq. 3.12).

$$L_p = \boldsymbol{F} L_p^0 \tag{3.12}$$

where L_p^0 is the initial length of L_p.

For a simple transfer to a multi-dimensional case, one can assume this deformation for each coordinate. This leads to a cuboid shape of Ω_p and neglects shear deformations. This approach is known as uniform GIMP or uGIMP [323] . In uGIMP, the grid crossing problem is an issue due to the fixed MP size as discussed in Section 3.2.

3.3   Dual domain material point method

The dual domain material point method (DDMP) is an approach to smooth the gradient of the nodal shape functions [342]. Unlike GIMP, no characteristic functions are used. Instead, an auxiliary stress σ_A (Eq. 3.13) is added to the stress tensor to improve the numerical properties. The numerical results are not affected as long as the additive term is of the same order as the numerical error. Such an approach is famous in the field of numerical mathematics and is often connected to the addition of viscosity to a system. Zhang et al. [342] proposed Eqs. 3.13 to 3.15 for the auxiliary stress [83].

$$\sigma_A = A(x,t) + \sum_j \frac{S_j(x)}{V_j} \sum_{p=1}^{N_p} v_p(t)\sigma_p(t)S_j(x_p) - \sigma(x,t) \qquad (3.13)$$

where

$$A(x,t) = \sum_{p=1}^{N_p} \alpha(x_p)v_p\sigma_p\delta(x-x_p) - \sum_{j=1}^{N} \frac{S_j(x)}{V_j} \sum_{p=1}^{N_p} \alpha(x_p)v_p\sigma_p S_j(x_p) \qquad (3.14)$$

and

$$\alpha(x) = 0.5 \left(\prod_{k=1}^{n_c} [n_c S_k(x)] \right)^{\frac{3}{2(n_c-1)d}} \qquad (3.15)$$

Eq. 3.13 proves to be of order two: $\sigma_A(x) = \mathcal{O}(\delta x^2)$. The addition of σ_A leads to the new internal force at the node i (Eq. 3.16).

$$f_i^{\text{int}} \approx - \int_\Omega (\sigma + \sigma_A) \cdot \nabla S_i \mathrm{d}\Omega = - \sum_{p=1}^{N_p} v_p\sigma_p \cdot \overline{\nabla S_i}(x_p) \qquad (3.16)$$

where $\overline{\nabla S_i}(x_p)$ is the modified gradient of the nodal shape function and can be written as Eq. 3.17.

$$\overline{\nabla S_i}(x_p) = \alpha(x)\nabla S_i(x) + (1-\alpha(x)) \sum_{j=1}^{N} \frac{S_j(x)}{V_j} \int S_j(y)\nabla S_i(y)\mathrm{d}\Omega \qquad (3.17)$$

Eq. 3.17 is continuous as the shape function S_j is continuous and the only discontinuous term $\alpha(x)\nabla S_i(x)$ is zero between the elements as $\alpha(x) = 0$. For a detailed explanation of all involved parameters the reader is referred to [342].

3.4 Spline grid shape function

Another way to get a smoother gradient for the nodal shape functions is to choose smoother shape functions. The problem with quadratic shape functions, for instance, is that they have negative values inside an element and can lead to singularities in the mass matrix. To overcome this issue, it is possible to use B-Splines [282], which is discussed in detail in Chapter 4. In one-dimensional cases, the use of a quadratic shape function is given by Eq. 3.18 and the use of a cubic shape function by Eq. 3.19.

$$N(x) = \begin{cases} \dfrac{1}{2h^2}x^2 + \dfrac{3}{2h}x + \dfrac{9}{8} & \text{for } -\dfrac{3}{2}h \leq x \leq -\dfrac{1}{2}h \\[2mm] -\dfrac{1}{h^2}x^2 + \dfrac{3}{4} & \text{for } -\dfrac{1}{2}h \leq x \leq \dfrac{1}{2}h \\[2mm] \dfrac{1}{2h^2}x^2 - \dfrac{3}{2h}x + \dfrac{9}{8} & \text{for } \dfrac{1}{2}h \leq x \leq \dfrac{3}{2}h \\[2mm] 0 & \text{else} \end{cases} \tag{3.18}$$

$$N(x) = \begin{cases} \dfrac{1}{6h^2}x^3 + \dfrac{1}{h^2}x^2 + \dfrac{2}{h}x + \dfrac{4}{3} & \text{for } -2h \leq x \leq -h \\[2mm] -\dfrac{1}{2h^3}x^3 - \dfrac{1}{h^2}x^2 + \dfrac{4}{3} & \text{for } -h \leq x \leq 0 \\[2mm] \dfrac{1}{2h^3}x^3 - \dfrac{1}{h^2}x^2 + \dfrac{4}{3} & \text{for } 0 \leq x \leq h \\[2mm] -\dfrac{1}{6h^2}x^3 + \dfrac{1}{h^2}x^2 - \dfrac{2}{h}x + \dfrac{4}{3} & \text{for } h \leq x \leq 2h \\[2mm] 0 & \text{else} \end{cases} \tag{3.19}$$

Eqs. 3.18 and 3.19 give a smooth gradient and a non-zero shape function as both are defined piecewise positive inside an element.

3.5 Adaptive material point method

For extremely large deformations within one time step, MPs can be separated by an empty element. For such cases, Ma et al. [185] proposed an adaptive

MP splitting algorithm when the accumulative strain of a particle exceeds the defined threshold (Fig. 3.5).

3.6   Mixed integration

The integration within one element in MPM is mostly performed by summing up the values of all MPs inside a given element. The quality of such integration depends on the location of the MPs and their quantities. In contrast to the MPM integration, the Gauss integration is based on a fixed number of integration points in an optimal location to achieve as high accuracy as possible. Beuth [31] proposed to use a combination of both integration schemes to mitigate the noise produced by grid crossing. Replacing the MPM integration by Gauss integration in Eq. 3.1 gives Eq. 3.20.

$$f_2^{\text{int}} \approx \sum_{q=1}^{N_q^e} \sigma_p \omega_p |J(\xi_q)| - \sum_{q=1}^{N_q^{e+1}} \sigma_p \omega_p |J(\xi_q)| \tag{3.20}$$

The Gauss integration is not suitable for partially filled elements as the whole element volume is used for the integration whereas the sum of the MP volumes can be smaller. Therefore, it is only recommended to apply Gauss integration if the total particle volume inside one element is at least 90% of the element volume [5]. The fastest Gauss integration is done by using just one integration point. The idea is to use an averaged stress in filled elements by averaging the MP stresses over MP volumes (Eq. 3.21), and are now used to integrate the internal forces.

$$\sigma = \frac{\sum_{q=1}^{N_q^e} \sigma_p \Omega_p}{\sum_{q=1}^{N_q^e} \Omega_p} \tag{3.21}$$

$$f_e^{\text{int}} = \sum_{q=1}^{N_q^e} \omega_q \boldsymbol{B} \sigma \det\left(\boldsymbol{J}(\xi_q)\right) \tag{3.22}$$

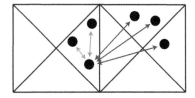

FIGURE 3.5: Adaptive splitting material point scheme.

The matrix $\boldsymbol{B}$ and $\boldsymbol{J}$ in Eq. 3.22 are the strain-displacement and the Jacobian matrices. Both matrices are given in Eq. 3.23 for a three-dimensional case.

$$
\boldsymbol{B} = \begin{pmatrix} \partial/\partial x & 0 & 0 \\ 0 & \partial/\partial y & 0 \\ \partial/\partial y & \partial/\partial x & 0 \\ 0 & \partial/\partial z & \partial/\partial y \\ \partial/\partial z & 0 & \partial/\partial x \end{pmatrix} \quad , \quad \boldsymbol{J} = \begin{pmatrix} \partial x/\partial \xi & \partial y/\partial \xi & \partial z/\partial \xi \\ \partial x/\partial \eta & \partial y/\partial \eta & \partial z/\partial \eta \\ \partial x/\partial \zeta & \partial y/\partial \zeta & \partial z/\partial \zeta \end{pmatrix} \quad (3.23)
$$

The Jacobian-matrix transforms local coordinates ξ, η and ζ to global coordinates x, y and z. Its determinant gives the transformed volume.

3.7 Integration schemes: explicit vs implicit

For dynamic problems, spatial discretisation on a gird results in a system of ordinary differential equations of the form of Eq. 3.24.

$$
M\vec{\dot{v}} = \vec{f}_{\text{ext}} - \vec{f}_{\text{int}} \tag{3.24}
$$

where $\boldsymbol{M}$, $\vec{\dot{v}}$, $\vec{f}_{\text{ext}}$ and $\vec{f}_{\text{int}}$ are the nodal values of the mass, acceleration, external force and internal force, respectively.

Eq. 3.24 represents an equilibrium between inertial, internal and external forces. Assuming this equilibrium holds true at time t, Eq. 3.24 becomes Eq. 3.25.

$$
\boldsymbol{M}^{\,t}\vec{\dot{v}}^{\,t} = \vec{f}^{\,t}_{\text{ext}} - \vec{f}^{\,t}_{\text{int}} \tag{3.25}
$$

where $\vec{\dot{v}}^{\,t}$ can be approximated by a finite difference scheme to calculate $\vec{v}^{\,t+1}$. Applying a forward difference scheme to Eq. 3.25 gives Eq. 3.26.

$$
\vec{v}^{\,t+1} = \Delta t \left(\boldsymbol{M}^{\,t}\right)^{-1} \left(\vec{f}^{\,t}_{\text{ext}} - \vec{f}^{\,t}_{\text{int}}\right) + \vec{v}^{\,t} \tag{3.26}
$$

where the right hand side of Eq. 3.26 depends only on the previous time steps. This is an example for an explicit integration scheme of order one. Commonly a lumped mass matrix is used to simplify the matrix inversion. To calculate $\vec{v}^{\,t+1}$ in Eq. 3.26, only a matrix vector multiplication has to be performed. Using the backward difference scheme to approximate the second derivative in Eq. 3.25 and assuming that the equilibrium holds at a time $t + 1$ gives an implicit formulation (Eq. 3.27).

$$
\boldsymbol{M}^{\,t+1}\Delta\vec{v} - \Delta t \vec{f}^{\,t+1} =: R(t+1) = 0 \tag{3.27}
$$

where $R(t+1)$ defines the residuum, $\Delta v = \vec{v}^{\,t+1} - \vec{v}^{\,t}$ and $\vec{f}^{\,t+1} = \vec{f}^{\,t+1}_{\text{ext}} - \vec{f}^{\,t+1}_{\text{int}}$. The residuum $R(t)$ is in general a non-linear function and to solve the equation $R(t+1) = 0$. The Newton-Raphson method can be applied to calculate $\Delta\vec{v}$.

3.8   Critical time step and Courant number

Explicit time integration schemes are conditionally stable. Assuming a linear elastic problem, the critical time step Δt_{cr} defines the time increment during which a wave with speed c crosses the smallest element length d (Eq. 3.28).

$$\Delta t_{cr} = \frac{d}{c} = \frac{d}{E/\rho} \tag{3.28}$$

where E is the Young modulus and ρ the density of the linear elastic material.

The critical time step defines the biggest time increment which can be used for a stable calculation. For non-linear problems it is often not possible to estimate the critical time step and, therefore, an estimation based on the linearised equations is made. To adjust the estimated time increment to the non-linear problem, an additional multiplier is used to reach stability (Eq. 3.29).

$$\Delta t = C_{NB}\Delta t_{cr} \tag{3.29}$$

where C_{NB} is the Courant number with $0 < C_{NB} < 1$.

3.9   Closure

Since the formulation of the standard MPM, many extensions and modifications have been developed. Most of them are dealing with the problem of MPs crossing the element boundary (grid crossing) or the accuracy of the quadrature using the scattered positions of the MPs. The choice of the suitable method depends on the application. If the accuracy of the numerical solution is of interest, then a B-Spline approach is a good choice. However, the standard MPM is still a good choice for short-duration simulations. The choice of the time-integration scheme also depends on the application and its efficiency. Most of the dynamic problems, including wave or shock propagation, can only be treated properly by an explicit integration scheme as an implicit one tends to smooth the solution in areas which might be of interest. On the other hand, the implicit integration is a good choice for long-duration simulations as the time increment can be large without compromising the stability condition.

4

Recent Developments to Improve the Numerical Accuracy

E. Wobbes, R. Tielen, M. Möller, C. Vuik and V. Galavi

4.1 Introduction

The material point method (MPM) has shown to be successful in simulating problems that involve large deformations. However, the standard algorithm suffers from many numerical shortcomings. While some of these shortcomings are inherited from the finite element method (FEM), other drawbacks are specific to MPM. The FEM-type inaccuracies include interpolation, time integration, and mass lumping errors [283]. On the other hand, MPM suffers from grid crossing errors [24] (see Chapter 3). The method typically projects the material point (MP) data to the background grid, also called mesh, and vice versa using piecewise-linear basis functions. The discontinuous gradients of these basis functions lead to unphysical oscillations in the solution when MPs cross element boundaries. In addition, MPM reconstructs scattered MP data using a low-order function-reconstruction technique that causes severe inaccuracies when large deformations are involved [286]. The most recent developments outlined in this chapter largely enhance the mathematical accuracy of the method, while keeping the physical qualities and computational efficiency close to those of original MPM.

B-spline material point method (BSMPM) provides a fundamental approach to smoothing the gradients of the basis functions. In essence, it replaces the piecewise-linear basis functions by higher-order B-spline basis functions that guarantee at least C_0-continuity of the gradients. BSMPM was originally introduced by Steffen et al. [282], but Tielen et al. [301] proposed a more general and straightforward implementation. The new approach uses the isogeometric analysis (IgA) formulation of B-splines based on the Cox-de-Boor formula [77]. Although BSMPM reduces the grid crossing, interpolation, and time stepping errors within MPM [283, 301], the accuracy of the solution can be further improved by replacing the direct mapping of the MP data to the background grid by more advanced techniques. Several function-reconstruction methods have been proposed for the projection of the scattered MP data [286, 301, 329].

In contrast with many standard techniques, which do not conserve physical quantities like mass and momentum, the Taylor Least Squares (TLS) function reconstruction [329] not only decreases the numerical errors but also ensures the conservation of the total mass and linear momentum. The TLS reconstruction is based on the least squares [163] approximation constructed from a set of Taylor basis functions [181]. It locally approximates quantities of interest, such as stress and density, allowing discontinuities across element boundaries. When used in combination with a suitable quadrature rule, the TLS technique preserves the mass and momentum after transferring the MP information to the background grid. Furthermore, the mapping of MP information involves the solution of a linear system that includes a mass matrix at every time step. While a consistent mass matrix can significantly increase the accuracy of BSMPM combined with a reconstruction technique [301], it can also lead to stability issues. Therefore, further research of this topic is needed.

Meanwhile, p-multigrid can be applied to solve linear systems resulting from IgA discretisations to ensure an efficient computation with a consistent mass matrix. Compared to the standard techniques, such as the Conjugate Gradient method, p-multigrid requires a considerably lower number of iterations. In addition, p-multigrid can be used to solve the momentum balance equation within BSMPM. Based on the research of Love and Sulsky [176], it is expected that a consistent mass matrix will conserve energy and angular momentum.

4.2 Most relevant concepts

This section explains the mass lumping procedure and presents the computational steps of the widely used Modified-Update-Stress-Last (MUSL) scheme [287]. For simplicity, only one-dimensional deformations of a one-phase continuum are considered.

4.2.1 Mass lumping

MPM solves the momentum balance equation in its weak form (see Chapter 2). After the spatial discretisation, the system solved on the background grid follows Eq. 4.1.

$$\boldsymbol{M}^{\mathrm{C}} \vec{a} = \vec{f}^{\,\mathrm{ext}} - \vec{f}^{\,\mathrm{int}} \tag{4.1}$$

where $\boldsymbol{M}^{\mathrm{C}}$ is the consistent mass matrix, $\vec{a}$ the acceleration vector, $\vec{f}^{\,\mathrm{ext}}$ the external force vector, and $\vec{f}^{\,\mathrm{int}}$ the internal force vector.

The consistent mass matrix is given by Eq. 4.2.

$$\boldsymbol{M}^{C} = \int_{\Omega} \rho \vec{\phi} \vec{\phi}^{T} \, d\Omega \tag{4.2}$$

where ρ is the density, $\vec{\phi}$ is the basis function vector, and Ω is the considered domain.

If the domain discretisation generates N_{n} degrees of freedom (DOFs), the basis function vector is denoted by $\vec{\phi}(x) = [\phi_{1}(x) \ \phi_{2}(x) \ \dots \ \phi_{N_{n}}(x)]^{T}$, where ϕ_{i} represents a basis function associated with the ith DOF.

Similar to FEM, this non-diagonal matrix can be transformed into a diagonal matrix by the lumping procedure when piecewise-linear or B-spline basis functions are used. Denoting element (i, j) of the consistent and lumped mass matrices by $\boldsymbol{M}^{C}_{(i,j)}$ and $\boldsymbol{M}_{(i,j)}$, respectively, the lumping procedure can be described by Eq. 4.3.

$$\boldsymbol{M}_{(i,j)} = \delta_{i,j} \sum_{j} \boldsymbol{M}^{C}_{(i,j)} \tag{4.3}$$

where $\delta_{i,j}$ is the Dirac delta function.

Since piecewise-linear and B-spline basis functions satisfy the partition of unity property (i.e. $\sum_{i=1}^{N_{n}} \phi_{i}(x) = 1 \ \forall \ x \in \Omega$), mass lumping can also be achieved variationally (Eq. 4.4).

$$\boldsymbol{M} = \begin{bmatrix} \diagdown \\ & \vec{m} \\ & & \diagdown \end{bmatrix} \quad \text{with} \quad \vec{m} = \int_{\Omega} \rho \vec{\phi} \, d\Omega \tag{4.4}$$

In the version of MPM considered in this chapter, Eq. 4.1 is typically used with the lumped mass matrix $\boldsymbol{M}$. Lumping of the mass matrix has a number of advantages. For example, it reduces the computational costs and improves convergence characteristics of the method [176]. However, generally a lumped mass matrix limits the spatial convergence to $\mathcal{O}(h^{2})$ [283] and hinders the conservation of energy and angular momentum [176].

4.2.2 Modified-Update-Stress-Last algorithm

Throughout this section, the superscript t denotes the time level, and Δt is the time-step size. The N_{mp} MPs are initialised at $t = 0$ s. Each MP carries a certain volume V_{mp}, density ρ_{mp}, position x_{mp}, displacement u_{mp}, velocity v_{mp}, and stress σ_{mp}. These values are time dependent, but the MP mass m_{mp} remains constant throughout the simulation. Assuming that all MP properties are known at time t, the computation for time $t + \Delta t$ proceeds as follows.

1. The data from the MPs is mapped to the DOFs of the background grid. For instance, the diagonal of the lumped mass matrix (Eq. 4.5) and the

internal forces $\vec{f}^{\,\text{int}}$ (Eq. 4.6) are computed.

$$\vec{m}^{\,\text{t}} = \sum_{mp=1}^{N_{\text{mp}}} m_{\text{mp}} \vec{\phi}\left(x_{\text{mp}}^{\text{t}}\right) \tag{4.5}$$

$$\left(\vec{f}^{\,\text{int}}\right)^{\text{t}} = \sum_{mp=1}^{N_{\text{mp}}} \sigma_{\text{mp}}^{\text{t}} \vec{\phi}'\left(x_{\text{mp}}^{\text{t}}\right) V_{\text{mp}}^{\text{t}} \tag{4.6}$$

In other words, Eq. 4.5 implies that for the ith DOF Eq. 4.7 holds.

$$m_{\text{i}}^{\text{t}} = \sum_{mp=1}^{N_{\text{mp}}} m_{\text{mp}} \phi_{\text{i}}\left(x_{\text{mp}}^{\text{t}}\right) \tag{4.7}$$

This direct mapping technique is typical for standard MPM. An improved approach is provided in Section 4.4.

2. The accelerations at the DOFs are obtained after combining the internal forces with any external forces.

$$\vec{a}^{\,\text{t}} = \left(\boldsymbol{M}^{\text{t}}\right)^{-1} \left[\left(\vec{f}^{\,\text{ext}}\right)^{\text{t}} - \left(\vec{f}^{\,\text{int}}\right)^{\text{t}}\right] \tag{4.8}$$

3. The velocity of each MP at time $t + \Delta t$ is determined.

$$v_{\text{mp}}^{\text{t}+\Delta\text{t}} = v_{\text{mp}}^{\text{t}} + \Delta t\, \vec{\phi}^{\,\text{T}}\left(x_{\text{mp}}^{\text{t}}\right) \vec{a}^{\,\text{t}} \quad \forall\, \text{mp} = \{1, 2, \ldots, N_{\text{mp}}\} \tag{4.9}$$

4. The velocities at the DOFs are subsequently obtained.

$$\vec{v}^{\,\text{t}+\Delta\text{t}} = \left(\boldsymbol{M}^{\text{t}}\right)^{-1} \sum_{mp=1}^{N_{\text{mp}}} m_{\text{mp}} \vec{\phi}\left(x_{\text{mp}}^{\text{t}}\right) v_{\text{mp}}^{\text{t}+\Delta\text{t}} \tag{4.10}$$

where $\vec{v}$ is the velocity vector consisting of the velocities at the DOFs v_{i}.

5. The incremental displacement vector $\Delta\vec{u}$ is computed.

$$\Delta\vec{u}^{\,\text{t}+\Delta\text{t}} = \Delta t\, \vec{v}^{\,\text{t}+\Delta\text{t}} \tag{4.11}$$

6. After these steps, the remaining part of the MP properties is updated.

$$u_{\text{mp}}^{\text{t}+\Delta\text{t}} = u_{\text{mp}}^{\text{t}} + \vec{\phi}^{\,\text{T}}\left(x_{\text{mp}}^{\text{t}}\right) \Delta\vec{u}^{\,\text{t}+\Delta\text{t}} \tag{4.12}$$

$$x_{\text{mp}}^{\text{t}+\Delta\text{t}} = x_{\text{mp}}^{\text{t}} + \vec{\phi}^{\,\text{T}}\left(x_{\text{mp}}^{\text{t}}\right) \Delta\vec{u}^{\,\text{t}+\Delta\text{t}} \tag{4.13}$$

$$\Delta\varepsilon_{\text{mp}}^{\text{t}+\Delta\text{t}} = \nabla\phi^{\text{T}}\left(x_{\text{mp}}^{\text{t}}\right) \Delta\vec{u}^{\,\text{t}+\Delta\text{t}} \tag{4.14}$$

where $\Delta\varepsilon_{\text{mp}}$ is the MP incremental strain.

The MP stress at time $t + \Delta t$ is computed from σ^t and $\Delta\varepsilon_{mp}^{t+\Delta t}$ using a constitutive model.

Considering only one-dimensional elastic deformations, it follows from [132, 189] that for small and large deformations, the stress is given by Eqs. 4.15 and 4.16, respectively.

$$\sigma_{mp}^{t+\Delta t} = \sigma_{mp}^t + E\,\Delta\varepsilon_{mp}^{t+\Delta t} \qquad \forall\,mp = \{1, 2, \dots, N_{mp}\} \qquad (4.15)$$

$$\sigma_{mp}^{t+\Delta t} = \sigma_{mp}^t + \left(E - \sigma_{mp}^t\right)\Delta\varepsilon_{mp}^{t+\Delta t} \qquad \forall\,mp = \{1, 2, \dots, N_{mp}\} \qquad (4.16)$$

where E is the Young modulus.

7. The volume and density of each MP are obtained from the volumetric strain increment ε_{vol} (Eqs. 4.17-4.18).

$$V_{mp}^{t+\Delta t} = \left(1 + \Delta\varepsilon_{vol,mp}^{t+\Delta t}\right) V_{mp}^t \qquad (4.17)$$

$$\rho_{mp}^{t+\Delta t} = \frac{\rho_{mp}^t}{\left(1 + \Delta\varepsilon_{vol,mp}^{t+\Delta t}\right)} \qquad (4.18)$$

4.3   B-spline Material Point Method

BSMPM [282,301] replaces the piecewise-linear basis functions by higher-order B-splines. The univariate B-spline basis functions can be introduced using a *knot vector*, a sequence of ordered non-decreasing points in $\mathbb{R}$ that are called *knots*. A knot vector is denoted by $\Xi = \{\xi_1, \xi_2, \dots, \xi_{N_n+p+1}\}$ with N_n and p being the number of basis functions and the polynomial order, respectively. The first and last knot are repeated $p+1$ times to make the resulting B-spline interpolatory at both end points. In contrast to linear basis functions, $\phi_{i,p}$ is not interpolatory, that is $\phi_{i,p}(x_j) \neq \delta_{i,j}$.

The Cox-de Boor formula [77] defines B-spline basis functions recursively. For $p = 0$, the basis functions are provided in Eq. 4.19.

$$\phi_{i,0}(\xi) = \begin{cases} 1 & \text{if } \xi_i \leq \xi < \xi_{i+1} \\ 0 & \text{otherwise} \end{cases} \qquad (4.19)$$

For $p > 0$, the basis functions are given by Eq. 4.20

$$\phi_{i,p}(\xi) = \frac{\xi - \xi_i}{\xi_{i+p} - \xi_i}\phi_{i,p-1}(\xi) + \frac{\xi_{i+p+1} - \xi}{\xi_{i+p+1} - \xi_{i+1}}\phi_{i+1,p-1}(\xi) \qquad \xi \in \hat{\Omega} \qquad (4.20)$$

where $\hat{\Omega}$ is the parametric domain.

Fig. 4.1 shows quadratic basis functions with $\Xi = \{0, 0, 0, 1/3, 2/3, 1, 1, 1\}$.

B-spline basis functions satisfy the following properties.

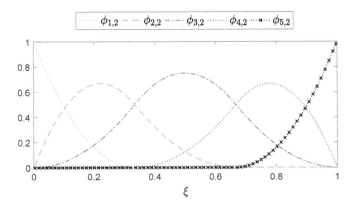

FIGURE 4.1: Example of quadratic B-spline basis functions.

1. They form a partition of unity:

$$\sum_{i=1}^{N_n} \phi_{i,p}(\xi) = 1 \quad \forall\, \xi \in \hat{\Omega} \tag{4.21}$$

2. Each $\phi_{i,p}$ has compact support $[\xi_i, \xi_{i+p+1}]$.

3. They are non-negative in their support:

$$\phi_{i,p}(\xi) \geq 0 \quad \forall\, \xi \in \hat{\Omega} \tag{4.22}$$

The gradients of the B-spline basis functions are defined as Eq. 4.23 [77].

$$\frac{\mathrm{d}\phi_{i,p}(\xi)}{\mathrm{d}\xi} = \frac{p}{\xi_{i+p} - \xi_i}\phi_{i,p-1}(\xi) - \frac{p}{\xi_{i+p+1} - \xi_{i+1}}\phi_{i+1,p-1}(\xi) \tag{4.23}$$

The gradients corresponding to the basis functions from Fig. 4.1 are provided in Fig. 4.2.

B-spline basis functions bring many advantages over the typically used linear ones. First of all, they guarantee at least C_0-continuity of the gradients and hence, significantly reduce the grid crossing error. Similarly to piecewise-linear basis functions, B-splines enable lumping of the mass matrix due to their non-negativity and partition of unity properties, which is an essential for many engineering studies. In addition, they decrease the interpolation and time-stepping errors [283].

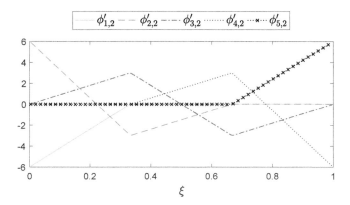

FIGURE 4.2: Example of the gradient of a quadratic B-spline basis function.

4.4 Mapping of material point data to background grid

In the MUSL algorithm, the computation of certain quantities involves a direct mapping of the MP data to the background grid. For example, the velocities at the DOFs are calculated from MP velocities and masses as shown in Eq. 4.10. This MPM mapping technique ensures the conservation of the mass $\mathcal{M}$ and linear momentum $\mathcal{P}$ of the system as follows.

$$\mathcal{M} = \sum_{i=1}^{N_n} m_i = \sum_{i=1}^{N_n} \left(\sum_{mp=1}^{N_{mp}} m_{mp}\phi_i(x_{mp}) \right) = \sum_{mp=1}^{N_{mp}} m_{mp} \sum_{i=1}^{N_n} \phi_i(x_{mp})$$

$$= \sum_{mp=1}^{N_{mp}} m_{mp} \tag{4.24}$$

$$\mathcal{P} = \sum_{i=1}^{N_n} m_i v_i = \sum_{i=1}^{N_n} m_i \left(\frac{1}{m_i} \sum_{mp=1}^{N_{mp}} m_{mp}\phi_i(x_{mp})v_{mp} \right)$$

$$= \sum_{mp=1}^{N_{mp}} m_{mp}v_{mp} \sum_{i=1}^{N_n} \phi_i(x_{mp}) = \sum_{mp=1}^{N_{mp}} m_{mp}v_{mp} \tag{4.25}$$

The above expressions are obtained from Eqs. 4.5 and 4.10, and by exploiting the partition of unity property of the piecewise-linear basis functions. Superscripts t and $t + \Delta t$ have been dropped to improve readability.

The MPM mapping can lead to significant numerical errors, especially when large deformations are considered [286]. For this reason, the TLS technique is discussed in this chapter. For each element, the TLS technique reconstructs quantities of interest, such as stress and density, from the MP data and evaluates them at the integration points. After that, a numerical quadrature rule is applied to determine the internal forces and velocities at the DOFs. If the integration is exact, the proposed mapping approach preserves the total mass and linear momentum. The ideas are introduced for one dimension (1D), but can be extended to multiple dimensions in a straightforward manner. The proof of the conservation properties of the TLS technique can be found in Wobbes et al. [329].

4.4.1   Taylor Least Squares reconstruction

For the Least Squares approximation, a set of N_{mp} distinct data points $\{x_{mp}\}_{mp=1}^{N_{mp}}$ is considered. The generic data values of these points are denoted by $\{f(x_{mp})\}_{mp=1}^{N_{mp}}$. It is assumed that $f \in F$, where F is a normed function space on $\mathbb{R}$, and $P = \text{span}\{\psi_i\}_{i=1}^{n_b} \subset F$ is a set of n_b basis functions. The Least Squares approximation at a point $x \in \mathbb{R}$ can be written as in Eq. 4.26.

$$\tilde{f}(x) = \sum_{i=1}^{n_b} \alpha_i \psi_i(x) = \vec{\psi}^{\,T}(x)\vec{\alpha} \quad \text{with} \quad \vec{\alpha} = \boldsymbol{D}^{-1}\boldsymbol{B}\vec{F} \qquad (4.26)$$

where $\vec{\alpha}$ is the vector of coefficients for the basis functions obtained using Eqs. 4.27, 4.28 and 4.29.

$$\boldsymbol{D} = \sum_{mp=1}^{N_{mp}} \vec{\psi}(x_{mp})\vec{\psi}^{\,T}(x_{mp}) \qquad (4.27)$$

$$\boldsymbol{B} = [\vec{\psi}(x_1) \quad \vec{\psi}(x_2) \quad \dots \quad \vec{\psi}(x_{N_{mp}})] \qquad (4.28)$$

$$\vec{F} = [f(x_1) \quad f(x_2) \quad \dots \quad f(x_{N_{mp}})]^{\mathrm{T}} \qquad (4.29)$$

The basis for P is formed by the local Taylor basis functions, which are defined using the concept of the volume average of a function f over Ω_e shown in Eq. 4.30.

$$\overline{f} := \frac{1}{|\Omega_e|} \int_{\Omega_e} f \, d\Omega_e \qquad (4.30)$$

where $|\Omega_e|$ is the volume of element e.

For example, if the element is one dimensional (i.e, $\Omega_e = [x_{min}, x_{max}]$ with $x_{max} > x_{min}$), then $|\Omega_e| = x_{max} - x_{min}$. The Taylor basis functions are then given by Eq. 4.31.

$$\psi_1 = 1 \quad , \quad \psi_2 = \frac{x - x_c}{\Delta x} \quad , \quad \psi_3 = \frac{(x - x_c)^2}{2\Delta x^2} - \overline{\frac{(x - x_c)^2}{2\Delta x^2}} \qquad (4.31)$$

$$\text{where} \quad x_c = \frac{x_{max} + x_{min}}{2} \quad \text{and} \quad \Delta x = \frac{x_{max} - x_{min}}{2}$$

An important quality of the Taylor basis that ensures the conservation property of the reconstruction technique is shown in Eq. 4.32 [181].

$$\int_{\Omega_e} \psi_i \, d\Omega_e = \begin{cases} |\Omega_e| & \text{if } i = 1, \\ 0 & \text{if } i \neq 1. \end{cases} \tag{4.32}$$

Suppose that a function f has to be reconstructed in such way that its integral over Ω_e, $\int_{\Omega_e} f(x) \, d\Omega_e = c$ ($c \in \mathbb{R}$), is preserved. The TLS approximation of f is equal to a linear combination of Taylor basis functions (Eq. 4.33).

$$f(x) \approx \tilde{f}(x) = \sum_{i=1}^{n_b} a_i \psi_i(x) \tag{4.33}$$

Using Eq. 4.32, the integral of $\tilde{f}$ can then be written as Eq. 4.34.

$$\int_{\Omega_e} \tilde{f}(x) \, d\Omega_e = \int_{\Omega_e} \sum_{i=1}^{n_b} a_i \psi_i(x) \, d\Omega_e = \sum_{i=1}^{n_b} a_i \int_{\Omega_e} \psi_i(x) \, d\Omega_e = \alpha_1 |\Omega_e| \tag{4.34}$$

Therefore, explicitly enforcing the condition shown in Eq. 4.35 conserves the integral.

$$\alpha_1 = \frac{c}{|\Omega_e|} \tag{4.35}$$

Consequently, the number of basis functions in the Least Squares approximation is reduced by one.

4.4.2 Example of Taylor Least Squares reconstruction

The outstanding properties of the TLS technique can be illustrated by reconstructing $f(x) = \sin(x) + 2$ on $[0, 4\pi]$. In this case, the integral that should be preserved is equal to 8π. The domain is discretised using four elements of size π and contains 11 data points. Two data points are located at the boundaries of the first element, (i.e., 0 and π). In $[2\pi, 3\pi]$, the data points are distributed uniformly in the interior of the domain. The remaining data points have random positions creating different types of data distribution within each element. The data point at π is used for both intervals, $[0, \pi]$ and $[\pi, 2\pi]$.

The TLS approximation is obtained using three Taylor basis functions. Fig. 4.3 visualises the data-point distribution for 10 integration points per element. The overall performance of the TLS technique is quantified by the Root-Mean-Square (RMS) error for function f and the relative error for its integral. Both errors are computed using 10 Gauss points within each element. The RMS error is equal to $3.8139 \cdot 10^{-2}$, while the relative error for the integral is equal to $7.0679 \cdot 10^{-16}$. It should be noted that the relative error for the integral computed with only two Gauss points per element is equal to $2.7903 \cdot 10^{-15}$. Thus, the TLS approach preserves the integral up to machine precision for this example.

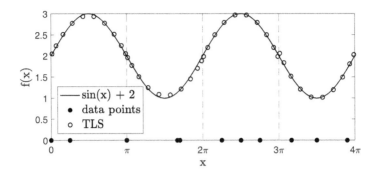

FIGURE 4.3: TLS reconstructions of $f(x) = \sin(x) + 2$ on $[0, 4\pi]$ for different types of data point distribution within an element.

4.4.3   Integration of Taylor Least Squares reconstruction

When the TLS reconstruction is considered as part of the MPM algorithm, MPs serve as data points. In order to ensure an accurate and conservative mapping of the information from MPs to the grid, the technique is combined with a Gauss quadrature rule with a suitable number of Gauss points. However, Gauss quadrature can be replaced by any numerical integration that provides an exact result.

The TLS technique is applied to replace the MPM integration in Eq. 4.6. Since in this case the conservation is not required, all unknown coefficients of the Taylor basis functions are obtained from the least-square approximation (i.e. coefficient a_1 is not enforced by Eq. 4.35). Thus, the internal forces at the DOFs are computed as follows.

1. Apply TLS approximation to reconstruct the stress field from the MP data within each active element without specifying the coefficient of the first Taylor basis function (Eq. 4.36).

$$\tilde{\sigma}_e = \sum_{i=1}^{n_b} s_i \psi_i \qquad (4.36)$$

where s_i is the coefficient corresponding to the ith Taylor basis function. Outside of Ω_e, $\tilde{\sigma}_e$ is zero.

The global approximation of the stress function, $\tilde{\sigma}$, is then given by Eq. 4.37.

$$\tilde{\sigma} = \sum_{e=1}^{N_e} \tilde{\sigma}_e \qquad (4.37)$$

2. Integrate the stress approximation using a Gauss quadrature (Eq. 4.38).

$$\vec{f}^{\text{int}} \approx \int_\Omega \tilde{\sigma}(x,t)\vec{\phi}' \, d\Omega = \sum_{g=1}^{N_g} \tilde{\sigma}\left(x_g\right) \vec{\phi}'\left(x_g\right) \omega_g \tag{4.38}$$

where N_g is the total number of Gauss points, x_g the global position of a Gauss point, and ω_g the weight of a Gauss point.

For exact integration of the approximated function, each active element should contain N_g/N_e Gauss points, where N_e is the total number of elements (knot spans) and N_g satisfies $n_b \leq 2N_g/N_e$. This implies that for a quadratic TLS approach the numerical integration requires at least two Gauss points per element.

The TLS technique is also used to map the MP velocities to the DOFs (i.e. it replaces Eq. 4.10). However, the coefficient of the first basis function is specified according to Eq. 4.35. The remaining coefficients are calculated from Eq. 4.26 explicitly excluding ψ_1 to maintain the integral value.

1. Apply TLS approximation to reconstruct the density field and momentum, which is given by the product of density and velocity from the MP data within each active element, while preserving the mass and momentum of the element (Eqs. 4.39 and 4.40).

$$\tilde{\rho}_e = \sum_{i=1}^{n_b} r_i\psi_i \text{ with } r_1 = \frac{1}{|\Omega_e|} \sum_{\{p|x_{\text{mp}}\in\Omega_e\}} m_{\text{mp}} \tag{4.39}$$

$$(\widetilde{\rho v})_e = \sum_{i=1}^{n_b} \gamma_i\psi_i \text{ with } \gamma_1 = \frac{1}{|\Omega_e|} \sum_{\{p|x_{\text{mp}}\in\Omega_e\}} m_{\text{mp}}v_{\text{mp}} \tag{4.40}$$

where r_i and γ_i are the coefficients corresponding to the i-th Taylor basis function. Outside of Ω_e, $\tilde{\rho}_e$ and $(\widetilde{\rho v})_e$ are zero.

The global approximations are given by Eq. 4.41.

$$\tilde{\rho} = \sum_{e=1}^{N_e} \tilde{\rho}_e \text{ and } (\widetilde{\rho v}) = \sum_{e=1}^{N_e} (\widetilde{\rho v})_e \tag{4.41}$$

2. Integrate the approximations using a Gauss quadrature to obtain the momentum vector $\vec{p}$ and the consistent mass matrix $\boldsymbol{M}^C$ (Eqs. 4.42-4.43).

$$\vec{p} = \sum_{g=1}^{N_g} (\widetilde{\rho v})\left(x_g\right) \omega_g \vec{\phi}\left(x_g\right) \tag{4.42}$$

$$\boldsymbol{M}^C = \sum_{g=1}^{N_g} \tilde{\rho}\left(x_g\right) \omega_g \vec{\phi}\left(x_g\right) \left(\vec{\phi}\left(x_g\right)\right)^T \tag{4.43}$$

As previously mentioned, the number of Gauss points per element should be specified so that exact integration is ensured.

3. Compute the velocity vector (Eq. 4.44).

$$\vec{v} = (\boldsymbol{M}^{C})^{-1}\vec{p} \tag{4.44}$$

The consistent mass matrix $\boldsymbol{M}^{C}$ can be replaced in Eq. 4.44 by a lumped mass matrix without losing the conservation properties of the algorithm. A consistent mass matrix typically provides more accurate results, but may lead to stability issues [329].

4.5   Application to vibrating bar problem

This section compares the performance of the standard MPM that follows the MUSL algorithm with its more advanced versions, such as BSMPM and BSMPM with the TLS reconstruction (TLS-BSMPM). The comparison is done based on an example that describes the vibration of a one-phase bar with fixed ends. A system of partial differential equations for the velocity and stress captures the motion (Eqs. 4.45 and 4.46).

$$\rho\frac{\partial v}{\partial t} = \frac{\partial \sigma}{\partial x} \tag{4.45}$$

$$\frac{\partial \sigma}{\partial t} = E\frac{\partial v}{\partial x} \tag{4.46}$$

The system is extended by a relation between the velocity v and displacement u given by Eq. 4.47.

$$v = \frac{\partial u}{\partial t} \tag{4.47}$$

The vibration is triggered by an initial velocity that varies along the bar leading to the following initial and boundary conditions (Eqs. 4.48 and 4.49).

$$u(x,0) = 0, \quad v(x,0) = v_0\sin\left(\frac{\pi x}{h}\right), \quad \sigma(x,0) = 0; \tag{4.48}$$

$$u(0,t) = 0, \quad u(h,t) = 0 \tag{4.49}$$

where H is the length of the bar and v_0 is the maximum initial velocity.

For small deformations, the analytical solution in terms of displacement, velocity, and stress is given by Eqs. 4.50 to 4.52.

$$u(x,t) = \frac{v_0 h}{\pi \sqrt{E/\rho}} \sin\left(\frac{\pi \sqrt{E/\rho}\, t}{h}\right) \sin\left(\frac{\pi x}{h}\right) \tag{4.50}$$

$$v(x,t) = v_0 \cos\left(\frac{\pi \sqrt{E/\rho}\, t}{h}\right) \sin\left(\frac{\pi x}{h}\right) \tag{4.51}$$

$$\sigma(x,t) = v_0 \sqrt{E\rho} \sin\left(\frac{\pi \sqrt{E/\rho}\, t}{h}\right) \cos\left(\frac{\pi x}{h}\right) \tag{4.52}$$

Table 4.1 gives the parameter values for the vibrating bar benchmark under small deformations. The contribution of the temporal errors to the overall error generated during the computation is minimised by selecting a small time-step size and short simulation time. To be more precise, the time-step size and total simulation time are set to $1 \cdot 10^{-7}$ s and $1.9 \cdot 10^{-6}$ s, respectively. Furthermore, the number of elements (knot spans) is varied from 5 to 40, while the initial number of MPs per cell or knot span (PPC) is fixed to 12. Grid crossing does not occur, and the maximal observed strain is equal to $5.3 \cdot 10^{-7}$ m.

The obtained results are illustrated in Fig. 4.4. As expected, MPM shows second-order convergence in the displacement and velocity. Since the stress is computed as a derivative of the displacement, its convergence rate is one. The use of BSMPM with the standard MPM mapping leads to a lower RMS error and third-order convergence for the velocity, but hinders the performance of the method for the displacement and stress. The poor performance of the method in terms of displacement and stress is caused by the large values of the error at the boundaries of the domain. This is illustrated in Fig. 4.5 for the stress distribution.

The TLS reconstruction in conjunction with a lumped mass matrix has little influence on the convergence behaviour of BSMPM. For this reason, the corresponding results are not shown in Fig. 4.4. However, the TLS technique eliminates the boundary issues due to the B-spline basis functions when a consistent mass matrix is used for the mapping of MP information. As a result, it significantly decreases the RMS error in the displacement and stress

TABLE 4.1
Parameters for small and large deformation vibrating bar.

Parameter	Symbol	Unit	Value
Height	h	m	1.00
Density	ρ	kg/m^3	$2.00 \cdot 10^3$
Young modulus (small def.)	E	kPa	$7.00 \cdot 10^3$
Young modulus (large def.)	E	kPa	40
Max. initial velocity (small def.)	v_0	m/s^2	0.28
Max. initial velocity (large def.)	v_0	m/s^2	0.80

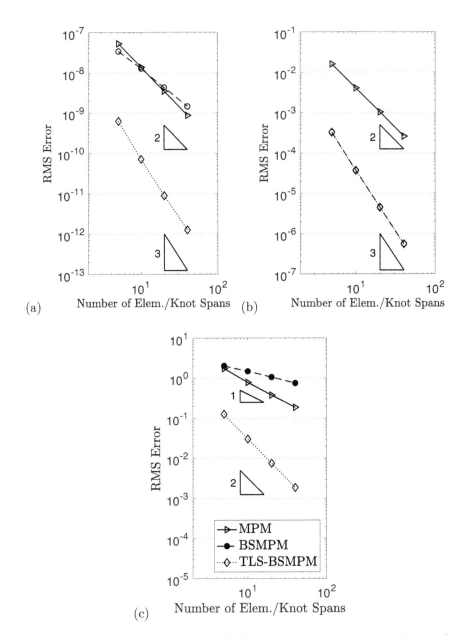

FIGURE 4.4: Spatial convergence of MP methods for the vibrating bar problem without grid crossing: (a) displacement, (b) velocity, and (c) stress.

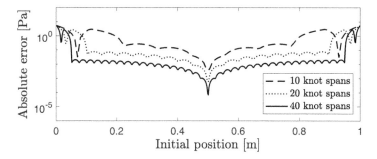

FIGURE 4.5: Absolute error obtained with BSMPM for stress distribution in the vibrating bar problem without grid crossing.

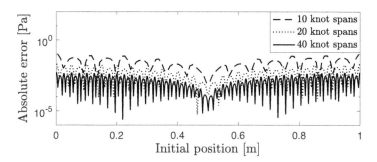

FIGURE 4.6: Absolute error obtained with TLS-BSMPM for stress distribution in the vibrating bar problem without grid crossing.

and leads to a higher convergence order for both quantities. The performance of the TLS function reconstruction technique in terms of the absolute error for the stress is depicted in Fig. 4.6. Moreover, the mapping with the TLS reconstruction preserves the relative error in the total mass and momentum under $7.5033 \cdot 10^{-15}$ and $2.1007 \cdot 10^{-16}$, respectively.

For large deformations, the parameters from Table 4.1 are used and the simulation time is increased to 0.1 s. The computations are performed with the time-step size of $1 \cdot 10^{-5}$ s, 20 elements (knot spans) and initially 8 MPs in each cell. Since the analytical solution is not available when large strains are considered, the results are compared to the solution generated with the Updated Lagrangian Finite Element Method (ULFEM) [25] using 160 DOFs.

Under large deformations, the MPs cross the element boundaries more than 450 times when the standard MPM is used. This results in unphysical oscillations in the solution for the stress as shown in Fig. 4.7. The use of the B-spline basis functions reduces the oscillatory behaviour, but still significantly deviates from the ULFEM solution. When the TLS reconstruction (with a consistent mass matrix) is applied as well, the solutions improve significantly

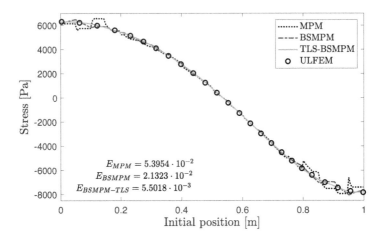

FIGURE 4.7: Stress distribution and corresponding relative errors in the L_2-norm in the vibrating bar with grid crossing.

leading to the reduction of the MPM error by a factor of 9.8. The method limits the error in the mass and linear momentum of the system to $2.9104 \cdot 10^{-14}$ and $5.7205 \cdot 10^{-15}$ during the simulation.

The vibrating bar example demonstrates that BSMPM with the standard MPM mapping considerably reduces the unphysical oscillations originating from grid crossing. However, its performance can be hindered by the issues at the boundaries of the domain. The application of the TLS reconstruction ensures an accurate solution at the boundaries leading to higher-order convergence. As a result, TLS-BSMPM significantly improves the solution of the standard MPM for small and large deformations. The obtained results also show that TLS-BSMPM conserves the mass and linear momentum of the system up to machine precision. Thus, the method preserves the physical properties of the standard MPM.

For this example, the use of a consistent mass matrix for the velocity computation in TLS-BSMPM is vital for the optimal performance of the method. Although not considered here, the consistent mass matrix in Eq. 4.1 can further improve the accuracy and physical properties of the algorithm. Therefore, it is important to ensure an efficient solution of the resulting linear systems.

4.6   Iterative solvers

When solving a linear system of equations, two general solution strategies can be distinguished. Direct solvers, such as Gaussian Elimination and Cholesky

Decomposition, determine the solution of a linear system of equations directly. However, since direct solvers require a high amount of computational resources, they are less preferable in the solution of large linear systems. Direct solvers can be used though in a different setting, for example as a preconditioner [254].

Alternatively, iterative solvers can be adopted, in which an initial guess is updated successively until a converged end-of-step solution has been reached. Hence, when considering Eq. 4.1, a sequence of approximations $\vec{a}^{(0)}, \vec{a}^{(1)}, \vec{a}^{(2)}, \ldots$ is constructed based on an initial guess $\vec{a}^{(0)}$. For each solution $\vec{a}^{(n)}$, the corresponding residual vector $\vec{r}^{(n)} = \vec{f} - \boldsymbol{M}^C \vec{a}^{(n)}$ is determined. Once the residual is smaller then a predefined tolerance, the method is said to have converged and the solution at the corresponding iteration is used. Different types of iterative methods can be distinguished. Basic iterative methods like the (damped) Jacobi or Gauss-Seidel method are easy to implement but require a relatively large number of iterations. The more advanced Krylov subspace methods like the Conjugate Gradient and Generalized Minimal Residual (GMRES) method show better convergence rates, which result in a smaller number of iterations needed to converge. Next to the choice of the iteration scheme, one or more stopping criteria have to be selected, such as Eq. 4.53.

$$\frac{||\vec{r}^{(k)}||_2}{||\vec{r}^{(0)}||_2} < \epsilon \tag{4.53}$$

Apart from the adopted stopping criterion and iterative method, the properties of the corresponding matrix play an important role in the performance of the iterative solver. In particular, the condition number influences the rate of convergence of the chosen iterative method. The condition number $\kappa(\boldsymbol{M})$ of a matrix $\boldsymbol{M} \in \mathbb{R}^{N \times N}$ in 2-norm is defined as given in Eqs. 4.54 and 4.55

$$\kappa(\boldsymbol{M}) := ||\boldsymbol{M}||_2 ||\boldsymbol{M}^{-1}||_2 \tag{4.54}$$

$$||\boldsymbol{M}||_2 := \sqrt{\lambda_{\max}(\boldsymbol{M}^\top \boldsymbol{M})} \tag{4.55}$$

where $\lambda_{\max}(\cdot)$ denotes the maximum eigenvalue of the corresponding matrix.

In case the matrix is symmetric and positive definite (SPD), the condition number is given by Eq. 4.56 [254].

$$\kappa(\boldsymbol{M}) = \frac{\lambda_{\max}(\boldsymbol{M})}{\lambda_{\min}(\boldsymbol{M})} \tag{4.56}$$

To illustrate the influence of the condition number on the convergence of iterative methods, the Conjugate Gradient method is considered. The approximated solution after n iterations, $\vec{a}^{(n)}$, obtained with the Conjugate Gradient method and the exact solution $\vec{a}$ satisfy the inequality given in Eq. 4.57 [114].

$$||\vec{a} - \vec{a}^{(n)}|| \leq 2 \left(\frac{\sqrt{\kappa(\boldsymbol{M})} - 1}{\sqrt{\kappa(\boldsymbol{M})} + 1} \right)^n ||\vec{a} - \vec{a}^{(0)}|| \tag{4.57}$$

Hence, the factor at which the initial difference between the approximation and the exact solution $\vec{a}$ decreases every iteration depends heavily on the condition number of the matrix. High values of $\kappa(M)$ imply a slow convergence, while low values of $\kappa(M)$ result in fast convergence.

Within MPM, the condition number of the mass matrix determines the rate of convergence when an iterative method is used to solve Eqs. 4.1 and 4.10. Therefore, the condition number of the mass matrix for different orders of the B-spline basis functions p and mesh widths h is presented in Table 4.2.

Although the numerical estimates are obtained on relatively coarse meshes, the strong dependence of the condition number on the approximation order p can be observed. Hence, a spatial discretisation with high-order B-spline basis functions leads to an ill-conditioned linear system of equations.

To illustrate the effect of the condition number on the performance of iterative solvers, Eq. 4.1 is considered and has to be solved in every time step with the mass matrix defined by Eq. 4.2, where the B-spline basis functions ϕ can be chosen of arbitrary order p.

The force vector $\vec{f}$ is chosen to be constant, simulating a gravitational force of 9.81 in the negative x- and y-direction. The resulting linear system is then solved with the Conjugate Gradient method for different approximation orders p. A tolerance of $\epsilon = 1 \cdot 10^{-8}$ is chosen combined with a zero initial guess. The number of iterations needed with the CG method, which can be adopted since M^C is SPD, are presented in Table 4.3.

TABLE 4.2
$\kappa(M)$ for different mesh widths and orders of B-spline basis functions.

$p \setminus h^{-1}$	8	16	32	64
1	$7.38 \cdot 10^0$	$8.55 \cdot 10^0$	$8.89 \cdot 10^0$	$8.97 \cdot 10^0$
2	$5.21 \cdot 10^1$	$5.51 \cdot 10^1$	$5.60 \cdot 10^1$	$5.62 \cdot 10^1$
3	$4.40 \cdot 10^2$	$4.24 \cdot 10^2$	$4.20 \cdot 10^2$	$4.22 \cdot 10^2$
4	$3.77 \cdot 10^3$	$3.36 \cdot 10^3$	$3.29 \cdot 10^3$	$3.30 \cdot 10^3$
5	$3.30 \cdot 10^4$	$2.61 \cdot 10^4$	$2.52 \cdot 10^4$	$2.52 \cdot 10^4$

TABLE 4.3
Number of iterations needed with the CG method for different values of p and h.

$p \setminus h^{-1}$	16	32	64	128
1	18	18	18	17
2	34	46	46	44
3	68	93	95	91
4	93	178	196	187
5	156	242	252	227

The effect of the condition number on the number of iterations needed before the CG method converges is clearly visible: a high condition number leads to a high number of CG-iterations. The use of high-order B-spline basis function therefore leads to a new challenge: the efficient solution of linear systems arising in IgA discretisations.

4.7   Efficient solvers for isogeometric analysis

Sangali and Tani [257] investigated the preconditioners based on the solution of the Sylvester equation and Collier et al. [71] the use of direct solvers. An alternative class of iterative solvers are multigrid methods. Multigrid methods aim to solve linear systems of equations by using a hierarchy of discretisations (Fig. 4.8). At each level of the hierarchy a basic iterative method (e.g. Jacobi, Gauss-Seidel) is applied (smoothing), whereas on the coarsest level a correction is determined (coarse grid correction). This correction is then transferred back to the finest level and used to update the solution. Information between different levels is transferred using prolongation and restriction operators.

Starting from the finest level, different strategies can be adopted to traverse the hierarchy, leading to different cycle types. Fig. 4.8 illustrates the most common cycle type, the V-cycle.

The hierarchy can be obtained in various ways. With h-multigrid, each level of the hierarchy corresponds to a discretisation with a certain mesh width h. Typically, one chooses h, $2h$, $4h$, ... to construct the hierarchy. h-Multigrid methods for IgA discretisations have been studied in [108,136]. An alternative solution strategy is the use of p-multigrid based methods, in which a hierarchy is constructed based on discretisations of different approximation orders. To show the structure of p-multigrid methods, the linear system given by Eq. 4.1

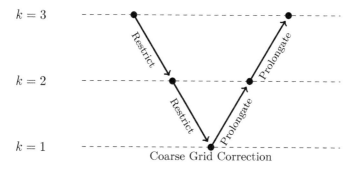

FIGURE 4.8: Description of a V-cycle

is reconsidered. To keep track of the mesh width h and approximation order p on the different grid levels, they are indicated as subscripts (Eq. 4.58).

$$M_{h,p}^C \vec{a}_{h,p} = \vec{f}_{h,p} \tag{4.58}$$

Furthermore, it is noted that B-spline basis functions are adopted for spatial discretisation. Starting from an initial guess $\vec{a}_{h,p}^0$, a single iteration of the (two-grid) p-multigrid method consists of the following steps [300].

1. Apply ν_1 pre-smoothing steps on Eq. 4.58 using Eq. 4.59.

$$\vec{a}_{h,p}^{(0,m+1)} = \vec{a}_{h,p}^{(0,m)} + \mathcal{S}(\vec{f}_{h,p} - M_{h,p}^C \vec{a}_{h,p}^{(0,m)}), \quad m = 0, \ldots, \nu_1 - 1 \tag{4.59}$$

   where $\mathcal{S}$ is a smoother (i.e. a basic iterative method like Jacobi or Gauss-Seidel).

2. Project the residual from level p onto level $p - 1$ using the restriction operator I_p^{p-1} and solve the residual equation given by Eq. 4.60 at level $p - 1$ to obtain the coarse grid correction.

$$M_{h,p-1}^C \vec{e}_{h,p-1} = \vec{r}_{h,p-1} \tag{4.60}$$

   where $\vec{e}_{h,p-1}$ denotes the error (or correction) at the coarse level.

3. Project the correction $\vec{e}_{h,p-1}$ onto level p using the prolongation operator I_{p-1}^p and update $\vec{a}_{h,p}^{(0,\nu_1)}$ using Eq. 4.61.

$$\vec{a}_{h,p}^{(0,\nu_1)} := \vec{a}_{h,p}^{(0,\nu_1)} + I_{p-1}^p(\vec{e}_{h,p-1}) \tag{4.61}$$

4. Apply ν_2 post-smoothing steps to Eq. 4.58 to obtain $\vec{a}_{h,p}^{(0,\nu_1+\nu_2)} =: \vec{a}_{h,p}^1$

The two-grid multigrid method can be applied recursively until level $p = 1$ has been reached, which results in a V-cycle. Alternatively, different schemes can be applied.

Eq. 4.60 is solved at level $p = 1$ using a Conjugate Gradient (CG) solver. Eq. 4.53 with $\epsilon = 10^{-4}$ is chosen as a stopping criterion for the CG method. A detailed description of the prolongation and restriction operator can be found in [300].

The mass matrices, which are needed at each level for the smoothing procedure, are obtained by rediscretisation. The solution $\vec{a}_{h,p}^1$ is used as an initial guess for the next cycle.

The advantage of p-multigrid methods is the fact that on the 'coarsest' level a linear system is solved, where the mass matrix has a more favourable condition number. Furthermore, since they coincide with B-spline basis functions for $p = 1$, established solution techniques for Lagrange Finite Elements can be used. The potential of p-multigrid is illustrated in the following paragraph by considering Eq. 4.1. The equation results from applying a force in

TABLE 4.4

Number of V-cycles needed
with p-multigrid for different
values of p and h.

$p \setminus h^{-1}$	8	16	32	64
1	1	1	1	1
2	4	4	4	3
3	15	12	10	9
4	22	17	13	11
5	58	38	27	20

the negative x- and y-direction. The number of pre- and post-smoothing steps is identical for all numerical experiments ($\nu_1 = \nu_2 = 8$).

The number of V-cycles required by p-multigrid is presented in Table 4.4 and can be compared with the ones presented in Table 4.3. Note that the number of cycles needed with p-multigrid is significantly lower compared to the iterations required by the CG method. Furthermore, p-multigrid, as standard h-multigrid methods, exhibits the h-independence property, implying that the number of cycles needed before the method has converged is independent of the mesh width h.

Numerical results indicate that p-multigrid methods can be used as an efficient solution technique to solve the equations arising in MPM.

4.8 Closure

The combined use of B-spline basis functions and a TLS reconstruction (TLS-BSMPM) significantly improves the solution of the standard MPM algorithm. For small strains, smaller errors and, in many cases, higher convergence rates can be obtained, while for large deformations, TLS-BSMPM reduces the oscillations originating from grid crossing. Furthermore, in contrast to many standard reconstruction techniques, the total mass and linear momentum are conserved provided a sufficiently accurate numerical quadrature method is adopted.

The use of the consistent mass matrix instead of the lumped one for the mapping of MP data to the background grid considerably improves the solution. However, the use of the consistent mass matrix within BSMPM requires efficient solution techniques for high order B-spline discretisations. Since the condition number grows exponentially with p in this case, (standard) iterative methods are inefficient. p-Multigrid methods, in contrast, have been shown to be an efficient solution approach that requires a significantly smaller number of iterations compared to CG in this case.

5

Axisymmetric Formulation

V. Galavi, F.S. Tehrani, M. Martinelli, A.Elkadi and D. Luger

5.1 Introduction

Many geotechnical problems are axisymmetric such as the cone penetration test or pile installations. So far, many MPM codes solved these problems as two-dimensional (2D) plane strain or three-dimensional (3D) problems. The numerical simulation of large deformation problems, such as the cone penetration testing and pile installation, can be carried out in a less expensive way if the effect of rotational symmetry of the pile and the surrounding soil is taken into consideration in the formulation. This results in a new MPM formulation, which is based on an axisymmetric formulation developed by Sulsky and Schreyer [284]. This MPM formulation can significantly reduce the number of elements and MPs compared to a three dimensional formulation. It also make the pre-processing and the post-processing much easier as these tasks are always tedious for 3D configurations.

This chapter presents the theory of the axisymmetric formulation of MPM and its use demonstrated with the numerical simulation of three problems, namely the loading of a circular shallow foundation, the granular column collapse and the cone penetration test.

5.2 Axisymmetric strains

Gens and Potts [112] described three geotechnical engineering examples which can be expressed by an axisymmetric formulation. These are (1) a pile under torsional load, (2) a pile under vertical load, and (3) hollow cylinder samples subject to torsion. This section covers the second case where the piles are subjected to a vertical load and in which the strain tensor has four non-zero components. These are the radial strain ε_r, the vertical strain ε_z, the shear strain γ, and the circumferential strain in the cylindrical coordinate system ε_θ, where θ refers to an angle and not the the Lode angle. This means that there is one more non-zero strain component in the axisymmetric compared to the plane strain. The displacements in the cylindrical coordinate system are denoted as u_r and u_z. The strain components are related to the displacements as given in Eqs. 5.1 and 5.2.

$$\varepsilon_r = \frac{\partial u_r}{\partial r} \quad , \quad \varepsilon_z = \frac{\partial u_z}{\partial z} \quad , \quad \varepsilon_\theta = \frac{\partial u}{\partial r} \tag{5.1}$$

$$\gamma_{rz} = 2\gamma = \left(\frac{\partial u_r}{\partial z} + \frac{\partial u_z}{\partial r} \right) \tag{5.2}$$

where r is the radial distance between the MP and the axis of symmetry.

The sign convention in mechanics sets tensile strains as positive, which differs from the geotechnical convention. In FEM, strains are related to the displacements as given in Eq. 5.3.

$$\varepsilon = B_i u_i = \begin{bmatrix} \dfrac{\partial N_i(x_p)}{\partial r} & 0 \\ 0 & \dfrac{\partial N_i(x_p)}{\partial z} \\ \dfrac{\partial N_i(x_p)}{\partial z} & \dfrac{\partial N_i(x_p)}{\partial r} \\ \dfrac{N_i(x_p)}{\partial r} & 0 \end{bmatrix} \tag{5.3}$$

where u_i is a vector containing nodal displacements of node i. The matrix N contains shape functions at the local location (x_p) of integration points (MPs) and the matrix B contains the gradient of shape functions.

By expanding Eq. 5.3, the strain components in the axisymmetric formulation can be expressed as given in Eqs. 5.4 to 5.7.

$$(\varepsilon_r)_p = \sum_{i=1}^{N_n} u_{r,i} \frac{\partial N_i(x_p)}{\partial r} \tag{5.4}$$

$$(\varepsilon_z)_p = \sum_{i=1}^{N_n} u_{z,i} \frac{\partial N_i(x_p)}{\partial z} \tag{5.5}$$

$$(\gamma_{rz})_p = \sum_{i=1}^{N_n} \left(u_{z,i} \frac{\partial N_i(x_p)}{\partial r} + u_{r,i} \frac{\partial N_i(x_p)}{\partial z} \right) \tag{5.6}$$

$$(\varepsilon_\theta)_p = \sum_{i=1}^{N_n} u_{r,i} \frac{N_i(x_p)}{r} \tag{5.7}$$

where N_n is the number of nodes of an element.

5.3 Internal forces

The main difference between the axisymmetric and the plane strain formulations lies in the matrix B (Eq. 5.3), which is used to calculate the internal forces and strains. Eq. 5.8 is used to get the internal force f_i^{int}.

$$f_i^{int} = \sum_{p=1}^{N_p} B_i^T (x_p) \sigma_p \Omega_p \qquad (5.8)$$

where N_p is the total number of MPs (integration points) within an element, Ω_p the integration weight (volume) of MP p, and σ_p the stress tensor of MP p. The components of the internal force in the axisymmetric formulation are written as given in Eqs. 5.9 and 5.10.

$$f_{r,i}^{int} = \sum_{p=1}^{N_p} \left\{ \left((\sigma_r)_p \frac{\partial N_i (x_p)}{\partial r} + (\sigma_{rz})_p \frac{\partial N_i (x_p)}{\partial z} + (\sigma_{\theta\theta})_p \frac{N_i (x_p)}{\partial r} \right) \Omega_p \right\} \qquad (5.9)$$

$$f_{z,i}^{int} = \sum_{p=1}^{N_p} \left\{ \left((\sigma_z)_p \frac{\partial N_i (x_p)}{\partial z} + (\sigma_{rz})_p \frac{\partial N_i (x_p)}{\partial r} \right) \Omega_p \right\} \qquad (5.10)$$

5.4 Procedure for one calculation step

The solution procedure for one calculation step of the axisymmetric MPM in drained condition is summarised as follows.

1. Lumped mass matrix at nodes is calculated from the MP mass (Eq 5.11).

$$m_i = \sum_{p=1}^{N_p} m_p N_i (x_p) \qquad (5.11)$$

where m_p is the mass of the MP p.

The initial mass of the MP is calculated with Eqs. 5.12-5.13.

$$m_p = \rho_p \Omega_p \qquad (5.12)$$

$$\Omega_p = r_p A_p \qquad (5.13)$$

where ρ_p, A_p, r_p and Ω_p are the initial density, area, radius and volume of MP p, respectively. The initial area of the MP, A_p, is set based on the Gauss quadrature or standard quadrature which can be considered as a fraction of the element area. The volume of a MP is therefore calculated as the area of a MP multiplied by its radius r_p (Eq. 5.13).

2. The momentum at nodes is calculated from the momentum at the MPs by a mass weighted mapping with Eq. 5.14.

$$m_i v_i = \sum_{p=1}^{N_p} m_p v_p N_i \left(x_p\right) \tag{5.14}$$

where v_i and v_p are the velocities at nodes and MPs, respectively.

3. The acceleration at nodes is calculated by solving the discrete system of equations given in Eq. 5.15.

$$m_i a_i = f_i^{\text{ext}} - f_i^{\text{int}} \tag{5.15}$$

where f_i^{int} and f_i^{ext} are the internal and external force vectors, respectively.

The external force vector consists of body forces and traction forces. The internal force is calculated from stresses in the MPs with Eq. 5.8. The components of the internal force vector are given by Eqs. 5.9 and 5.10.

4. The acceleration at nodes is used to update the velocities of MPs using the shape functions (Eq. 5.16).

$$v_p^{t+\Delta t} = v_p^{\text{T}} + \sum_{i=1}^{N_n} \Delta t N_i \left(x_p\right) a_i^{\text{T}} \tag{5.16}$$

5. The nodal velocities are then calculated from the updated MP velocities by solving (Eq. 5.14).

6. The nodal incremental displacements can be calculated by integrating the nodal velocities (Eq. 5.17).

$$\Delta u_i^{t+\Delta t} = \Delta t v_i^{t+\Delta t} \tag{5.17}$$

7. The strains and stresses are calculated at MPs (Eq. 5.18).

$$\Delta \varepsilon_p^{t+\Delta t} = B \left(x_p\right) \Delta u_e^{t+\Delta t} \tag{5.18}$$

The subscript e stands for element which means that the nodal incremental displacements of the element are used. The calculated strains can then be used for calculation of stresses using a constitutive model. The components of the strain vector in the axisymmetric formulation are given by Eqs. 5.4 and 5.7.

8. The volume and density of MPs are updated using the volumetric strain $\Delta \varepsilon_v$.

$$\Omega_p^{t+\Delta t} = \left(1 + \Delta \varepsilon_{v,p}^{t+\Delta t}\right) \Omega_p^t \tag{5.19}$$

$$\rho_p^{t+\Delta t} = \frac{\rho_p^{\text{T}}}{\left(1 + \Delta \varepsilon_{v,p}^{t+\Delta t}\right)} \tag{5.20}$$

where the subscript p refers to the MP p and not the plastic component.

9. Displacements and positions of MPs are updated with Eqs. 5.21 and 5.21.

$$u_\text{p}^{t+\Delta t} = u_\text{p}^\text{T} + \sum_{i=1}^{N_\text{n}} N_i\left(x_\text{p}\right) \Delta u_i^{t+\Delta t} \tag{5.21}$$

$$x_\text{p}^{t+\Delta t} = x_\text{p}^\text{T} + \sum_{i=1}^{N_\text{n}} N_i\left(x_\text{p}\right) \Delta u_i^{t+\Delta t} \tag{5.22}$$

10. At this step, the mesh is reset and a space search is performed to find new positions of MPs in the background mesh.

A similar procedure can be developed for the two-phase formulation [309] (see Chapter 14), which includes the generation and dissipation of excess pore water pressure. However, this extension falls out of the scope of this present chapter.

5.5   Examples

Three examples are considered to verify the implementation of the axisymmetric formulation and are (1) a circular surface load on an elastic ground, (2) a granular column collapse, and (3) a cone penetration test (CPT).

5.5.1   Circular surface load on an elastic ground

An analytical solution for circular surface load on an elastic ground is given and then the solution is compared with results of MPM. According to Boussinesq, the vertical stress at a point (r, z) under a point load can be found from Eq. 5.23.

$$\sigma_z = \frac{3q}{2\pi z^2} \frac{1}{\left(1 + \dfrac{r^2}{z}\right)^{3/2}} \tag{5.23}$$

By integrating it over a circular area with the radius of R, the vertical stress below the centre point at any depth z can be found from Eq. 5.24.

$$\sigma_z = q\left(1 - \frac{z^3}{\left(r^2 + z^2\right)^{3/2}}\right) \tag{5.24}$$

Similarly, the horizontal radial stress is derived as given in Eq. 5.25.

$$\sigma_r = \frac{q}{2}\left(1 + 2\nu - \frac{2\left(1 + \nu\right)z}{\left(R^2 + z^2\right)^{1/2}} + \frac{z^3}{\left(R^2 + z^2\right)^{3/2}}\right) \tag{5.25}$$

where ν is the Poisson ratio. For the centre point right below the circular load, the vertical and horizontal radial stresses can be found by setting the depth z to zero in Eqs. 5.24 and 5.25. This gives Eq. 5.26.

$$z = 0: \quad \sigma_z = q \quad \text{and} \quad \sigma_r = \frac{q}{2}\left(1 + 2\nu\right) \tag{5.26}$$

The vertical strain ε_z, due to the triaxial load under the centre point is given by Eq. 5.27.

$$\varepsilon_z = \frac{1}{E}\left(\sigma_z - 2\nu\sigma_r\right) \tag{5.27}$$

where E is the Young modulus.

By substituting σ_z and σ_r from Eqs. 5.24 and 5.25 in Eq. 5.27, and integrating from $z = z$ to $z = \infty$, the elastic deformation D under the centre point at any z can be found (Eq. 5.28).

$$D = \frac{q}{E}\left(\left(2 + 2\nu^2\right)\left(R^2 + z^2\right)^{1/2} - \frac{(1+\nu)z^2}{\left(R^2 + z^2\right)^{1/2}}\left(\nu + 2\nu^2 - 1\right)z\right) \tag{5.28}$$

For $z = 0$, total displacement below the surface load can be found (Eq. 5.29).

$$z = 0: \quad D = \frac{qR}{E}\left(2 - 2\nu^2\right) \tag{5.29}$$

The circular footing is simulated using the axisymmetric MPM. A surface pressure of -1,000 kPa with a radius of 1 m is applied on an elastic medium with a stiffness of $E = 10^4$ kPa and Poisson ratio of $\nu = 0.3$. From the analytical solutions (Eqs. 5.26 and 5.29), the vertical and horizontal stresses and the vertical settlement at the centre point below the surface load are $\sigma_z = -1,000$ kPa, $\sigma_r = -800$ kPa and $D = 0.182$ m. It should be noted that the analytical solutions are based on the small deformation assumption.

The simulated vertical and horizontal stresses are depicted in Fig. 5.1 (a) and (b). The vertical stresses at the centre of the footing are $\sigma_z = -1144$ kPa, $\sigma_r = -832$ kPa, respectively, which are in good agreement with the analytical solution. The predicted vertical displacement at the centre of the footing (Fig. 5.1 c) is 17.46 cm.

By comparing the results of the numerical simulations with the analytical solutions, it can be concluded that the axisymmetric MPM formulation is capable of predicting the stresses and deformation of the circular footing problem.

5.5.2   Granular column collapse

The granular column collapse is a well-established experiment in the field of granular mechanics, which is discussed in detail in Chapter 9. Two experimental layouts exist – the plane strain and axisymmetric layouts. Most numerical simulations are carried with plane strain or 3D models as the experimental

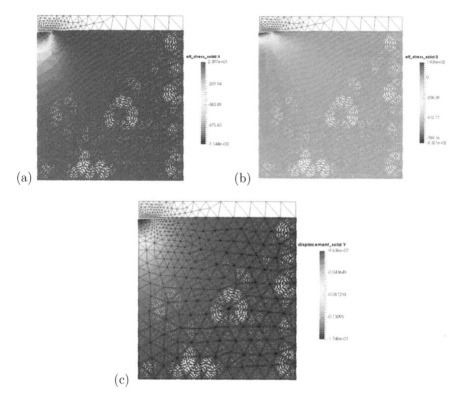

FIGURE 5.1: (a) Vertical stress, (b) horizontal stress, and (c) vertical displacement field obtained from the axisymmetric MPM simulation.

results suggest little differences between both configurations. In this example, the axisymmetric MPM is compared against the 3D MPM [5] in simulating a dry granular column collapse problem in the rotational symmetric condition. The 3D model is reduced in size by considering a quarter of the model. The 3D element and MP discretisation are shown in Fig. 5.2 (a). The mesh is composed of 24,460 tetrahedral elements for 5,230 MPs. The axisymmetric model is shown in Fig. 5.2 (b) and includes the MP discretisation. The mesh consists of 540 triangle elements and 312 MPs.

The Mohr-Coulomb material model is used to describe the mechanical behaviour of the soil and the material parameters are given in Table 5.1. Fig. 5.3 shows results of the 3D model and the 2D model at different steps; the simulations are superimposed in order to facilitate the comparison. The results obtained from the three dimensional model are the same for both cases in terms of failure mode and post-failure behaviour, which is consistent with the MPM theory and the experimental data. However, the computational time required to simulate the 3D problem using one core is 28 s whereas the axisymmetric simulation is 0.47 s on the same machine, which is almost 60 times faster calculation. This is due to the fact that the axisymmetric formulation needs fewer elements and MPs.

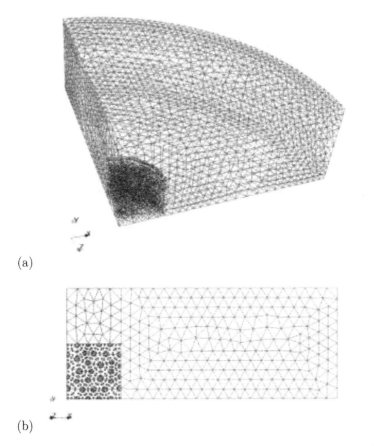

(a)

(b)

FIGURE 5.2: Element and MP discretisation for (a) the 3D model and (b) the axisymmetric model.

TABLE 5.1

Material parameters used for the granular column collapse.

Parameter	Symbol	Unit	Value
Young modulus	E	kPa	1,000
Poisson ratio	ν	-	0.2
Cohesion	c	kPa	0
Friction angle	φ'	°	30
Dilatancy angle	ψ	°	0

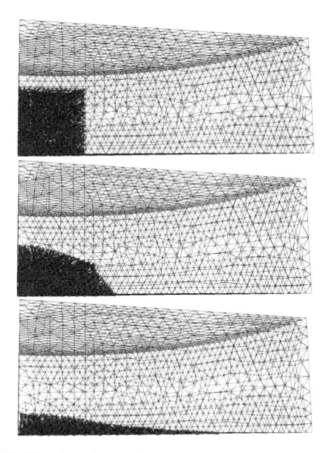

FIGURE 5.3: Granular column collapse: 3D model (red dots) versus axisymmetric model (blue dots).

5.5.3 Cone penetration testing in dry sand

The cone penetration test (CPT) is one of the most used methods for site characterisation because of its simplicity in execution. Many design methods are developed based on the cone resistance correlations. Numerical simulation of the CPT has always been a challenging task due to the large deformations involved in the problem. Different methods have been used in the past for numerical simulation of the CPT, such as the arbitrary Lagrangian-Eulerian finite element technique (e.g. [303]), the particle finite element method (PFEM) (e.g. [204]), the cavity expansion method (e.g. [239]) and the discrete element method (e.g. [15]). A more extensive discussion on the CPT simulations is given in Chapter 18.

Tehrani and Galavi [294] studied the effectiveness of the numerical spherical cavity expansion method and MPM in simulating the cone penetration testing in sand described in this section.

The cone has a diameter of $\phi = 0.036$ m and is initially embedded at a depth of $z = 0.36$ m ($= 10\phi$) below the soil surface. A constant vertical velocity of 0.02 m/s is applied to the cone, which is modelled as a rigid body. A frictional contact condition (see Chapter 6) is considered at the interface of the cone and the soil and is based on the work of Bardenhagen et al. [22]. This contact condition allows frictional sliding and separation but prevents interpenetration. The friction between the soil and the steel cone is assumed to be $\mu = \tan 20° = 0.36$, which is 2/3 of the critical state friction angle of the soil ($\varphi'_{cs} = 30°$).

To increase the accuracy in the contact formulation, a moving mesh (Chapter 6) is used to ensure that the contact nodes of the soil and the cone are always at the same elevation. In addition, the fine mesh defined always remains around the cone. During the penetration, the top part of the background mesh moves together with the cone penetrator while the bottom part is compressed.

Fig. 5.4 shows the MPM model, which has 5,950 triangle elements and 28,782 MPs. The elements are refined in the vicinity of the cone penetrator in order to increase the accuracy.

The sand is a silica sand called Baskarp sand. Simulations are carried out for three different relative densities ($I_D = 30\%$, 50% and 90%). The initial stress in the ground is set based on the unit weight of the soil and a surcharge. The unit weight of the soil is kept constant and equal to 16 kN/m^3 for

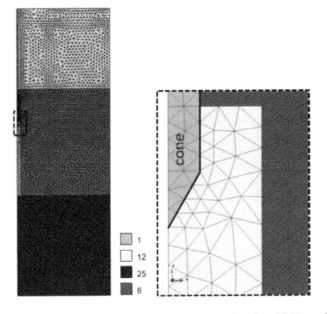

FIGURE 5.4: Mesh and MP discretisation for the CPT model.

TABLE 5.2
Material parameters of the soil in the CPT.

Parameter	Symbol	Unit	Value		
			$I_D = 30\%$	$I_D = 50\%$	$I_D = 90\%$
Young modulus	E	kPa	7,894	14,000	43,659
Poisson ratio	ν	-	0.2	0.2	0.2
Cohesion	c	kPa	1	1	1
Friction angle	φ'	°	32.8	36.5	43.4
Dilatancy angle	ψ	°	2.8	6.5	13.4

all the relative densities, although this is not physically correct but it ensures that the initial stresses are the same for all cases. This offers convenience when studying the effect of different strength and stiffness properties of the soil on final results. A thin layer with higher density is placed on the ground surface in order to apply a uniform surcharge of 25 kPa, which results in an additional horizontal effective stress of almost 12.5 kPa in the model. According to Salgado and Prezzi [256], the cone resistances q_c can be computed as given in Eq. 5.30.

$$q_c = 1.64 p_A \exp\left(0.1041\varphi'_c + (2.64 - 0.02\varphi'_c)I_D\right)\left(\frac{\sigma'_h}{p_A}\right)^{(0.841-0.47I_D)} \tag{5.30}$$

where σ'_h is the horizontal effective stress, p_A the reference stress, which is $p_A = 100$ kPa, and I_D the relative dilatancy index (see Chapter 7).

The cone resistance q_c is obtained using Eq. 5.30 for a reference point 1 m below the surcharge layer as 2.26 MPa, 3.95 MPa and 12.1 MPa for relative densities of 30%, 50% and 90%, respectively.

The Mohr-Coulomb model is used to describe the mechanical behaviour of the sand. The stiffness parameters are determined using the equations given by [180] for normally consolidated and uncemented predominantly silica sands (Eq. 5.31).

$$E_{oed} = 4q_c \tag{5.31}$$

The peak friction angle φ'_p is determined by considering the friction angle to be the sum of the critical state friction angle and the dilatancy angle ($\varphi' = \varphi'_{cs} + \psi$). The dilation angle ψ is calculated using the relative dilatancy index [40]. Table 5.2 summarises the material parameters of Baskarp sand.

The values of the cone resistance q_c are obtained from the MPM simulations and are shown in Fig. 5.5. These are compared with the q_c values obtained with Eq. 5.30 and the results show good agreement with the predicted cone resistance.

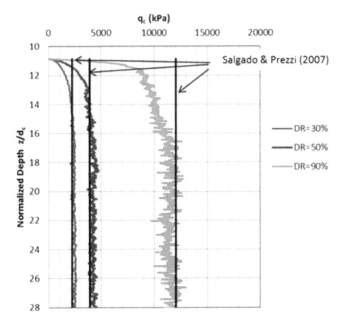

FIGURE 5.5: Empirical and numerically predicted cone resistance (averaged stresses at the cone tip).

5.6 Closure

This chapter presents the axisymmetric MPM formulation, which is computationally more efficient than 3D simulations as it requires a smaller computational domain and significantly fewer MPs. The axisymmetric MPM formulation is then validated by simulating the loading of a circular footing and the results are validated with Boussinesq's theoretical solution. The axisymmetric MPM formulation is then used to simulate the granular column collapse and the results are compared with 3D MPM simulations. It is shown that the axisymmetric formulation gives the same results as the 3D simulation but for a lower computational cost. Finally, a cone penetration test is simulated and validated by comparing the numerical prediction with the empirical prediction.

6

Numerical Features Used in Simulations

F. Ceccato and P. Simonini

6.1 Introduction

Numerical methods solve predefined governing equations of a given boundary value problem in a specific sequence. Chapter 2 explains how these equations are solved with MPM. Additional algorithms are required in order to obtain a numerically stable or computationally efficient solution as well as for defining specific conditions necessary for finding the correct solution to the problem. This chapter presents some of the most common numerical features used in MPM simulations.

The application of boundary conditions along moving boundaries in large deformation MPM simulations requires special attention, especially for non-zero tractions. Dynamic phenomena with waves propagating throughout the medium might require the introduction of absorbing boundaries in order to avoid reflections at the boundaries. Moreover, artificial bulk viscosity damping can mitigate the oscillations present across shock waves in explicit integration schemes. Simulation of quasi-static problems with a dynamic formulation requires the introduction of local damping in order to enable the convergence to a static equilibrium state. Mass scaling improves the computational efficiency by introducing non-physical mass to the system in order to achieve larger explicit time steps. Contact conditions between different bodies and the boundary may require dedicated conditions. Volumetric locking can cause severe perturbation to the solution, especially for nearly-incompressible materials, and can be overcome with a dedicated algorithm.

6.2   Boundary conditions

Two kinds of boundary conditions can be distinguished – essential and natural. Essential boundary conditions are imposed directly on the solution, and the degrees of freedom are directly eliminated from the system of equations. These are typically fixities or prescribed velocities or displacements. Natural boundary conditions are imposed on a secondary variable, such as stresses or pressures. Calculated velocities are directly overwritten by the prescribed values, while prescribed traction is included in the weak form of the governing equations.

In classical FEM, the application of prescribed boundary conditions is simple as these can be specified directly on the boundary nodes, which coincide with the boundary of the continuum body and are well defined throughout the computation. However, the computational mesh in MPM does not necessarily align with the boundary of the material making the application of the prescribed boundary conditions more challenging. The difficulty arises when dealing with non-zero boundary conditions, including non-zero tractions and non-zero kinematic boundary conditions. Zero kinematic boundary conditions, also called fixities, can be applied at the degree of freedom of the computational grid in a similar way as in FEM. However, those conditions have to be applied to boundary nodes that might become active during computations. This is when a material point is located in the element connected to the node. Applying essential boundary conditions that are not aligned with element boundaries is not trivial. In some cases, the problem is overcome by workarounds such as introducing stiff regions or applying a moving mesh (see Section 6.8). Recently, the implicit boundary method (IBM), which was originally developed for FEM, has been extended to MPM [72]. It consists in defining a trial and test function space that will implicitly lead to the enforcement of the essential boundary conditions.

Zero-traction boundary conditions are automatically enforced to be satisfied by the solution of equations of motion (see Chapter 2). Dealing with non-zero traction boundary conditions is more complex. Traction loads can be applied on either the element boundaries or the material points (MPs). The first option is applicable only when the boundary of the body remains aligned with loaded element boundaries throughout the computation. The second option consists in storing the load on selected MPs that move through the mesh. When the traction is assigned at the element boundary, the nodal traction force is integrated like in FEM applying Gauss quadrature and then used in the momentum balance equation (Fig. 6.1 a). The applied load is thus integrated accurately and the traction nodal force is non-zero only for the nodes belonging to the loaded surface. When the traction is applied on the MPs, this is mapped from MPs to all nodes of the element where it is located by means of the shape functions (Fig. 6.1 b). Such mapping procedure is carried

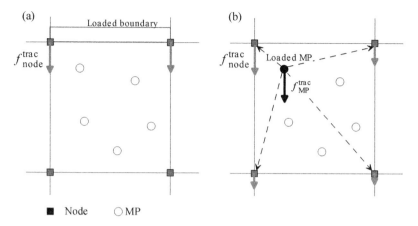

FIGURE 6.1: (a) Traction is integrated from element boundary to adjacent nodes. (b) Traction is mapped from a loaded MP to the nodes of the element.

out according to Eq. 6.1.

$$\vec{f}^{\text{trac}} = \sum_{\text{el}=1}^{n_{\text{el}}} \sum_{\text{MP}=1}^{n_{\text{MP,el}}} \mathbf{N}^{\text{T}}(\vec{\xi}_{\text{MP}}) \vec{f}_{\text{MP}}^{\text{trac}} \tag{6.1}$$

The disadvantage of this way of dealing with surface tractions is that the effect of the surface force is distributed across the layer of elements that borders the boundary. If a MP is affected by a moving boundary condition, the element containing the point becomes part of the contour. Then, the contour has an equivalent thickness of the size of the element and the boundary condition is spread affecting all MPs within the cell, even those which are not at the boundary. To minimise numerical errors, the thickness of the elements along the boundary should be small and the use of a fine mesh is recommended.

Another difficulty with moving boundaries appears when an external traction does not have a prescribed direction. This means that if the shape of the contour changes, the direction of the applied condition has to be updated during the calculation. This is computationally expensive as the normal vector must be calculated for every time step.

6.3 Absorbing boundaries

The use of finite boundaries produces wave reflections that do not characterise a naturally unbounded domain. Different strategies have been proposed in literature to treat wave reflections (see e.g. [113] for an overview). One of the most common approaches in dynamic codes is the use of absorbing boundaries. This method [183] applies viscous damping forces (dashpot) along the artificial boundary. Al-Kafaji [5] observed that the boundary that is supported by dashpots tends to creep continuously, particularly under sustained loading. In order to control creep, dashpots are replaced by Kelvin-Voigt elements, which consider the parallel combination of a spring and a dashpot as shown in Fig. 6.2.

The traction vector corresponding to the absorbing boundary has three components: a component normal to the boundary τ_n^{ab}, and two components in tangential directions $\tau_{t_1}^{ab}$ and $\tau_{t_2}^{ab}$. The response of the Kelvin-Voigt element is described by Eq. 6.2.

$$\tau_n^{ab} = -a\rho c_p v_n - k_p u_n \tag{6.2a}$$

$$\tau_{t_1}^{ab} = -b\rho c_s v_{t_1} - k_s u_{t_1} \tag{6.2b}$$

$$\tau_{t_2}^{ab} = -b\rho c_s v_{t_2} - k_s u_{t_2} \tag{6.2c}$$

where a and b are dimensionless parameters, v_n, v_{t_1} and v_{t_2} the velocities, u_n, u_{t_1} and u_{t_2} the displacements, ρ the unit mass, c_p and c_s the velocities of the compression and shear waves, respectively. k_p and k_s represent the stiffness per unit area associated to the elastic component.

Eq. 6.2 can be rewritten in a matrix form as given by Eq. 6.3.

$$\vec{\tau}^{ab} = -\eta^{ab}\vec{v} - \mathbf{K}^{ab}\vec{u} \tag{6.3}$$

In an isotropic linear elastic material, the wave velocities (Eq. 6.4) are functions of the constrained modulus E_c, the elastic shear modulus G and the mass density ρ.

$$c_p = \sqrt{\frac{E_c}{\rho}} \quad \text{and} \quad c_s = \sqrt{\frac{G}{\rho}} \tag{6.4}$$

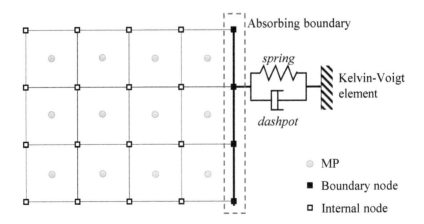

FIGURE 6.2: Absorbing boundary with Kelvin-Voigt element.

The first term in the right-hand side of Eq. 6.2 represents the traction given by the dashpot, which is proportional to the velocity. The second term represents the traction given by the spring, which is proportional to the displacement. The matrix form of the discretised equation of motion, including the force at the absorbing boundary, reads as Eq. 6.5.

$$\mathbf{M}\vec{a} = \vec{f}^{\text{trac}} - \vec{f}^{\text{ab}} + \vec{f}^{\text{grav}} - \vec{f}^{\text{int}} \tag{6.5}$$

where the contribution of the absorbing boundary $\vec{f}^{\text{ab}}$ is computed by integrating $\vec{\tau}^{\text{ab}}$ over the surface $\partial\Omega_{\text{ab}}$ wherever the absorbing boundary is applied. It can then be written in matrix form as Eqs. 6.6 to 6.8.

$$\vec{f}^{\text{ab}} = \mathbf{C}^{\text{ab}}\vec{v} + \mathbf{K}^{\text{ab}}\vec{u} \tag{6.6}$$

$$\mathbf{C}^{\text{ab}} = \int_{\partial\Omega_{\text{ab}}} \mathbf{N}\eta^{\text{ab}}\mathbf{N}^{\text{T}}\mathrm{d}S \tag{6.7}$$

$$\mathbf{K}^{\text{ab}} = \int_{\partial\Omega_{\text{ab}}} \mathbf{N}k^{\text{ab}}\mathbf{N}^{\text{T}}\mathrm{d}S \tag{6.8}$$

where the dashpot matrix $\mathbf{C}^{\text{ab}}$ contains the dashpot coefficients (Eq. 6.7), and the spring matrix $\mathbf{K}^{\text{ab}}$ contains the spring coefficients (Eq. 6.8).

The coefficients k_{p} and k_{s} can be expressed as functions of the elastic moduli of the material adjacent to the absorbing boundary and a virtual thickness δ as Eq. 6.9.

$$k_{\text{p}} = \frac{E_{\text{c}}}{\delta} \quad \text{and} \quad k_{\text{s}} = \frac{G}{\delta} \tag{6.9}$$

The virtual thickness δ can be interpreted as the thickness of a virtual layer, which extends outside the boundary. Note that for $\delta \to 0$ the absorbing boundary reduces to a rigid boundary. It reduces to a dashpot boundary for $\delta \to \infty$.

Since MPs move through the mesh, the scheme must be implemented in an incremental form such that displacement and velocity increments of a MP are only accounted for when it enters the considered element (Eqs. 6.10 and 6.11).

$$\vec{f}^{\text{ab,t}} = \vec{f}^{\text{ab,t}-1} + \Delta\vec{f}^{\text{vb,t}} \tag{6.10}$$

$$\Delta\vec{f}^{\text{vb,t}} = \mathbf{C}^{\text{t}}\Delta\vec{v} + \mathbf{K}^{\text{t}}\Delta\vec{u} \tag{6.11}$$

For the two-phase formulation, two sets of Kelvin-Voigt elements need to be defined. For the liquid phase, the traction of the absorbing boundary has only the normal component and is given by Eq. 6.12.

$$p_{\text{L}}^{\text{ab}} = -a_{\text{L}}\rho_{\text{L}}c_{\text{L}}v_{\text{L,n}} - k_{\text{L}}u_{\text{L,n}} \tag{6.12}$$

where $v_{\text{L,n}}$ and $u_{\text{L,n}}$ are the normal component of liquid velocity and displacement, respectively. The velocity of the compression wave in the liquid is given by Eq. 6.13.

$$c_{\text{L}} = \sqrt{\frac{K_{\text{L}}}{\rho_{\text{L}}}} \tag{6.13}$$

The stiffness coefficient k_L can be expressed as a function of the bulk modulus of the liquid K_L and the virtual thickness δ_L as Eq. 6.14.

$$k_\mathrm{L} = \frac{K_\mathrm{L}}{\delta_\mathrm{L}} \tag{6.14}$$

The traction in the matrix form on the absorbing boundary nodes can be written as Eq. 6.15.

$$\vec{f}_\mathrm{L}^{\,\mathrm{ab}} = \mathbf{C}_\mathrm{L}^{\mathrm{ab}} \vec{w} + \mathbf{K}_\mathrm{L}^{\mathrm{ab}} \vec{u}_\mathrm{L} \tag{6.15}$$

where $\mathbf{C}_\mathrm{L}^{\mathrm{ab}}$ and $\mathbf{K}_\mathrm{L}^{\mathrm{ab}}$ are the dashpot and spring matrix for the liquid phase respectively. The traction for the solid part at the absorbing boundary represents the traction applied on the soil skeleton and is, therefore, proportional to the solid velocity $\vec{v}_\mathrm{S}$ and displacement $\vec{u}_\mathrm{S}$ (Eq. 6.16).

$$\vec{f}_\mathrm{S}^{\,\mathrm{ab}} = \mathbf{C}_\mathrm{S}^{\mathrm{ab}} \vec{v}_\mathrm{S} + \mathbf{K}_\mathrm{s}^{\mathrm{ab}} \vec{u}_\mathrm{S} \tag{6.16}$$

The dry density ρ_dry and the effective constrained modulus E_c' should be used to estimate dashpot and spring coefficients in Eq. 6.16.

In order to illustrate the effect of introducing an absorbing boundary, the problem of wave propagation in a linear elastic saturated porous medium is considered. Although the example is one-dimensional, it is simulated with a three-dimensional MPM implementation (*Anura3D* [14]). A 2.5 m long column is discretised with $1,000$ rows of six tetrahedral elements. One MP is initially placed inside each element. Roller boundaries are prescribed at the lateral surfaces. An absorbing boundary is applied at the bottom of the column. The material parameters are listed in Table 6.1. The dashpot coefficients and the virtual thickness of Eqs. 6.15 and 6.16 can differ for the solid and the liquid phase but in the following example identical values are used for both phases. The dashpot coefficients a and b are assumed equal to 1.0. Lysmer and Kuhlemeyer [183] showed that this choice gives the maximum absorption for both compression and shear waves for a wide range of incidence angles. The column is instantly loaded with $\sigma_\mathrm{y} = 10$ kPa at the top surface. The propagation of the undrained wave is studied following the pore pressure of a MP located 0.675 m below the top surface.

In undrained conditions, the total load is distributed between pore pressure p_L and effective stress depending on the stiffness of the phases. The expected normalised pressure is given by Eqs. 6.17 and 6.18.

$$\frac{p_\mathrm{L}}{\sigma_\mathrm{y}} = \frac{K_\mathrm{L}/n}{E_\mathrm{c,u}} = 0.5 \tag{6.17}$$

$$E_\mathrm{c,u} = E' \frac{1-\nu}{(1+\nu)(1-2\nu)} + \frac{K_\mathrm{L}}{n} \tag{6.18}$$

where $E_\mathrm{c,u}$ is the undrained confined compression modulus. The speed of the undrained wave can be calculated with Eq. 6.19.

$$c_\mathrm{p,u} = \sqrt{\frac{E_\mathrm{c,u}}{\rho_\mathrm{sat}}} \tag{6.19}$$

where $\rho_{\text{sat}} = (1 - n)\rho_{\text{s}} + n\rho_{\text{L}}$. For the considered problem $c_{\text{p,u}} = 2{,}236$ m/s. The wave travels through the soil and is expected to reach the considered MP at $t = 3.02 \cdot 10^{-4}$ s. At the bottom it is reflected and starts travelling upwards reaching again the MP at $t = 1.93 \cdot 10^{-3}$ s. If the bottom boundary is rigid and impermeable the pore pressure increases by a factor 2, but if there is an absorbing boundary the amplitude of the reflected wave is decreased, thus proving the efficiency of the implemented absorbing boundary. Increasing δ/h, where h is the length of the column, the amplitude of the reflected wave decreases (Fig. 6.3).

TABLE 6.1
Material parameters for the one-dimensional wave propagation problem with absorbing boundary.

Parameter	Symbol	Unit	Value
Effective Young modulus	E'	kPa	$5 \cdot 10^6$
Poisson ratio	ν	-	0
Liquid bulk modulus	K_{L}	kPa	$2 \cdot 10^6$
Porosity	n	-	0.4
Dry density	ρ_{dry}	kg/m^3	1 600
Liquid density	ρ_{L}	kg/m^3	1 000
Permeability	k	m/s	$1.0 \cdot 10^{-5}$

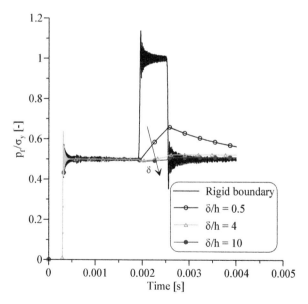

FIGURE 6.3: Performance of absorbing boundary for two-phase formulation. Normalised pore pressure for a MP at 0.675 m below the top surface (after [52]).

6.4 Stress initialisation

As for all numerical methods, the initial stress state has to be defined or computed prior to the analysis. One way of initialising the stresses is to carry out a phased analysis in which in the first phase the initial stresses are computed. This is done by increasing the gravity until the solution converges to quasi-static equilibrium. This is done by evaluating the normalised kinetic energy E_{kin}, and the normalised out-of-balance force falls below a predefined tolerance (tol_{E} or tol_{F}) as given by Eqs. 6.20 and 6.21.

$$\frac{E_{\text{kin}}}{W_{\text{ext}}} < tol_{\text{E}} \tag{6.20}$$

$$\frac{\|\vec{f}^{\text{ext}} - \vec{f}^{\text{int}}\|}{\|\vec{f}^{\text{ext}}\|} < tol_{\text{F}} \tag{6.21}$$

where W_{ext} is the work of the external forces, $\vec{f}^{\text{ext}}$ the external force, $\vec{f}^{\text{int}}$ the

, and $\|\vec{f}^{\text{ext}} - \vec{f}^{\text{int}}\|$ the norm of the out-of-balance force.

   In the cases of a horizontal ground surface and groundwater table, the initial stress can be initialised with the K_0-procedure, which is based on the effective stress principle. With this approach, the vertical stress at MP increases linearly with depth and can be computed with Eqs. 6.22 and 6.23.

$$\sigma'_{\text{v,0}} = \rho g z_{\text{p}} - \rho_{\text{w}} g z_{\text{w}} \tag{6.22}$$

$$\sigma'_{\text{h0}} = K_0 \sigma'_{\text{v,0}} \tag{6.23}$$

where z_{p} is the depth of the considered MP below a horizontal surface, ρ_{w} the density of water z_{w} the depth of the considered MP below the groundwater table, and K_0 the earth pressure coefficient at rest.

6.5 Local damping

Although dynamic MPM formulations are developed to solve problems that evolve relatively fast and in which inertia plays an important role, they can be also applied to processes in which the steady state solution is of interest. To accelerate the convergence to quasi-static equilibrium, an artificial local damping can be included in the formulation in order to introduce energy dissipation in the dynamic momentum conservation. Artificial local damping can also be applied to mimic natural damping in soils and rocks, which often is not adequately represented by most of the constitutive models applied in geomechanics. In this case the damping in the numerical simulation should reproduce in magnitude and form the energy losses in the natural system when subjected to a dynamic loading.

Local damping consists in adding a force to a node. This force is proportional to the out-of-balance force and acts in the opposite direction of the velocity. The discretised momentum balance (Eq. 2.23) is modified as Eqs. 6.24 and 6.25.

$$\mathbf{M}\vec{a} = \vec{f} + \vec{f}^{\,\mathrm{damp}} \tag{6.24}$$

$$\vec{f}^{\,\mathrm{damp}} = -\alpha \|\vec{f}\| \frac{\vec{v}}{\|\vec{v}\|} \tag{6.25}$$

where $\vec{f}^{\,\mathrm{damp}}$ is the damping force and $\vec{f} = \vec{f}^{\,\mathrm{ext}} - \vec{f}^{\,\mathrm{int}}$ the out-of-balance force. The dimensionless parameter α is the local damping factor.

In the two-phase formulation, the liquid and the solid phase are damped separately. The momentum balance for the liquid (Eq. 2.68) takes the form of Eq. 6.26.

$$\mathbf{M}_{\mathrm{L}}\vec{a_{\mathrm{L}}} = \vec{f}_{\mathrm{L}}^{\,\mathrm{ext}} - \vec{f}_{\mathrm{L}}^{\,\mathrm{int}} + \vec{f}_{\mathrm{L}}^{\,\mathrm{damp}} - \vec{f}^{\,\mathrm{drag}} \tag{6.26}$$

The first two terms on the right-hand side represent the out-of-balance force for the liquid phase (Eq. 6.27).

$$\vec{f}_{\mathrm{L}} = \vec{f}_{\mathrm{L}}^{\,\mathrm{ext}} - \vec{f}_{\mathrm{L}}^{\,\mathrm{int}} \tag{6.27}$$

The damping force for the liquid can be written as Eq. 6.28.

$$\vec{f}_{\mathrm{L}}^{\,\mathrm{damp}} = -\alpha_{\mathrm{L}} \|\vec{f}_{\mathrm{L}}\| \frac{\vec{v}_{\mathrm{L}}}{\|\vec{v}_{\mathrm{L}}\|} \tag{6.28}$$

The momentum equation for the mixture is given by Eqs. 6.29 and 6.30.

$$\mathbf{M}_{\mathrm{S}}\vec{a}_{\mathrm{S}} = -\mathbf{M}_{\mathrm{L}}\vec{a}_{\mathrm{L}} + \vec{f}^{\,\mathrm{ext}} - \vec{f}^{\,\mathrm{int}} + \vec{f}^{\,\mathrm{damp}} \tag{6.29}$$

$$\vec{f}^{\,\mathrm{damp}} = \vec{f}_{\mathrm{S}}^{\,\mathrm{damp}} + \vec{f}_{\mathrm{L}}^{\,\mathrm{damp}} \tag{6.30}$$

where $\vec{f} = \vec{f}^{\,\mathrm{ext}} - \vec{f}^{\,\mathrm{int}}$ is the out-of-balance force of the mixture. The damping force for the solid can be written as Eq. 6.31.

$$\vec{f}_{\mathrm{S}}^{\,\mathrm{damp}} = -\alpha_{\mathrm{S}} \|\vec{f} - \vec{f}_{\mathrm{L}}\| \frac{\vec{v}_{\mathrm{S}}}{\|\vec{v}_{\mathrm{S}}\|} \tag{6.31}$$

Local damping is originally developed to achieve quasi-static solutions. However, it has some characteristics that make it attractive to use in dynamic simulations if proper values for the damping coefficient α are used. For quasi-static problems high values of α between 0.5 and 0.75 can be used to accelerate the convergence to static equilibrium. In dynamic problems, a small value of α between 0.05 and 0.15 can simulate natural energy dissipation of the material, if it is not taken into account by the constitutive model. Local damping

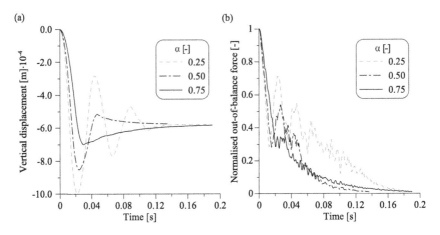

FIGURE 6.4: Effect of local damping factor on the (a) displacement of MP and (b) out-of-balance force.

should not be used in highly dynamic problems, where wave propagation is important.

To understand the effect of the local damping factor, a column of 1 m under gravity loading is considered. The column is discretised with a structured mesh of 10 rows of 6 tetrahedral elements. 4 MPs are located inside each element. The material is linear elastic with a Young modulus of $E = 10$ MPa, a Poisson ratio $\nu = 0.3$, and a unit mass of $\rho_s = 1,590$ kg/m^3. Initial stresses are zero and gravity is instantly applied at the beginning of the analysis.

Fig. 6.4 (a) shows the displacement in time of a MP located at the top of the column. After application of gravity the point oscillates around the quasi-static solution. The amplitude of these oscillations decreases with time and this reduction increases with the increase of the local damping factor. Fig. 6.4 (b) shows that the local damping accelerates the decrease of the out-of-balance forces, thus, accelerating the convergence to the static equilibrium.

6.6 Artificial bulk viscosity damping

Artificial bulk viscosity damping [321] has the purpose to improve the simulation of shock propagation. A shock front mathematically represents a travelling surface of discontinuity; explicit integration schemes tend to oscillate in the vicinity of shocks leading to violation of the conservation properties. Using viscosity damping these oscillations will be mitigated, but the shock front will be smeared across several elements.

The concept of artificial bulk viscosity damping is to add a viscosity term q_{vis} to the stress tensor in case of compression, and can be regarded as a viscous contribution to the conservation law. No damping is applied for relaxation. The viscosity term reads as Eq. 6.32.

$$q_{\text{vis}} = \begin{cases} \rho\,(c_1 l\,\text{tr}(\dot{\varepsilon}))^2 - \rho c_0 l c\,\text{tr}(\dot{\varepsilon}) & \text{if} \quad \text{tr}(\dot{\varepsilon}) < 0 \\ 0 & \text{if} \quad \text{tr}(\dot{\varepsilon}) > 0 \end{cases} \tag{6.32}$$

where c_0 and c_1 are dimensionless constants, ρ is the unit mass of the material, l is the characteristic length of an element, and $\text{tr}(\dot{\varepsilon})$ is the volumetric strain rate.

The quadratic term in Eq. 6.32 is present in the original formulation [321] and the linear term is added. Both terms disappear in the limit case $q_{\text{vis}}(l \rightarrow 0) = 0$, such that the amount of viscous damping decreases as the element size shrinks.

The introduction of the artificial viscosity term into the system of discrete equations reduces the critical time step size for explicit time integration schemes, which can be computed with Eqs. 6.33 and 6.34.

$$\Delta t_{\text{crit}} = \frac{l}{c}\left(\sqrt{1 + \xi^2} - \xi\right) \tag{6.33}$$

$$\xi = c_0 - c_1^2\left(\frac{l}{c}\right)\text{tr}(\dot{\varepsilon}) \tag{6.34}$$

The effect of artificial bulk viscosity damping is illustrated with an example of stress wave propagation along a column. A column of 2 m is discretised with 1,000 rows of 6 tetrahedral elements. One MP is placed inside each element. A vertical load of 10 kPa is instantly applied at the top of the column. The material is isotropic linear elastic and the velocity of the compression wave is 200 m/s. Fig. 6.5 shows the stress at a MP located at a distance of 1 m below the column top for different values of the parameters c_0 and c_1. If no viscosity damping is applied ($c_0 = c_1 = 0$), the response is oscillating but the amplitude of the oscillations decreases significantly by introducing a small amount of artificial viscosity.

6.7 Contact between bodies

In geotechnical engineering, problems involving sliding contacts at interfaces are common in soil-structure problems such as pile installation, retaining structures, or landslides. MPM is capable of handling non-slip contact between different bodies because their velocities belong to the same vector field. To model the relative motion at the interface between the contacting bodies a specific contact algorithm is required.

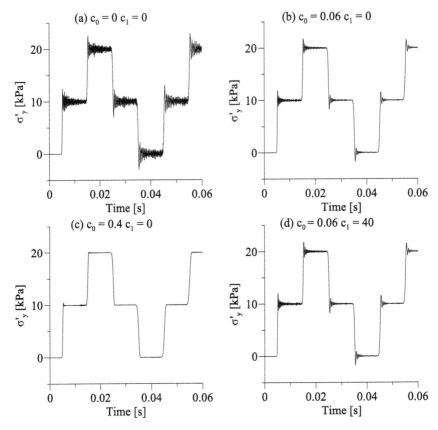

FIGURE 6.5: Evolution of stress in time at a fixed location for different viscosity coefficients c_0 and c_1.

The first examples of contact algorithms were efficient in simulating separation between bodies, but frictional contact forces were not considered [141, 339]. A frictional contact algorithm [22, 143] models the interaction between grains in granular materials. The algorithm allows sliding and rolling with friction as well as separation between grains, and correctly prohibits interpenetration. The algorithm automatically detects the contact nodes. Thus, a predefinition of the contact surface is not required. The contact formulation can also be used for adhesion problems [5, 56] as well as for the two-phase MPM formulation [58].

The contact algorithm can also be implemented in GIMP [184] but it requires the definition of a penalty function to avoid non-physical oscillations of contact forces due to numerical errors in the definition of the contact surface. The key concept of the penalty approach is to allow limited interpenetration between the contacting materials. The method is able to reduce numerical oscillations in the contact force. Moreover, the interpenetration is limited to

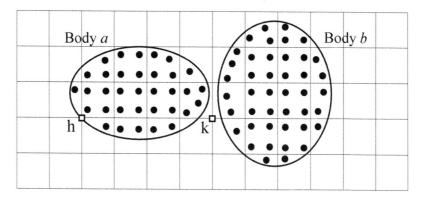

FIGURE 6.6: Example of two bodies in contact.

a very low level with an optimal selection of the penalty function properties, while the accuracy of the computation is effectively improved.

In FEM, contact problems are solved with interface elements, which can also be used in MPM [316, 325].

An alternative approach is the use of level-set based contact algorithms [13, 173] to simulate soil-penetration problems such as the installation of off-shore foundations. The idea is to describe the soil with two independent and overlapping domains. The use of a distributed Lagrangian multiplier and a level-set function provide the necessary contact interaction. This approach is specific to penetration problems and the extension to other types of applications is not straightforward.

6.7.1 Formulation

The contact algorithm presented in this section follows [23, 56], which is developed at the nodal velocity level. The algorithm can be considered as a predictor-corrector scheme in which the nodal velocity is predicted from the solution of the governing equation of each body separately and then corrected using the velocity of the combined system and the contact rule.

The contact algorithm is explained by considering two bodies a and b (Fig. 6.6), which are in contact at time t. The momentum balance equations of each body and of the combined system of bodies are assembled separately (Eq. 6.35).

$$\mathbf{M}_a^t \vec{a}_a^t = \vec{f}_a^t \tag{6.35a}$$

$$\mathbf{M}_b^t \vec{a}_b^t = \vec{f}_b^t \tag{6.35b}$$

$$\mathbf{M}^t \vec{a}^t = \left(\mathbf{M}_a^t + \mathbf{M}_b^t \right) \vec{a}^t = \left(\vec{f}_a^t + \vec{f}_b^t \right) = \vec{f}^t \tag{6.35c}$$

The nodal accelerations of each body and the combined system are calculated solving the momentum equations (Eq. 6.35) and then used to predict the nodal velocities at time $t + \Delta t$ (Eq. 6.36).

$$\vec{v}_a^{t+\Delta t} = \vec{v}_a^t + \Delta t \vec{a}_a^t \tag{6.36a}$$

$$\vec{v}_b^{t+\Delta t} = \vec{v}_b^t + \Delta t \vec{a}_b^t \tag{6.36b}$$

$$\vec{v}^{t+\Delta t} = \vec{v}^t + \Delta t \vec{a}^t \tag{6.36c}$$

Contact nodes are detected by comparing the predicted velocity of a single body $(\vec{v}_{a,I}^{t+\Delta t}, \vec{v}_{b,I}^{t+\Delta t})$ to that of the combined system $(\vec{v}_I^{t+\Delta t})$ for each node I. If the velocities differ, then the node is identified as a contact node (Eqs. 6.37).

$$\vec{v}_{a,I}^{t+\Delta t} = \vec{v}_I^{t+\Delta t} \quad \Rightarrow \quad \text{it is not a contact node} \tag{6.37a}$$

$$\vec{v}_{a,I}^{t+\Delta t} \neq \vec{v}_I^{t+\Delta t} \quad \Rightarrow \quad \text{it is a contact node} \tag{6.37b}$$

For example, considering nodes h and k of body a (Fig. 6.6), it is clear that only the MPs of body a contribute to the momentum equation of the combined system at node h. Thus, $\vec{v}_{a,h}^{t+\Delta t} = \vec{v}_h^{t+\Delta t}$, and therefore it is not a contact node. In contrast, at node k, which is shared between two bodies, the combined velocity differs from the single body velocity. Consequently, this node is defined as a contact node. When a contact node is detected, the algorithm proceeds checking whether the bodies are approaching or separating. This is done by comparing the normal component of the single body velocity with the normal component of the the the velocity of the combined bodies. Hence, the two cases are possible (Eq. 6.38).

$$(\vec{v}_{a,k}^{t+\Delta t} - \vec{v}_k^{t+\Delta t}) \cdot \vec{n}_{a,k}^t > 0 \quad \Rightarrow \quad \text{approaching} \tag{6.38a}$$

$$(\vec{v}_{a,k}^{t+\Delta t} - \vec{v}_k^{t+\Delta t}) \cdot \vec{n}_{a,k}^t < 0 \quad \Rightarrow \quad \text{separating} \tag{6.38b}$$

where $\vec{n}_{a,k}^t$ is the unit outward normal to body a at node k and time t.

The normal unit vector is provided by the gradient of mass of the considered body (Eq. 6.39).

$$\tilde{\vec{n}}_{\beta,k}^t = \frac{\sum_{p\beta} m_{p\beta} \mathbf{G}_{kp}}{\|m_{p\beta} \mathbf{G}_{kp}\|} \quad \beta = a, b \tag{6.39}$$

where $\mathbf{G}_{kp}$ is the gradient of the shape function of node k at the location of MP p, and $m_{p\beta}$ is the mass of MP p of body β ($\beta = a, b$)

Eq. 6.39 does not satisfy the collinearity condition (i.e. $\vec{n}_{a,k}^t = -\vec{n}_{b,k}^t$), and can lead to non-conservation of momentum, including penetration. A collinear unit normal can be obtained by averaging the two unit normals [143] (Eqs. 6.39 and 6.40).

$$\vec{n}_k = \vec{n}_{a,k} = -\vec{n}_{b,k} = \frac{\tilde{\vec{n}}_{a,k} - \tilde{\vec{n}}_{b,k}}{\|\tilde{\vec{n}}_{a,k} - \tilde{\vec{n}}_{b,k}\|} \tag{6.40}$$

If one of the bodies (e.g. body a) is much stiffer than the other (e.g. body b) or if the surface of a body is flat/convex but the surface of the other body is concave, the unit normal can be chosen as given in Eq. 6.41.

$$\vec{n}_k^t = \vec{n}_{a,k}^t = -\vec{n}_{b,k}^t = \tilde{n}_{a,k}^t \tag{6.41}$$

If the boundary of a body coincides with element boundaries, the normal can be defined more easily and accurately using the geometrical properties of the mesh. This is only possible if the mesh moves with the body throughout the simulation and does not deform excessively. The moving mesh procedure (Section 6.8) can be applied in some cases to keep the element boundaries aligned with the boundary of the body, thus identifying very accurately the interface nodes and the normals.

The algorithm allows for free separation. If the bodies are separating, then no correction is required and each body moves with the single body velocity $\vec{v}_a^{t+\Delta t}$. If the bodies are approaching, the algorithms checks whether sliding occurs by introducing the contact law.

The predicted relative normal and tangential velocities at an approaching contact node can be respectively written as given in Eqs. 6.42 and 6.43.

$$\hat{\vec{v}}_{a,k,\text{norm}}^{t+\Delta t} = \left[\left(\vec{v}_{a,k}^{t+\Delta t} - \vec{v}_k^{t+\Delta t}\right) \cdot \vec{n}_k^t\right]\vec{n}_k^t \tag{6.42}$$

$$\hat{\vec{v}}_{a,k,\text{tan}}^{t+\Delta t} = \vec{n}_k^t \cdot \left[\left(\vec{v}_{a,k}^{t+\Delta t} - \vec{v}_k^{t+\Delta t}\right) \cdot \vec{n}_k^t\right] \tag{6.43}$$

where $\cdot$ indicates the cross product of two vectors.

These components can be used to predict the contact forces at the node (Eqs. 6.44 and 6.45).

$$\vec{f}_{\text{norm}}^{t+\Delta t} = \frac{m_{a,k}^t \hat{\vec{v}}_{a,k,\text{norm}}^{t+\Delta t}}{\Delta t} \tag{6.44}$$

$$\vec{f}_{\text{tan}}^{t+\Delta t} = \frac{m_{a,k}^t \hat{\vec{v}}_{a,k,\text{tan}}^{t+\Delta t}}{\Delta t} \tag{6.45}$$

where $m_{a,k}^t$ is the nodal mass integrated from the material points of body a.

The maximum tangential force is given in Eq. 6.46.

$$f_{\text{tan,max}}^{t+\Delta t} = f_{\text{adh}}^{t+\Delta t} + \mu\|\vec{f}_{\text{norm}}^{t+\Delta t}\| \tag{6.46}$$

where $f_{\text{adh}}^{t+\Delta t}$ is the adhesive force at the contact node and μ is the friction coefficient.

Depending on the magnitude of the predicted contact forces, stick and slip contacts can be distinguished (Eqs. 6.47).

$$\text{If} \quad \|\vec{f}_{\text{tan}}^{t+\Delta t}\| < f_{\text{tan}}^{\text{max},\,t+\Delta t} \quad \Rightarrow \quad \text{stick contact} \tag{6.47a}$$

$$\text{If} \quad \|\vec{f}_{\text{tan}}^{t+\Delta t}\| > f_{\text{tan}}^{\text{max},\,t+\Delta t} \quad \Rightarrow \quad \text{slip contact} \tag{6.47b}$$

No correction is required for sticking bodies (Eq. 6.47a). The single body velocity is equal to the system velocity ($\vec{v}_{a,k}^{t+\Delta t} = \vec{v}_k^{t+\Delta t}$). Eq. 6.47b is for sliding bodies. The velocity needs to be corrected. The correction is applied to the normal component in order to avoid interpenetration between bodies and to the tangential component in order to satisfy the contact rule (Eq. 6.46). The predicted single body velocity $\vec{v}_{a,k}^{t+\Delta t}$ is corrected to a new velocity $\tilde{\vec{v}}_{a,k}^{t+\Delta t}$ (Eq. 6.48).

$$\tilde{\vec{v}}_{a,k}^{t+\Delta t} = \vec{v}_{a,k}^{t+\Delta t} + \vec{c}_{norm}^{t+\Delta t} + \vec{c}_{tan}^{t+\Delta t} \tag{6.48}$$

Eq. 6.48 is equivalent to applying the contact forces given in Eq. 6.49.

$$\tilde{\vec{v}}_{a,k}^{t+\Delta t} = \vec{v}_{a,k}^{t+\Delta t} + \frac{\Delta t}{m_{a,k}^t} \left(\tilde{\vec{f}}_{norm}^{t+\Delta t} + \tilde{\vec{f}}_{tan}^{t+\Delta t} \right) \tag{6.49}$$

The correction for the normal component $\vec{c}_{norm}^{t+\Delta t}$ is obtained imposing that the normal component of the corrected single body velocity coincides with the normal component of the the velocity of the combined bodies (Eq. 6.50).

$$\tilde{\vec{v}}_{a,k}^{t+\Delta t} \cdot \vec{n}_{k^t} = \vec{v}_{k\,t+\Delta t} \cdot \vec{n}_{k^t} \tag{6.50}$$

Eq. 6.50 gives Eq. 6.51.

$$\vec{c}_{norm}^{t+\Delta t} = -\left[\left(\vec{v}_{a,k}^{t+\Delta t} - \vec{v}_{k\,t+\Delta t} \right) \cdot \vec{n}_k^t \right] \vec{n}_k^t \tag{6.51}$$

Therefore, the normal contact force is as given in Eq. 6.52.

$$\tilde{\vec{f}}_{norm}^{t+\Delta t} = \frac{m_{a,k}^t}{\Delta t} \vec{c}_{norm}^{t+\Delta t} \tag{6.52}$$

When sliding occurs, the maximum tangential contact force assumes Eq. 6.53.

$$\tilde{\vec{f}}_{tan}^{t+\Delta t} = f_{tan}^{max,\,t+\Delta t} \vec{t} \tag{6.53}$$

where $\vec{t}$ is the unit vector tangential to the contact surface, which can be evaluated with Eq. 6.54 where $\hat{\vec{v}}_{a,k,tan}^{t+\Delta t}$ is defined Eq. 6.42.

$$\vec{t} = \frac{\hat{\vec{v}}_{a,k,tan}^{t+\Delta t}}{\|\hat{\vec{v}}_{a,k,tan}^{t+\Delta t}\|} \tag{6.54}$$

Substituting Eq. 6.46 into Eq. 6.53 gives in Eq. 6.55.

$$\tilde{\vec{f}}_{tan}^{t+\Delta t} = \left(f_{adh}^{t+\Delta t} + \mu \| \tilde{\vec{f}}_{norm}^{t+\Delta t} \| \right) \vec{t} \tag{6.55}$$

The correction for the tangential component can be written as Eq. 6.56.

$$\vec{c}_{tan}^{t+\Delta t} = \frac{\Delta t}{m_{a,k}^t} \left(f_{adh}^{t+\Delta t} + \mu \| \tilde{\vec{f}}_{norm}^{t+\Delta t} \| \right) \vec{t} \tag{6.56}$$

The adhesive term can be expressed as Eq. 6.57.

$$f_{\text{adh}}^{t+\Delta t} = a_f A_k^t \tag{6.57}$$

where A_k^t is the contact area associated with the node k, which is integrated from the contact elements that share node k, and a_f is an adhesion factor that represents the adhesive force for unit of surface.

The equation for the corrected velocity is obtained by combining Eqs. 6.49, 6.52, 6.56 and 6.57.

$$\tilde{\vec{v}}_{a,k}^{t+\Delta t} = \vec{v}_{a,k}^{t+\Delta t} - \left[\left(\vec{v}_{a,k}^{t+\Delta t} - \vec{v}_k^{t+\Delta t} \right) \cdot \vec{n}_k^t \right] \vec{n}_k^t$$
$$- \left\{ \left[\left(\vec{v}_{a,k}^{t+\Delta t} - \vec{v}_k^{t+\Delta t} \right) \cdot \vec{n}_k^t \right] \mu + \frac{a A_k^{t+\Delta t}}{m_{a,k}^t} \right\} \vec{t} \tag{6.58}$$

Fig. 6.7 illustrates with a flow chart the main steps of the presented contact algorithm. Having calculated the velocity of the contact node k at time $t+\Delta t$, the corrected acceleration vector at the node must be recalculated as Eq. 6.59.

$$\tilde{\vec{a}}_{a,k}^t = \frac{\tilde{\vec{v}}_{a,k}^{t+\Delta t} - \vec{v}_{a,k}^t}{\Delta t} \tag{6.59}$$

This corrected acceleration is used to update the MP velocity. The same procedure explained here for body a must be applied to body b.

The contact algorithm is applied between the Lagrangian phase and the convective phase. The nodal velocities are first predicted in the Lagrangian phase. Then, the corrected nodal velocities and accelerations are computed with the contact algorithm and these new values of nodal accelerations are used to compute the velocities of MPs and update their positions, strains and stresses in the convective phase.

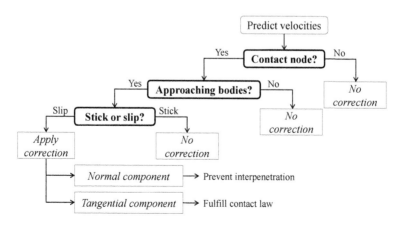

FIGURE 6.7: Flow chart illustrating the contact algorithm (after [57]).

6.7.2   Example

A 1m-wide cube of a linear elastic material ($E = 20$ MPa, $\nu = 0.33$, $\rho = 2{,}000$ kg/m^3) is initially positioned at the top of a sliding base inclined $\alpha = 26.53°$ and with the same elastic parameters. The domain is discretised with 8,133 tetrahedral elements forming an unstructured mesh with 4 MPs located in the initial active elements.

Stresses are initialised with a gravity loading (see Section 6.4) and the contact formulation is turned off in order to have a full contact between bodies. At $t = 0$, the block starts sliding down the inclined plane driven by gravity. The friction coefficient at the interface between the block and the sliding plane is $\mu = 0.2$. From the momentum balance equation, the displacement along the plane can be expressed as a function of time by assuming a rigid body motion (Eq. 6.60).

$$s(t) = 0.5 \left(\sin \alpha - \mu \cos \alpha \right) g t^2 \qquad (6.60)$$

where g is the gravity.

Fig. 6.8 shows the evolution of the displacement of the block with time and compares the MPM results with the analytical solution (Eq. 6.60). There is an excellent agreement between the numerical and the theoretical solution.

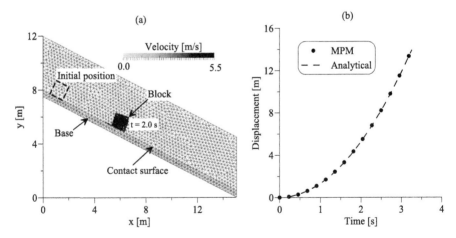

FIGURE 6.8: (a) Geometry of the problem and (b) evolution of block displacement along time.

6.8 Moving mesh

The moving mesh procedure exploits the fact that the computational grid in MPM does not store any permanent information. Therefore, it can be kept fixed or be freely modified at the end of each time step. This approach is useful when applying non-zero traction and kinematic boundary conditions because it ensures that the computational mesh aligns with the surface where tractions (kinematics) are prescribed [30] (Section 6.2). It is particularly convenient when simulating the movement of a body inside the soil such as in pile driving or anchor installation where it is necessary to keep a refined mesh always near the structure [55, 59, 224, 325].

The positions of the grid nodes are modified according to predefined rules after having updated stress, strain, velocity and position of MPs at the end of the convective phase. The MPs are assigned to the elements and the computation proceeds with a new time step.

The discretised domain is divided as follows with the moving mesh procedure.

Moving mesh portion of the grid whose elements do not deform along the simulation but move rigidly with certain predefined rules.

Deforming mesh portion of the grid located between the moving mesh and the fix boundary of the domain that modifies its dimensions with time. Thus, its elements deform accordingly.

All the elements of the discretised domain must be associated to only one of the aforementioned parts. In general, the moving mesh portion can translate in one or more directions and rotate rigidly following predefined rules.

The moving mesh usually contains a rigid body and it moves with the same displacement of this body, thus keeping the boundary of the elements aligned to the boundary of this structure. This is convenient when the contact algorithm (see Section 6.7.1) is applied because it prevents the need for identifying the contact surface and computing the normal directions at each time increment. It is also possible to attach the moving mesh to a surface over which a traction is applied and that moves with a prescribed velocity.

Because of the movement of the moving mesh zone, the elements of the deforming mesh zones change their aspect ratio with time. It is important to choose a discretisation that ensures a reasonable aspect ratio of all the elements during the entire simulation. For extremely large deformations, the domain can be remeshed if the mesh gets distorted.

To illustrate this procedure, the problem of a cantilever wall sliding over the foundation base is considered (Fig. 6.9). The material parameters are shown in Table 6.2. The wall retains a soil with a friction angle $\varphi = 30°$ over which a load of 20 kPa is applied. The friction coefficient between the wall and the foundation layer is $\mu = 0.2$.

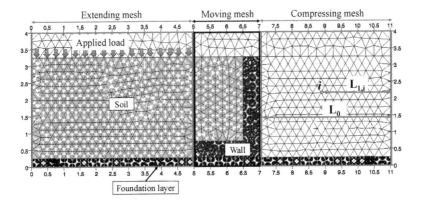

FIGURE 6.9: Geometry and discretisation of the cantilever wall problem.

TABLE 6.2
Material parameters for the retaining wall example.

Parameter	Symbol	Unit	Wall	Foundation	Soil
Young modulus	E	MPa	150	40	20
Poisson ratio	ν	-	0.33	0.33	0.33
Density	ρ	kg/m^3	2500	1590	1590
Friction angle	φ	°	-	-	30

A moving mesh zone is defined in a volume that includes the cantilever wall. In this example, the moving mesh translates horizontally with a displacement u, which is the average horizontal displacement of the MPs discretising the structure. The compressing mesh is located between the moving mesh part and the right boundary. Conversely, the extending mesh is located between the moving mesh and the left boundary.

The horizontal coordinates of the nodes belonging to the deforming mesh regions are updated as given in Eq. 6.61.

$$x'_i = x_i + \frac{L_{1,i}}{L_0} u \qquad (6.61)$$

where x_i and x'_i are the initial and updated coordinate, respectively, L_0 the initial length of the compressing or extending mesh, and $L_{i,1}$ the distance of the i-th node from the fix boundary.

In the first phase of the considered example, gravity is applied and the external traction is not active. In this condition, the system reaches the static equilibrium because the frictional force at the base of the wall is sufficient for equilibrating the soil active thrust. In the second phase, the load on the top soil surface is applied and the wall starts sliding because the soil thrust exceeds the maximum contact force at the wall-foundation interface. Fig. 6.10

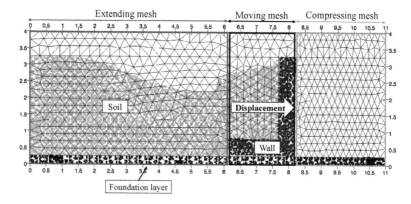

FIGURE 6.10: Computational grid at the end of the simulation with the use of moving mesh procedure.

shows the position of the retaining wall and the deformed computational grid after 1.45 s from the load application. Note that the elements belonging to the moving mesh zone do not deform and the boundary of the cantilever wall remains aligned with the element boundaries. The shape of the elements belonging to the compressing or the extending mesh appears deformed.

6.9 Anti-locking algorithms

MPM can suffer from the use of low-order elements, which reduce the computational cost but decrease the numerical stability as in FEM. Locking occurs when an element wants to deform but cannot because it is constrained by neighbouring elements, making this element nearly incompressible. This leads to large errors in the solution. The bulk modulus is very large and small errors in strain will yield to large errors in stress.

Fig. 6.11 depicts an example of locking. Element e_1 (Fig. 6.11 a) is defined by nodes 1 and 2 on the x-axis, and node 3 on the y-axis. The area of the triangular element remains constant when the element is incompressible. If nodes 1 and 2 are fixed, y_3 must remain in the same place giving $u_y^3 = 0$. Therefore, the remaining degree of freedom is the horizontal displacement u_x^3. Similarly, the only remaining degree of freedom in the element e_2 (Fig. 6.11 b) defined by nodes 4, 1, and 3, is the vertical displacement u_y^3. Two triangles can be assembled (Fig. 6.11 c). Since incompressibility for element e_1 requires $u_y^3 = 0$ and incompressibility for element e_2 requires $u_x^3 = 0$, node 3 cannot move and both elements are completely locked up. With nodes 1 through 4 locked up, the nodes of elements 3, 4, 5, 6, 7 and 8 will also be locked

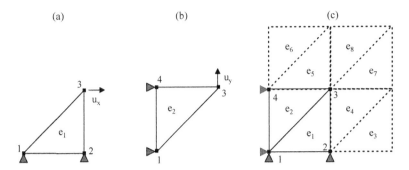

FIGURE 6.11: A pathological case of volumetric locking in triangular elements.

(Fig. 6.11 c). Such locking usually propagates throughout the entire mesh yielding an unrealistic stiff response and an erroneous velocity field.

Several anti-locking algorithms have been suggested to mitigate the instability associated with the linear shape functions [81, 196]. A solution to the problem is the Nodal Mixed Discretisation (NMD) technique for linear tetrahedral elements [81]. In this procedure, the element volumetric behaviour is averaged over the elements sharing its nodes via a least squares smoothing process. The effect of applying the NMD scheme is to increase the number of degrees of freedom per element. The strain rate of an element $\dot{\varepsilon}$ is divided into its deviatoric and volumetric components. The volumetric strain rate for a node i is defined as a weighted average of the surrounding element values $\dot{\varepsilon}_{\mathrm{vol,el}}$ with Eq. 6.62.

$$\dot{\varepsilon}_{\mathrm{vol,i}} = \frac{\sum\limits_{el=1}^{no_{\mathrm{el,i}}} \dot{\varepsilon}_{\mathrm{vol,el}}\Omega_{\mathrm{el}}}{\sum\limits_{el=1}^{no_{\mathrm{el,i}}} \Omega_{\mathrm{el}}} \qquad (6.62)$$

where $no_{\mathrm{el,i}}$ are the elements surrounding the node i, and Ω_{el} is the volume of the element el.

A mean value for the element, $\bar{\dot{\varepsilon}}_{\mathrm{vol}}$ is calculated by taking the average of nodal quantities (Eq. 6.63).

$$\bar{\dot{\varepsilon}}_{\mathrm{vol}} = \frac{1}{no_{\mathrm{node,el}}} \sum\limits_{i=1}^{no_{\mathrm{node,el}}} \dot{\varepsilon}_{\mathrm{vol,i}} \qquad (6.63)$$

where $no_{\mathrm{node,el}}$ is the number of nodes in an element.

The element strain rates are redefined by superposition of the deviatoric part, which is unchanged, and the volumetric part modified as illustrated above. This technique is often called strain smoothing.

6.10   Mass scaling

The simulation of quasi-static or slow-process problems with a dynamic explicit code can require high computational effort because the time step size is bounded by the stability condition. However, if the inertia effect is negligible, the time step size can be artificially increased by scaling the density with the mass scaling factor β. The critical time step size increases by a factor $\sqrt{\beta}$ (Eq. 6.64).

$$\Delta t^{\beta}_{\text{crit}} = \frac{l_{\text{e}}}{\sqrt{\dfrac{E_{\text{c}}}{\beta\rho}}} = \sqrt{\beta}\Delta t^1_{\text{crit}} \tag{6.64}$$

where Δt^1_{crit} is the critical time step for $\beta = 1$ (no mass scaling is applied).

Mass scaling is a useful technique to improve computational efficiency of dynamic codes in simulating quasi-static and slow-process motion. However, sensitivity analyses are necessary to calibrate the mass scaling factor in slow process problems. High values of β can lead to wrong results. For instance, it has been shown that $\alpha > 0.15$ generates an overestimation of the tip resistance when simulating CPT [58].

6.11   Closure

This chapter presents the theoretical framework and implementation of some of the numerical techniques adopted in MPM to deal with specific problems such as application of boundary conditions, MP initialisation, contact between bodies and so on. Numerical examples are also provided to illustrate the effect of these features in MPM simulations. In many cases, these procedures have been originally developed for other numerical methods, but they have been successfully extended for MPM.

Absorbing boundary, local damping, viscous damping, mass scaling, and the NMD technique are examples of numerical features proposed for Finite Elements or Finite Difference methods that are well suited also for MPM. The topics discussed in this chapter are still under development within the research community especially regarding the treatment of boundary conditions, contact problems and the use of high-order elements to prevent volumetric locking.

7

An Introduction to Critical State Soil Mechanics

E.J. Fern and K. Soga

7.1 Introduction

The central theory of critical state soil mechanics [261] is the critical state theory [250]. It states that any soil under sustained uniform shearing will reach a state called the critical state. It will deform continuously without any changes in volume and sustain a constant stress state, although there is a dependency to the mean effective stress. The theory states that there is a unique relationship between the shear strength, the mean pressure and the element volume at critical state, and this relationship can be predicted. This is a powerful concept in soil mechanics because it permits the evaluation of the ultimate shear strength for a given mean effective stress, the amount of volume change for any soil at a given initial state. Although the actual mathematical form of the stress-strain relationship varies with grain characteristics, the theory appears to work with a wide range of granular materials ranging from fine clay to coarse sand and acts as reference point for the stress-strain relationship, also called constitutive models. These models are necessary for any MPM simulation in order to obtain the stress fields and give a stiffness to the material.

7.2 The critical state

Fig. 7.1 illustrates the stress-strain relationship of loosely and densely packed granular systems confined at an initial confining pressure and then sheared. The densely packed system shows a peak shear resistance and then softens to a state called the critical state. The shearing is accompanied by volumetric change. Initially the system contracts (volume or void ratio decreases) but

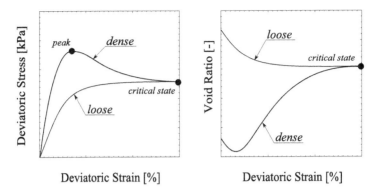

FIGURE 7.1: Stress-strain curves of loose and dense soils sheared at the same confining stress.

then starts to dilate (volume or void ratio increases). The rate of dilatancy increases, reaches its maximum and then decreases until reaching zero at the critical state. The decrease of the dilatancy rate results in a decrease of the shear resistance, which is called strain-softening. In contrast, when a loosely packed system is sheared at the same initial confining pressure, the shear resistance monotonically increases and reaches the critical state. The soil volume decreases with increasing shear deformation.

The critical state is marked with a dot in Fig. 7.1. The shear resistance at the critical state of the loosely packed system coincides with the one of the densely packed system for a given pressure. This implies that the initial arrangement of particle packing (dense or loose) is lost by shearing and the granular system has a unique packing arrangement at the critical state.

At a microscopic level, a granular system consists of grains. Fig. 7.2 shows a schematic description of the grain packing at different states. These packets of grains are rearranged when the granular is sheared. The densely packed system (Fig. 7.2-left) has grains which jam into one another by deformation and these grains have to move apart in order to allow packets of grains to flow by shear. Grains need to override each other resulting in volume increase. However, discrete element method (DEM) simulations and microscopic observations through computer tomography (CT) scans show that the grains group together to form chains, which results in larger pore space inside the chains. The shear-induced dilatancy is a product of complex granular interactions.

In the loosely packed system (Fig. 7.2-middle), the large voids decrease their volumes upon shearing. At the same time, the shear resistance increases with a better overall packing arrangement. The strength and density of the granular then vary until reaching the critical state when the packets of grains can mobilize over each other. At the critical state (Fig. 7.2-right), the shear resistance is constant and the granular structure continuously changes by shearing but the average void space remains the same to have the constant void ratio state.

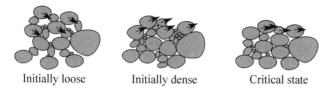

Initially loose   Initially dense   Critical state

FIGURE 7.2: Schematic description of the grain packing at different states.

At critical state, the granular material will be continuously deformed in shear without any changes in volume or stresses. Such conditions can be expressed mathematically as given in Eq. 7.1.

$$\frac{\mathrm{d}\eta'}{\mathrm{d}\varepsilon_{\mathrm{dev}}} = \frac{\mathrm{d}\varepsilon_{\mathrm{vol}}}{\mathrm{d}\varepsilon_{\mathrm{dev}}} = \frac{\mathrm{d}^2\varepsilon_{\mathrm{vol}}}{\mathrm{d}\varepsilon_{\mathrm{dev}}^2} = 0 \tag{7.1}$$

where $\eta' = q/p'$ is the effective stress ratio. In this chapter, $\varepsilon_{\mathrm{dev}}$ is the deviatoric (shear) strain and $\varepsilon_{\mathrm{vol}}$ is the volumetric strain. The dilatancy rate is defined as $D = \mathrm{d}\varepsilon_{\mathrm{vol}}/\mathrm{d}\varepsilon_{\mathrm{dev}}$ and the rate of the dilatancy rate as $\mathrm{d}D/\mathrm{d}\varepsilon_{\mathrm{dev}} = \mathrm{d}^2\varepsilon_{\mathrm{vol}}/\mathrm{d}\varepsilon_{\mathrm{dev}}^2$.

The stress state and the density of any granular are constant and uniquely defined when the soil is subjected to sufficient shearing [250]. This state is defined by a critical state line (CSL) as shown in Eq. 7.2 and in Fig. 7.3.

$$\eta'_{\mathrm{cs}} = M \quad \text{and} \quad e_{\mathrm{cs}} = f(p') \tag{7.2}$$

where M is the critical state stress ratio for a given value of the Lode angle θ, and e_{cs} is the critical state void ratio.

7.2.1 Critical state of clay

A normally consolidated clay has never been previously loaded and, hence, it has a loose structure at a given confining stress (Fig. 7.3 a-b, point A). It exhibits contractive behaviour by shearing in drained conditions before reaching the critical state at point B. The initial void ratio is larger than the one at critical state for a given confining stress p' and the specimen must reduce its volume in order to reach the critical state. If a shearing test is conducted in undrained conditions (i.e. pore water cannot move in or out of the clay), the constant volume constraint due to the large bulk stiffness of the pore water (i.e. incompressible conditions) will bring the effective stress state of the clay to the position shown in C. A positive excess pore pressure develops and the difference between point A and point C is the magnitude of the excess pore pressure.

An overconsolidated clay undergoes a loading-unloading cycle in terms of confining stress. If the degree of unloading stress relative to the loading stress is large (i.e. heavily overconsolidated), the void ratio after unloading can be

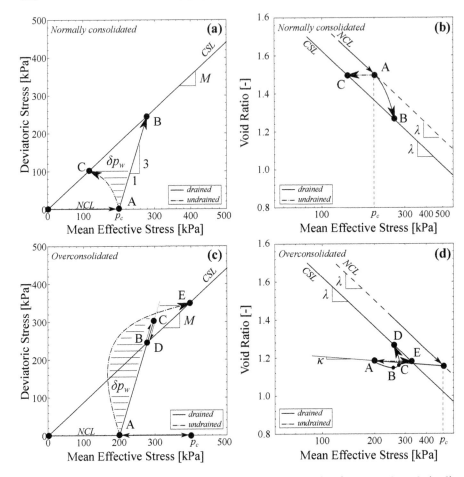

FIGURE 7.3: Triaxial compression stress paths in (a-b) drained and (c-d) undrained conditions.

below the critical state void ratio at a given confining pressure (Fig. 7.3 c-d, point A). When the clay is sheared in drained conditions, it dilates and increases its volume. During this process, the clay develops a peak strength at point C. The clay then softens until the stresses reach the critical state stress ratio at point D. When the clay is sheared in undrained conditions, the mean effective stress moves to the right in the $(\ln p', e)$ plane and reaches the critical state at point E. The mean effective stress at failure is greater than the mean effective stress prior to shearing. This means that negative excess pore pressure is generated due to the dilative tendency of the heavily overconsolidated clay.

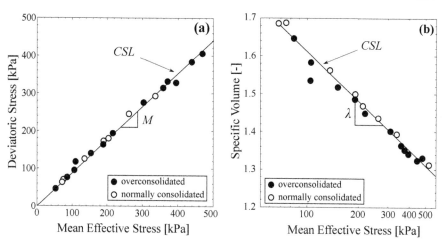

FIGURE 7.4: CSL in the (a) (p', q) plane and (b) $(\ln p', v)$ plane for drained triaxial compression tests on Weald clay (data from [250]).

There is experimental evidence of the existence of the critical state for clays. Fig. 7.4 shows the final stress and strain state for a series of triaxial compression tests of Weald clay for both normally consolidated and over-consolidated specimens. It can be seen that all tests have reached the same ultimate critical state irrespective of their initial state. Fig. 7.4 (a) shows the final stresses in the (p', q) plane and (b) in the $(\ln p', e)$ plane for both normally consolidated specimens (white markers) and overconsolidated specimens (black markers). The final stresses are set on a line with a linear relationship between the mean effective stress and the deviatoric stress. The final density of the clay, expressed in terms of specific volume v or void ratio e, has a logarithmic relation with the mean effective stress p' at critical state (Eq. 7.3). The normal compression line (NCL) was found to be parallel to the CSL for clays in the $(\ln p', e)$ plane [250].

$$\nu_{cs} = \Gamma - \lambda \ln p'_{cs} \qquad (7.3)$$

where $\nu_{cs} = 1 + e_{cs}$ is the critical state specific volume with e_{cs} the critical state void ratio, Γ the critical state specific volume at $p' = 1$ kPa and λ the slope of the CSL. M, Γ and λ are material properties determined from experiments.

7.2.2 Critical state of sand

Similar to what is observed for clays, there is experimental evidence of the existence of a critical state for sands. Fig. 7.5 shows the results of triaxial compression tests of Toyoura Sand with different sample preparations to create different initial void ratios at different confining stresses [315]. The left figure shows the final strength of the tests along with the CSL, giving a linear

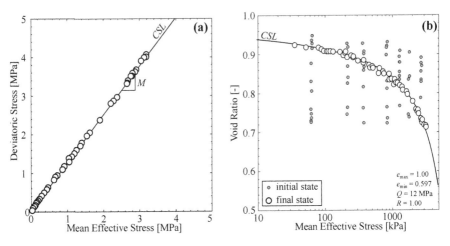

FIGURE 7.5: CSLs in the (a) (p', q) plane and (b) (e, p') plane for Toyoura Sand (data from [315]).

relationship. The right figure shows the initial (black markers) states prior to shearing and the final (white markers) states at critical state. It shows that all tests, no matter whether the specimens are dense or loose, reach a non-linear CSL in the $(\ln p', e)$ plane, which can only be approximated with Eq. 7.3 over a small range of pressure. An alternative equation is presented in Section 7.3.3.

7.2.3 Critical state stress ratio in the generalised stress space

In Terzaghi's soil mechanics [296], the effective friction angle φ'_{cs} is used to characterise the strength of soils with a nil cohesion ($c' = 0$).

$$\sin \varphi'_{cs} = \frac{\sigma'_1 - \sigma'_3}{\sigma'_1 + \sigma'_3} \tag{7.4}$$

where σ'_1 is the major principal effective stress and σ'_3 is the minor principal effective stress. This means that the classical Mohr-Coulomb model assumes that the magnitude of intermediate principal effective stress σ'_2 does not contribute to its failure.

Experimental data indicates that the critical state stress ratio M is different in triaxial compression ($\sigma'_1 > \sigma'_2 = \sigma'_3$) than in triaxial extension ($\sigma'_1 = \sigma'_2 > \sigma'_3$). If it is assumed that the friction angle φ'_{cs} is the same for both triaxial compression and extension, a relationship between M and φ'_{cs} can be obtained (Eqs. 7.5 and 7.6).
Triaxial compression ($\sigma'_1 > \sigma'_2 = \sigma'_3$):

$$M_{tc} = \frac{6 \sin \varphi'_{cs}}{3 - \sin \varphi'_{cs}} \tag{7.5}$$

Triaxial extension ($\sigma_1' = \sigma_2' > \sigma_3'$):

$$M_{\text{te}} = \frac{6\sin\varphi_{\text{cs}}'}{3 + \sin\varphi_{\text{cs}}'} \tag{7.6}$$

It is common to use an invariant called the Lode angle θ (Eq. 7.7) for the general principal stress space ($\sigma_1' > \sigma_2' > \sigma_3'$).

$$\tan\theta = \sqrt{3}\left[\frac{(\sigma_2' - \sigma_3')}{(\sigma_1'\sigma_2') + (\sigma_1' - \sigma_3')}\right] \quad 0° < \theta < 60° \tag{7.7}$$

The Lode angle θ is 0° in triaxial compression and 60° in triaxial extension. In order to use the critical state theory in wider geotechnical problems, it is necessary to extend the prediction of the critical state stress ratio M to any stress state. Since M is often evaluated from a triaxial compression test, M_{tc} can be used as a basis and then extended to other stress states with the Lode angle θ. Three alternatives are presented in this chapter. The most common is to estimate M with the Mohr-Coulomb criterion [295] (Eq. 7.8). The second alternative is to use the Matsuoka-Nakai criterion [197] (Eq. 7.9). However, it is an implicit function which requires an iterative process to solve. The third alternative is an explicit approximation of the Matsuoka-Nakai criterion [152] (Eq. 7.10).

Mohr-Coulomb:

$$M = \frac{3\sqrt{3}}{\cos\theta\left(1 + \dfrac{6}{M_{\text{tc}}}\right) - \sqrt{3}\sin\theta} \tag{7.8}$$

Matsuoka-Nakai:

$$\frac{27 - 3M^2}{3 - M^2 + \dfrac{8}{9}M^3\sin\theta\left(\dfrac{3}{4} - \sin^2\theta\right)} = \frac{27 - 3M_{\text{tc}}{}^2}{3 - M_{\text{tc}}{}^2 + \dfrac{2}{9}M_{\text{tc}}{}^3} \tag{7.9}$$

Jefferies-Shuttle:

$$M = M_{\text{tc}} - \frac{M_{\text{tc}}{}^2}{3 + M_{\text{tc}}}\cos\left(\frac{3\theta}{2}\right) \tag{7.10}$$

7.3 State indices

Critical state soil mechanics can be used to evaluate whether the granular material will contract or dilate upon shearing, depending on the stress and void ratio state. If the state is above the CSL, then the material has a contractive tendency during shearing as it reaches the critical state. If the state is below the CSL, then the material has a dilative tendency during shearing as it reaches to the critical state. The distance between the current state and the critical state is proportional to the degree of contraction or dilatancy. Various normalized state indices are available and are often used in constitutive models.

7.3.1   Equivalent liquidity index

For saturated clay, the critical state specific volume or void ratio can be expressed in terms of gravimetric water content w using the following equation $w = e/G_s$, where G_s is the specific gravity. It is therefore possible to express the critical state specific volume v_{cs} in terms of critical state water content e_{cs} (Eq. 7.11). Two points of this curve are known – the liquid limit (w_L, p_L') and plastic limit (w_P, p_P') which can be quantified by laboratory tests. Fig. 7.6 (a) illustrates the CSL in terms of water content.

$$w_{cs} = \Gamma_w - \lambda \ln p_{cs}' \qquad (7.11)$$

where w_{cs} is the critical state water content and Γ_w the critical state water intercept at $p' = 1$ kPa. The liquidity index I_L [296] is a normalised index which indicates the state of the clay. The liquidity index is $I_L = 1$ at the liquidity limit $(w = w_L)$ and $I_L = 0$ at the plastic limit $(w = w_P)$. As these two points are known as well as the CSL, it is possible to obtain the liquidity index at critical state I_L^{cs} as shown in Eq. 7.12 and illustrated in Fig. 7.6 (b).

$$I_L = \frac{w - w_P}{w_L - w_P} \qquad (7.12)$$

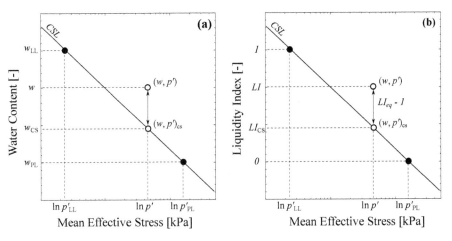

FIGURE 7.6: CSL expressed in terms of (a) water content and (b) liquidity index.

$$I_{\mathrm{L}}^{\mathrm{cs}} = \frac{w_{\mathrm{cs}} - w_{\mathrm{P}}}{w_{\mathrm{L}} - w_{\mathrm{P}}} = \frac{\ln\left(\dfrac{p_{\mathrm{P}}'}{p'}\right)}{\ln\left(\dfrac{p_{\mathrm{P}}'}{p_{\mathrm{L}}'}\right)} \tag{7.13}$$

The equivalent liquidity index I_L^{eq} [261] indicates the normalised distance from the current state to the critical state (Eq. 7.14).

$$I_L^{\mathrm{eq}} = I_{\mathrm{L}} - I_{\mathrm{L}}^{\mathrm{cs}} + 1 \tag{7.14}$$

The equivalent liquidity index is a measurement of how much a clay must contract or dilate to achieve the critical state and its value is an indicator of the magnitude of volume change. When the equivalent liquidity index is $I_L^{\mathrm{eq}} > 1$, then the clay must contract to achieve the critical state. When the equivalent liquidity index is $I_L^{\mathrm{eq}} < 1$, soil must dilate in order to achieve the CSL. The equivalent liquidity index is $I_L^{\mathrm{eq}} = 1$ at critical state.

7.3.2 State parameter

The state parameter [28] measures the distance between the current void ratio and the critical state void ratio (Eq. 7.15) for a given mean effective stress p' (Fig. 7.7). Note that the state parameter is denoted with a capital Ψ in order to avoid confusion with the dilatancy angle ψ.

$$\Psi = e - e_{\mathrm{cs}} \tag{7.15}$$

The state parameter is an indicator of the direction and magnitude of the volumetric behaviour of the sand upon shearing. When it is positive ($\Psi > 0$),

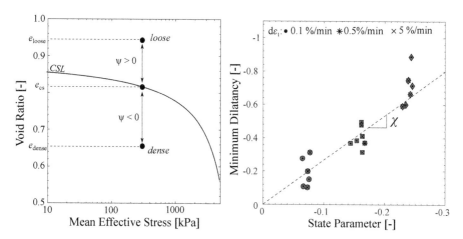

FIGURE 7.7: (a) definition of the state parameter, (b) correlation between the minimum dilatancy rate and the state parameter (data from [101]).

the sand will contract. When it is negative ($\Psi < 0$), it will dilate. When it is nil ($\Psi = 0$), the sand is at the critical state density and no changes in volume will take place while shearing is taking place. The state parameter is an equivalent expression of the equivalent liquidity index I_L^{eq} (Eq. 7.16).

$$I_L^{\text{eq}} = \frac{e - e_{\text{cs}}}{G_s(w_{\text{L}} - w_{\text{P}})} = \frac{\Psi}{G_s I_{\text{p}}} + 1 \tag{7.16}$$

where $I_{\text{p}} = w_{\text{L}} - w_{\text{P}}$ is the plasticity index.

It is possible to relate the state parameter Ψ to dilatancy rate with a dilatancy coefficient χ [148, 151].

$$D = \chi \frac{M}{M_{\text{tc}}} \Psi \tag{7.17}$$

As for the relative dilatancy index, the conversion is stress state and soil fabric dependent and may somewhat differ from the suggested value of $\chi = 3.5$ [149, 150].

7.3.3 Relative dilatancy index

The relative dilatancy index I_{R} [40] characterises the dilatancy characteristic of sand. It is a function of the relative density index I_{D} and the relative crushability index I_{C} (Eq. 7.18).

$$I_{\text{R}} = I_{\text{D}} I_{\text{C}} - R \tag{7.18}$$

The relative density index I_{D} (Eq. 7.19) defines the density of the packing by comparing the current void ratio state with the loosest and densest states possible which are characterised by the maximum void ratio e_{min} and the minimum void ratio e_{max}. These values are material properties and are obtained by laboratory experiments [157].

$$I_{\text{D}} = \frac{e_{\text{max}} - e}{e_{\text{max}} - e_{\text{min}}} \tag{7.19}$$

The relative crushability I_{C} (Eq. 7.20) defines the influence of pressure p' and the crushing pressure Q on the ability of a sand to dilate. If a sand is subjected to a high confining pressure, it will be more difficult to dilate. This pressure is normalised with the crushing pressure Q, which is a function of the mineralogy of the sand, typically 10 MPa for silica sand and 20 MPa for silica silt.

$$I_{\text{C}} = \ln\left(\frac{Q}{p'}\right) \tag{7.20}$$

When the relative dilatancy index is positive ($I_{\text{R}} > 0$), the sand is dense and has to dilate in order to reach the critical state. When the index is negative ($I_{\text{R}} < 0$), the sand is loose and has to contract in order to reach the critical state. When the index is nil ($I_R = 0$), the sand is at critical state. The magnitude of the index is proportional to changes in volume required to reach the critical state [40].

The relative dilatancy index I_R relates to the dilatancy rate with a dilatancy coefficient α (Eq. 7.21).

$$D_{1,\max} = \max\left(-\frac{d\varepsilon_{\text{vol}}}{d\varepsilon_1}\right) = \alpha I_R \qquad (7.21)$$

where α is the dilatancy coefficient with a default value of 0.3. It is important to note that the definition of dilatancy by Bolton is different from the dilatancy defined in this chapter. The dilatancy coefficient α appears to be a function of the soil structure [291].

The relative dilatancy index is nil at critical state and can, therefore, be used to formulate a CSL [41, 201] as given in Eq. 7.22. This dilatancy-based CSL is non-linear and takes into account the crushability of the sand grains. Fig. 7.5 shows the data from triaxial compression tests on Toyoura sand.

$$I_R = 0 \rightarrow e_{\text{cs}} = e_{\max} - \frac{e_{\max} - e_{\min}}{\ln(Q/p')}R \qquad (7.22)$$

7.4   Stress-dilatancy theories

Taylor [292] carried out a series of direct shear tests on dry Ottawa Sand and noticed that the development of strength changed with initial density and that this change in behaviour was associated with the volume change of the sand during shearing. A loose sand contracted and did not reach a peak strength whilst a dense sand dilated and exhibited a peak strength. He explained this phenomenon as a consequence of grains interlocking. Taylor [292] proposed a stress-dilatancy theory concept based on the work balance equation (Eq. 7.23) but did not explicitly write the equations. The external work corresponds to the product of the measured displacements and forces (assuming that the elastic deformation is negligible). The internal work corresponds to the frictional force. Fig. 7.8 (a) illustrated the direct shear tests.

$$\tau dx - \sigma'_N dy = \mu \sigma'_N dx \qquad (7.23)$$

where τ is the shear stress, σ'_N the normal effective stress, μ the friction coefficient, dx and dy the incremental displacements in the horizontal and vertical planes, respectively. The work balance equation (Eq. 7.23) can be rearranged to express the strength as a function of the friction coefficient and the incremental displacements characterising the interlocking (Eq. 7.24).

$$\frac{\tau}{\sigma'_N} = \mu + \frac{dy}{dx} \qquad (7.24)$$

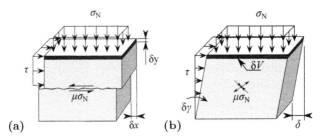

FIGURE 7.8: Schematic description of the (a) direct shear and (b) simple shear tests (after [101]).

The strength can then be express as the mobilised strength ($\tan \varphi'_m = \tau/\sigma'_N$), the friction coefficient as the critical state strength ($\tan \varphi'_{cs} = \mu$) and interlocking as the dilatancy rate ($\tan \psi = dy/dx$). The work balance equation then gives Taylor's stress-dilatancy theory (Eq. 7.25).

$$\tan \varphi'_m = \tan \varphi'_{cs} + \tan \psi \tag{7.25}$$

Roscoe and Schofield [249] followed the same path but with the simple shear test conditions for which the internal work equation can be expressed as shown in Eq. 7.26 and illustrated in Fig. 7.8 (b).

$$\sigma'_N d\varepsilon_{vol} + \tau d\gamma = \mu \sigma'_N d\gamma \tag{7.26}$$

The total strains can be decomposed into a reversible (elastic) and an irreversible (plastic) part. The work balance for the irreversible and plastic part gives Eq. 7.27 which is known as the Cam-Clay work equation.

$$\sigma'_N d\varepsilon^p_{vol} + \tau d\gamma^p = \mu \sigma'_N d\gamma^p \tag{7.27}$$

where the superscript p stands for plastic.

Eq. 7.27 can be rearranged in order to express the strength as a function of the critical state strength and dilatancy (Eq. 7.28).

$$\frac{\tau}{\sigma'_N} = \mu + \frac{d\varepsilon^p_{vol}}{d\gamma^p} \tag{7.28}$$

Newland and Allely [213] analysed the dilative behaviour of a granular system at a grain-size scale. They presented the theory for perfectly spherical particles but pointed out that the theory was independent of particle shapes. The stress-dilatancy theory then becomes Eq. 7.29.

$$\varphi' = \varphi'_{cs} + \psi \tag{7.29}$$

Rowe [253] recognised that the principal stress directions play a major role and established a new stress-dilatancy relationship. Bolton [40] stated that the strength by Rowe's theory is overestimtaed by 20% in plane strain shear conditions and the following corrected expression was suggested (Eq. 7.30).

$$\varphi' = \varphi'_{cs} + 0.8\psi \qquad (7.30)$$

It is possible to express Eq. 7.28 with the stress and strain variables defined earlier [249]. Stresses are expressed in terms of effective stress ratio η', the critical state stresses in terms of the critical state effective stress ratio M and the volumetric behaviour in terms of dilatancy rates D as shown in Eq. 7.31.

$$\eta' = M + D \qquad (7.31)$$

Eq. 7.31 is the basis for the development of the original Cam-Clay model, which will be presented in the next Chapter. Nova [216] included an additional parameter in order to offer more flexibility (Eq. 7.32).

$$\eta' = M + (N - 1)D \qquad (7.32)$$

where N is the dilatancy parameter which is often 0.2 [149, 150]. Eq. 7.32 is the basis for the development of the Nor-Sand model, which will be presented in the next chapter.

There have been many other suggestions for the stress-dilatancy relationship (i.e. [208, 217]). However, they all have in common the expression of the stress and strain variables as invariants and include the idea that the development of strength is the consequence of grains interlocking. These equations form the base to develop Cambridge-type constitutive models.

7.5 Closure

The critical state theory [250] is a simple theory which predicts with acceptable accuracy the ultimate state of any granular material sheared sufficiently. The critical state is a uniquely defined line in the (p', q, e) space. When a granular material is at critical state failure, the stress and void ratio have to be on this line. They provide a reference state from which it is possible to understand and predict the behaviour of the granular in any condition or state. It is also possible to formulate state indices in which the critical state is used as a reference (e.g. the equivalent liquidity index for clay or the state parameter for sand) and a normalised state is defined as a quantified distance to the critical state. These state indices are of interest for constitutive modelling. Because of these features, the critical state theory has become the most widely used theory in soil mechanics and is a cornerstone in assessing the behaviour of any soil.

The critical state theory is more than just a set of mathematical expressions which characterise the mechanical behaviour of granular material. It is a theory based on experimental observations and provides a physical explanation and relationship between the development of strength with respect to the development of strains. It enables reliable prediction on the ultimate strength of any soil subjected to shearing [247]. It therefore provides a unique reference state from which all critical state constitutive models are derived. As shown in Chapter 8, it is possible to develop powerful constitutive models for both clay and sand in conjunction with the critical state framework [262]. The success of the theory was driven by the establishment of a constitutive model called Cam-Clay [248, 249] and Nor-Sand [148], which are presented in Chapter 8.

8

An Introduction to Constitutive Modelling

E.J. Fern and K. Soga

8.1 Introduction

MPM is a continuum mechanics methods for which all materials require a mathematical idealisation of the development of stresses and strains. This mathematical idealisation is known as the constitutive model. So far, the choice of the constitutive model was mainly based on the experience from small deformation finite element (FE) simulations. However, the constitutive models play an important role in large deformation simulations as they define the failure at small strain and control the dissipation of energy at large deformation [103]. Although there has been limited research on the role of constitutive models in large deformation simulations, it is a necessary to understand the origin of constitutive models in order to understand their behaviour at large deformation.

The prediction of the shear strength of soil is a centrepiece of soil mechanics and has been achieved with failure criteria, such as Mohr-Coulomb [295]. These failure criteria predict the shear strength by assuming an elastic hardening of the soil until failure occurs (first generation of models, see Chapter 1). Therefore, the predicted shear strength is solely based on the current stress state. It is only with the emergence of the critical state soil mechanics (CCSM) [250, 262] that the elasto-plastic stress path was included in a constitutive model with the development of Original Cam-Clay [249]. This was achieved by formulating the energy balance equations and the stress-dilatancy rules (see Chapter 7). They were subsequently transformed into a constitutive model (second generation of models, see Chapter 1). However, this model was restricted to shearing behaviours of normally and lightly overconsolidated soils, and a second model was suggested to include isotropic compression stress paths – Modified Cam-Clay [248]. It is only in recent years that these two models were extended for overconsolidated clays [124] and sands [148] as the Cam-Clay models suffered from limitations with their hardening rule [247] (third generation of models, see Chapter 1).

8.2 General formulation

Many of the classical elasto-plastic constitutive models are based on the decomposition of deformation into reversible (elastic) and irreversible (plastic) components (Eq. 8.1).

$$d\vec{\varepsilon} = d\vec{\varepsilon}^{e} + d\vec{\varepsilon}^{p} \tag{8.1}$$

where $d\vec{\varepsilon}$, $d\vec{\varepsilon}^{e}$ and $d\vec{\varepsilon}^{p}$ are the total, elastic and plastic strain increment tensors expressed as a vector, respectively.

The models use the concept of effective stresses (Eq. 8.2), which states that all changes in effective stresses result in changes in strains and vice versa. This can be expressed by Eq. 8.3 for small deformations.

$$\vec{\sigma}' = \vec{\sigma}^{tot} - p_{w}\vec{I} \tag{8.2}$$

$$d\vec{\sigma}' = \boldsymbol{D}d\vec{\varepsilon} \tag{8.3}$$

where $\vec{\sigma}'$ is the effective stress vector, $\vec{\sigma}^{tot}$ the total vector , p_{w} the pore water pressure, $\vec{I} = [1, 1, 1, 0, 0, 0]^{T}$ an identity vector and $\boldsymbol{D}$ the stiffness matrix.

Stresses are objective when they remain the same irrespective of the referential. However, Eq. 8.3 only holds true for small deformations and the effective stress tensor needs to be updated for large deformations to be objective using, for example, the Jaumann rate of stress (see Chapter 2).

8.2.1 Elasticity

The elastic stiffness is defined by Eq. 8.4 and marked with the superscript e for elasticity.

$$d\vec{\sigma}' = \boldsymbol{D}^{e}d\vec{\varepsilon^{e}} \tag{8.4}$$

The elastic stiffness matrix $\boldsymbol{D}^{e}$ is defined by Eq. 8.5

$$\boldsymbol{D}^{e} = \begin{pmatrix} K + 4/3G & K - 2/3G & K - 2/3G & 0 & 0 & 0 \\ K - 2/3G & K + 4/3G & K - 2/3G & 0 & 0 & 0 \\ K - 2/3G & K - 2/3G & K + 4/3G & 0 & 0 & 0 \\ 0 & 0 & 0 & G & 0 & 0 \\ 0 & 0 & 0 & 0 & G & 0 \\ 0 & 0 & 0 & 0 & 0 & G \end{pmatrix} \tag{8.5}$$

where K is the bulk modulus and G the shear modulus.

8.2.2 Plasticity

The direction of the strain increments is given by the potential function P and is known as the flow rule (Eq. 8.6).

$$d\vec{\varepsilon}^{p} = \Lambda \frac{\partial P}{\partial \vec{\sigma}'} \tag{8.6}$$

where Λ is the plastic multiplier.

Constitutive models that adopt different potential and yield functions $(P \neq F)$ are said to be *non-associative*, and those which adopt the same function $(P = F)$ are said to be *associative*. Although non-associative flow rules provide more flexible behaviour in terms of stress-strain relationships of certain soils, and perhaps more realistic behaviours, the associated flow models provide essential soil behaviour understanding and modelling opportunities.

The stress space in which the soil behaves elastically is defined by the yield surface F and is a function of the stress state and the stress history of the soil (Eq. 8.7).

$$F(\vec{\sigma}', W_{\mathrm{p}}) \tag{8.7}$$

where W_{p} is a history variable which stores information of the past deformation and can be a scalar, a vector or a tensor. It is often related to the plastic strain – $W_{\mathrm{p}}(\vec{\varepsilon}^{\mathrm{p}})$.

The consistency condition guarantees that the stress state satisfies the yield criteria during plastic deformation (Eq. 8.8) as well as the effective stress concept.

$$\mathrm{d}F(\vec{\sigma}', W_{\mathrm{p}}) = \frac{\partial F}{\partial \vec{\sigma}'}\mathrm{d}\vec{\sigma}' + \frac{\partial F}{\partial W_{\mathrm{p}}}\frac{\partial W_{\mathrm{p}}}{\partial \vec{\varepsilon}^{\mathrm{p}}}\mathrm{d}\vec{\varepsilon}^{\mathrm{p}} = 0 \tag{8.8}$$

An incremental stress-strain equation (Eq. 8.3) can be developed by substituting Eqs. 8.1, 8.4 and 8.6 into 8.8, which gives Eq. 8.9.

$$\mathrm{d}\vec{\sigma}' = \left[\boldsymbol{D}^{\mathrm{e}} - \frac{\boldsymbol{D}^{\mathrm{e}}\dfrac{\partial F}{\partial \vec{\sigma}'}\left(\dfrac{\partial P}{\partial \vec{\sigma}'}\right)^{\mathrm{T}}\boldsymbol{D}^{\mathrm{e}}}{\left(\dfrac{\partial F}{\partial \vec{\sigma}'}\right)^{\mathrm{T}}\boldsymbol{D}^{\mathrm{e}}\dfrac{\partial P}{\partial \vec{\sigma}'} - \dfrac{\partial F}{\partial W_{\mathrm{p}}}\left(\dfrac{\partial W_{\mathrm{p}}}{\partial \vec{\varepsilon}^{\mathrm{p}}}\right)^{\mathrm{T}}\dfrac{\partial P}{\partial \vec{\sigma}'}} \right]\mathrm{d}\vec{\varepsilon} \tag{8.9}$$

Eq. 8.9 is expressed in terms of the yield function F and the energy dissipation mechanism $\partial W_{\mathrm{p}}/\partial \vec{\varepsilon}^{\mathrm{p}}$, which are intrinsic constituents of a constitutive model.

The Cam-Clay and Nor-Sand models are formulated in terms of stress invariants and these are the mean effective stress p', the deviatoric stress q and the Lode angle θ. Therefore, the partial derivative of the yield function F with respect to $\vec{\sigma}'$ gives Eq. 8.10.

$$\frac{\partial F}{\partial \vec{\sigma}'} = \frac{\partial F}{\partial p'}\frac{\partial p'}{\partial \vec{\sigma}'} + \frac{\partial F}{\partial q}\frac{\partial q}{\partial \vec{\sigma}'} + \frac{\partial F}{\partial \theta}\frac{\partial \theta}{\partial \vec{\sigma}'} \tag{8.10}$$

The partial derivatives of the yield function and history variables are specific to each constitutive model. For instance, Cam-Clay uses the consolidation pressure p_{c} as a history variable and Nor-Sand uses the image pressure p_{i}. The size of the yield surface is determined by the magnitude of these history variables, and are a function of plastic volumetric strain $\varepsilon_{\mathrm{vol}}^{\mathrm{p}}$, plastic deviatoric strain $\varepsilon_{\mathrm{dev}}^{\mathrm{p}}$ or both.

The hardening/softening rule (Eq. 8.11) determines how the yield surface grows or shrinks and, subsequently, defines the stress state at which the soil behaves elastically.

$$\frac{\partial F}{\partial W_{\mathrm{p}}} = \frac{\partial F}{\partial p_{\mathrm{c\ or\ i}}} \frac{\partial p_{\mathrm{c\ or\ i}}}{\partial \bar{\varepsilon}^{\mathrm{p}}} = \frac{\partial F}{\partial p_{\mathrm{c\ or\ i}}} \left(\frac{\partial p_{\mathrm{c\ or\ i}}}{\partial \varepsilon^{\mathrm{p}}_{\mathrm{vol}}} \frac{\partial \varepsilon^{\mathrm{p}}_{\mathrm{vol}}}{\partial \bar{\varepsilon}^{\mathrm{p}}} + \frac{\partial p_{\mathrm{c\ or\ i}}}{\partial \varepsilon^{\mathrm{p}}_{\mathrm{dev}}} \frac{\partial \varepsilon^{\mathrm{p}}_{\mathrm{dev}}}{\partial \bar{\varepsilon}^{\mathrm{p}}} \right) \quad (8.11)$$

The energy dissipating mechanism is an assumption of the constitutive model. For instance, Cam-Clay assumes that the energy is dissipated by volumetric plastic strain $\varepsilon^{\mathrm{p}}_{\mathrm{vol}}$, whereas Nor-Sand assumes that the energy is dissipated by plastic deviatoric strain $\varepsilon^{\mathrm{p}}_{\mathrm{dev}}$.

8.3   Cam-Clay

Cam-Clay is a simple model initially developed for normally consolidated and lightly overconsolidated clays[1]. It is the first model to capture the different mechanical behaviours from various stress and strain states. Two versions of Cam-Clay exist – Original Cam-Clay [249] and Modified Cam-Clay [248]. Both models follow the same principles but are formulated in different ways.

8.3.1   Yield function

Original Cam-Clay finds its roots in Taylor's stress-dilatancy theory [292] (see Chapter 7) from which it is possible to formulate a flow rule in terms of stress invariants (Eq. 8.12).

$$p'\mathrm{d}\varepsilon^{\mathrm{p}}_{\mathrm{vol}} + q\mathrm{d}\varepsilon^{\mathrm{p}}_{\mathrm{dev}} = p'M\mathrm{d}\varepsilon^{\mathrm{p}}_{\mathrm{dev}}$$

$$\rightarrow \quad \eta' = M - D \quad (8.12)$$

where $\eta' = q/p'$ is the effective stress ratio, M the critical state stress ratio and $D = d\varepsilon^{\mathrm{p}}_{\mathrm{vol}}/d\varepsilon^{\mathrm{p}}_{\mathrm{dev}}$ the dilatancy rate.

The original formulation [249] assumes normality between the stress and strain increments [87] and gives Eq. 8.13.

$$\frac{\mathrm{d}q}{\mathrm{d}p'} \frac{\mathrm{d}\varepsilon^{\mathrm{p}}_{\mathrm{dev}}}{\mathrm{d}\varepsilon^{\mathrm{p}}_{\mathrm{vol}}} = -1 \quad (8.13)$$

It is then possible to substitute Eq. 8.13 into Eq. 8.12, which gives Eq. 8.14.

$$q = Mp' + p'\frac{\mathrm{d}q}{\mathrm{d}p'} \quad (8.14)$$

---

[1]Cam-Clay is elasto-plastic for normally and lightly overconsolidated soils and contains plasticity in the hardening phase. This is not true when used for heavily overconsolidated soils for which the hardening phase is solely elastic. The model is then said to be elastic-plastic.

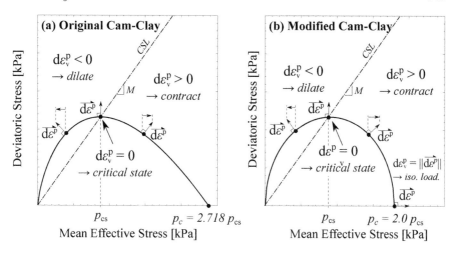

FIGURE 8.1: Yield surfaces of (a) Original Cam-Clay and (b) Modified Cam-Clay.

The deviatoric stress q can be expressed as $q = \eta' p'$ and substituted in Eq. 8.14, which gives Eq. 8.15.

$$\rightarrow \eta' p' = M p' + p' \frac{\mathrm{d}(\eta' p')}{\mathrm{d}p'} \tag{8.15}$$

The development of Eq. 8.15 gives Eq. 8.16.

$$\rightarrow \eta' = M + \frac{\mathrm{d}\eta'}{\mathrm{d}p'} p' + \eta' \rightarrow \mathrm{d}\eta' = -M \frac{\mathrm{d}p'}{p'} \tag{8.16}$$

The integration of Eq. 8.16 with the known point $(p' = p_\mathrm{c},\ q = 0,\ \eta' = 0)$ gives Eq. 8.17.

$$\eta' = -M \ln \left(\frac{p'}{p_\mathrm{c}} \right) \tag{8.17}$$

Eq. 8.17 forms a surface on which the soil yields and is shown in Fig. 8.1 (a). This yield surface can be expressed as a function as shown in Eq. 8.18.

$$F = q - M p' \ln \left(\frac{p_\mathrm{c}}{p'} \right) \tag{8.18}$$

The stresses are necessarily located within (elastic, $F < 0$) or on the yield surface (elasto-plastic, $F = 0$) due to the consistency condition.

The relationship between the history variable p_c and the corresponding critical state pressure p_cs is $\ln(p_\mathrm{c}/p_\mathrm{cs}) = 1$.

The yield function of Original Cam-Clay has a singularity at the tip of the yield surface $(p' = p_\mathrm{c},\ q = 0)$; the strain increment is undefined along the isotropic stress path (increasing p' with $q = 0$). This led to the development of Modified Cam-Clay [249], for which the yield surface (Eq. 8.19, Fig. 8.1 b) is chosen to avoid singularity points and the energy dissipating mechanism is subsequently back calculated (Eq. 8.20).

$$F = \left(\frac{\eta'}{M}\right)^2 + (p' - p_c) \tag{8.19}$$

$$p'\mathrm{d}\varepsilon^p_{vol} + q\mathrm{d}\varepsilon^p_{dev} = p'\sqrt{(\mathrm{d}\varepsilon^p_{vol})^2 + (M\mathrm{d}\varepsilon^p_{dev})^2} \tag{8.20}$$

The relationship between the history variable p_c and the corresponding critical state pressure p_{cs} is $p_c/p_{cs} = 2$.

The yield surfaces are defined by p' and q and hence the partial derivatives of $\partial F/\partial p'$ and $\partial F/\partial q$ for Original Cam-Clay (Eq. 8.18) and Modified Cam-Clay (Eq. 8.19) can be computed and are given in Eqs. 8.21 and 8.26.

Original Cam-Clay:

$$\frac{\partial F}{\partial p'} = M\left[1 - \ln\left(\frac{p_c}{p'}\right)\right] \tag{8.21}$$

$$\frac{\partial F}{\partial q} = 1 \tag{8.22}$$

$$\frac{\partial F}{\partial p_c} = -M\frac{p'}{p_c} \tag{8.23}$$

Modified Cam-Clay:

$$\frac{\partial F}{\partial p'} = 2p' - p_c \tag{8.24}$$

$$\frac{\partial F}{\partial q} = 1 \tag{8.25}$$

$$\frac{\partial F}{\partial p_c} = -p' \tag{8.26}$$

As discussed in Chapter 7, the critical state stress ratio M is a function of the Lode angle θ and its partial derivative $\partial F/\partial \theta$ is given by Eq. 8.27.

$$\frac{\partial F}{\partial \theta} = \frac{\partial F}{\partial M}\frac{\partial M}{\partial \theta} \tag{8.27}$$

The partial derivative $\partial F/\partial M$ is specific to the constitutive model, and is given in Eqs. 8.28 and 8.29 for original and Modified Cam-Clay, respectively.

Original Cam-Clay:

$$\frac{\partial F}{\partial M} = -\ln\left(\frac{p'}{p_c}\right) \tag{8.28}$$

Modified Cam-Clay:

$$\frac{\partial F}{\partial M} = 2\frac{\eta'^2}{M^3} \tag{8.29}$$

8.3.2 Volumetric behaviour and dilatancy

The normality condition between the strains and the stresses implies that the strain increment is normal to the yield surface as illustrated in Fig. 8.1. The critical state theory [250] suggests that a logarithmic relationship exists between the mean critical state effective stress p'_{cs} and the specific volume at critical state (Eq. 8.30), as is shown in Fig. 8.2.

$$v_{cs} = \Gamma - \lambda \ln(p'_{cs}) \tag{8.30}$$

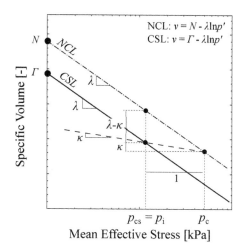

FIGURE 8.2: Critical state and normal consolidation lines as proposed in both Cam-Clay models.

where $v_{cs} = 1 + e_{cs}$ is the critical state specific volume, Γ the critical state specific volume at 1 kPa pressure and λ the critical state slope.

When the stress ratio state on the yield surface is smaller than the critical state stress ratio ($\eta' < M$), the volumetric strain increment is positive for a given deviatoric strain increment. The soil is contractive and the state of the soil is located above the critical state line (CSL) (Eq. 8.30) in the (v, $\ln p'$) space. When the stress ratio state on the yield surface is larger than the critical state stress ratio ($\eta' > M$), the volumetric strain increment is negative from the stress dilatancy relationship and the soil dilates. When the stress state is located at the summit of the yield function ($\eta' = M$), the soil is at critical state and the volumetric strain increment is nil (that is, the plastic strain vector in Fig. 8.1 (a) is in the vertical direction). Therefore, no changes in volume occur and the dilatancy is nil as the soil is on the CSL.

Both the Original and Modified Cam-Clay models assume that the plastic (or virgin) consolidation by isotropic compression is parallel to the CSL, as shown in Fig. 8.2. This is often called the normal compression line (NCL), which is defined by Eq. 8.31.

$$v = N - \lambda \ln(p') \tag{8.31}$$

During an isotropic compression test, the stress and strain state evolve along the NCL. The size of the yield surface increases and the isotropic stress path gives $p' = p_c$ as shown in Fig. 8.2. When the soil is unloaded, it exhibits elastic behaviour and expands. This expansion is characterised by elastic swelling, and is modelled in Cam-Clay as shown in Eq. 8.32.

$$v = v_c - \kappa \ln \left(\frac{p'}{p_c} \right) \tag{8.32}$$

where v_c is the specific volume prior to unloading. κ is the swelling modulus, which is the slope of the swelling line in the $(v, \ln p')$ space.

The position of the NCL in relation to the CSL is given by the yield function and the values of λ and κ. This gives Eq. 8.33 and Eq. 8.34.

Original Cam-Clay:

Modified Cam-Clay:

$$N = \Gamma + \lambda - \kappa \qquad (8.33) \qquad N = \Gamma + (\lambda - \kappa)\ln 2 \qquad (8.34)$$

The material hardens and the yield surface grows during isotropic compression. Both Cam-Clay models assume that the rate of increase of the yield surface follows the NCL. Therefore, the hardening is driven by plastic volumetric strain only (i.e. $\partial p_c / \partial \varepsilon_{dev}^P = 0$) and the hardening law is expressed by Eq. 8.35, which can be derived geometrically from Fig. 8.2.

$$\frac{\partial p_c}{\partial \varepsilon_{vol}^P} = \frac{\nu}{\lambda - \kappa} p_c \qquad (8.35)$$

8.3.3 Elasticity for clay

It is assumed that the mechanical behaviour is elastic when the stress state is inside the yield surface. In this chapter, the elastic behaviour is considered to be isotropic and hence two parameters are needed – the bulk modulus K and Poisson ratio ν [2]. It can be shown that the bulk modulus is dependent on the swelling modulus κ, the mean pressure p' and the specific volume v (Eq. 8.36), by using the κ-line swelling model and differentiating it.

$$K = \frac{\Delta p'}{\Delta \varepsilon_{vol}^e} = \frac{\Delta p'}{\Delta v} v = \frac{v p'}{\kappa} \qquad (8.36)$$

8.3.4 Simulations of drained triaxial compression tests

Fig. 8.3 shows simulations of triaxial compression tests with both the Original and the Modified Cam-Clay models. The model parameters are given in Table 8.1. The initial confinement pressure is $p_0' = 20$ kPa and the consolidation pressure is $p_c = 20$ kPa for the normally consolidated specimens, and $p_c = 100$ kPa for the overconsolidated specimens.

The triaxial compression test is a conventional laboratory test in which a cylindrical soil sample is consolidated by isotropic pressure and is then sheared by increasing the axial stress but keeping the radial stress constant. A drained test implies that the pore water pressure is kept constant during axial compression and the specimen volume can change by allowing the pore water to

---

[2] Poisson ratio is denoted by the Greek letter "nu" and the specific volume denoted with the Roman letter "v".

TABLE 8.1

Cam-Clay parameters for simulations. The CSL intercept Γ and the NCL intercept N are related (Eqs. 8.33 and 8.34).

Parameter	Symbol	Unit	Value
Swelling modulus	κ	-	0.062
Poisson ratio	ν	-	0.2
CS stress ratio	M_{tc}	-	0.89
CSL slope	λ	-	0.161
NCL intercept	N	-	2.76
Consolidation pressure	p'_c	kPa	20 or 100

move in or out of the specimen. Hence, the total stress increment is equal to the effective stress increment; the excess pore pressure is zero.

A normally consolidated state implies that the sample has never been subjected to a higher pressure. The initial mean effective stress is the consolidation pressure $p' = p_c$ as shown by point A in Figs. 8.3 (a) and 8.4 (a). It is on the "wet" side of the CSL as its initial water content is higher than its critical state value. The drained triaxial compression ($d\sigma'_1 > 0$, $d\sigma'_3 = 0$) imposes an effective stress path with a stress ratio rate of $dq/dp' = d\sigma'_1/(d\sigma'_1/3) = 3$ as shown in the figures.

As the material is sheared with the given stress path, the mobilised shear resistance increases in a non-linear manner until reaching the CSL, and the specific volume decreases from the NCL to the CSL. This phase is known as the hardening phase, in which the yield surface grows until the stress state is located at the summit of the yield surface and on the CSL. Accordingly, the consolidation pressure p_c increases. For a given set of model parameters, Modified Cam-Clay predicts a stiffer response than Original Cam-Clay and this is due to the difference in yield surface shape.

Overconsolidated samples have been previously isotropically loaded and then unloaded before any shearing takes place. Therefore, the yield surface expands during the loading phase with $p' = p_c$. The sample is then unloaded and behaves elastically along the κ-line. That is, the yield surface remains at its position by keeping the value of p_c but the unloaded stress state is located within the yield surface ($p' < p_c$). This state corresponds to point A in Figs. 8.3 (b) and 8.4 (b). The overconsolidation densifies the material and its initial specific volume prior to shearing can be located below the CSL if the unloaded pressure is much smaller than the consolidation pressure (i.e. $\ln(p_c/p') > 1$ for Original Cam-Clay and $p_c/p' > 2$ for Modified Cam-Clay). The state of the soil is on the "dry" side of the critical state.

When the heavily overconsolidated sample is sheared in drained conditions with the triaxial compression stress path of $dq/dp' = 3$, the sample initially exhibits elastic behaviour from point A to point B crossing the CSL. It does not reach a critical state because the soil is elastic and the specific volume state is not on the CSL. The soil contracts during this elastic phase due to increase in mean pressure p'.

The soil then yields when the stress state reaches the yield surface as shown in Figs. 8.3 (a) and 8.4 (a) and this corresponds to the peak shear resistance (point B). Original Cam-Clay (Fig. 8.3 b) predicts a lower peak strength than Modified Cam-Clay (Fig. 8.4 b) for the same consolidation pressure p_c. This is due to the different shapes of the yield function. From this point, the soil behaves as an elastic-plastic material. As the stress ratio state at yield is greater than the critical state ($\eta' > M$) for heavily overconsolidated clay, it dilates and at the same time softens to reach the critical state by keeping the stress path of $dq/dp' = 3$. Therefore, the shear resistance decreases as shearing continues and then finally reaches the critical state when the stress ratio becomes $\eta' = M$ and the specific volume is on the CSL in the $(v, \ln p')$ plane. The yield surface shrinks and the consolidation pressure p_c decreases (i.e. softening).

8.3.5 Simulations of undrained triaxial compression tests

Triaxial tests can also be run in undrained conditions. It means that no pore water can flow in or out of the sample. The water is relatively incompressible and hence the sample volume is kept constant throughout the test during shearing. Triaxial compression tests will give the total stress path to be $dq/dp' = 3$, but the effective stress path will not follow $dq/dp' = 3$ due to excess pore pressure generation. The magnitude of excess pore pressure depends on the dilative or contractive nature of the sample.

Figs. 8.3 (c) and 8.4 (c) illustrate the triaxial compression case of a normally consolidated clay sample in undrained conditions. The constant volume condition implies that the mean effective stress state of the clay needs to move from point A to point B in the horizontal direction of the $(v, \ln p')$ plane in order to reach the critical state of $p_{cs(B)}$. Positive excess pore pressure develops during shearing and the shear resistance at the critical state becomes $q_{cs(B)}$ as shown in the figure. The effective stress path moves to the upper left direction in the (p', q) space and the path can be evaluated from the incremental stress-strain relationship with the constant volume constraint ($d\varepsilon_{vol} = 0$), as shown by Eq. 8.37.

$$d\varepsilon_{vol} = d\varepsilon_{vol}^e + d\varepsilon_{dev}^p = \frac{dp'}{K} + d\varepsilon_{dev}^p = 0 \qquad (8.37)$$

$$\rightarrow dp' = -K d\varepsilon_{vol}^p \qquad (8.38)$$

Eq. 8.38 shows that the contractive tendency of the clay ($d\varepsilon_{vol}^p > 0$) decreases the mean effective stress.

The soil hardens and the consolidation pressure increases from $p_{c(A)}$ to $p_{c(B)}$. However, the degree of hardening is much smaller than the drained case in Figs. 8.3 (a) and 8.4 (a). Hence, the undrained shear strength of

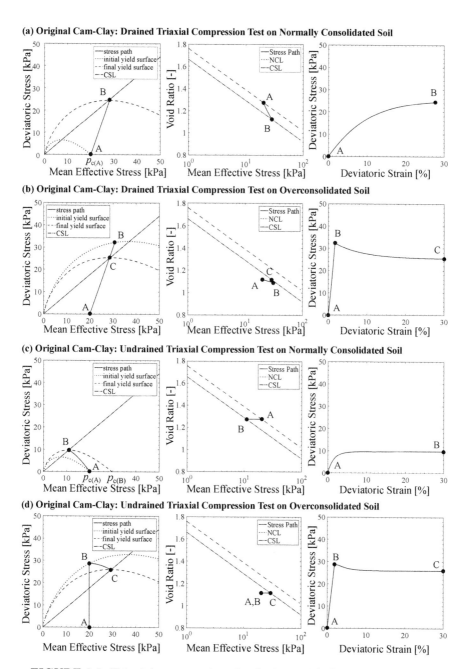

FIGURE 8.3: Triaxial compression simulations with Original Cam-Clay.

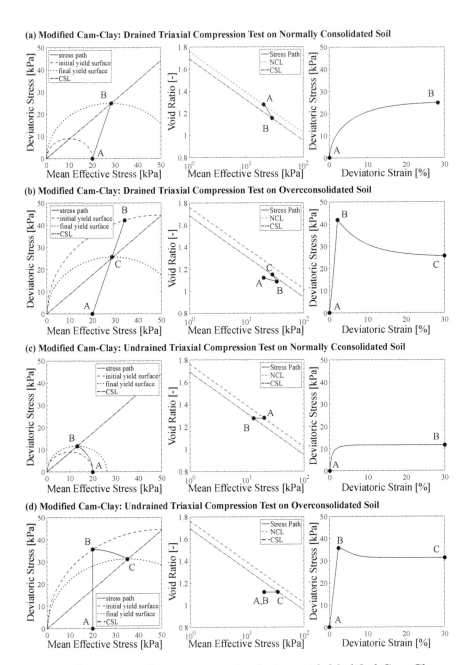

FIGURE 8.4: Triaxial compression simulations with Modified Cam-Clay.

normally consolidated clay is smaller than the drained shear strength in n triaxial compression condition.

Figs. 8.3 (d) and 8.4 (d) illustrate the triaxial compression case of a heavily overconsolidated sample in undrained conditions. The stress-strain path differs largely from the one observed for the normally consolidated sample. The overconsolidation increases the size of the yield surface and the sample initially undergoes elastic deformation from point A to point B. In undrained conditions, the mean effective stress remains constant ($dp' = 0$) as the deviatoric stress q increases during elastic deformation ($d\varepsilon_{vol} = d\varepsilon^e_{vol} = 0$). This is because $dp' = Kd\varepsilon^e_{vol} = 0$.

Once the stress state reaches the yield surface at point B, the soil yields and becomes elasto-plastic. As the soil is on the dry side of the CSL, the effective stress state needs to move to the right direction from Point B to Point C in the $(v, \ln p')$ plane so that the constant volume condition ($dv = 0$) can be kept. The clay has a dilative tendency ($d\varepsilon^p_{vol} < 0$) and hence the effective mean pressure increases with shearing ($dp' > 0$) according to Eq. 8.38. The mean pressure at the critical state in undrained conditions is greater than that in drained conditions. Hence, the undrained shear strength of heavily overconsolidated clay is greater than the drained shear strength in the triaxial compression case.

8.4   Nor-Sand

Nor-Sand [148] is a simple elastic-plastic model for sand. It can be viewed as a bounding surface model with two surfaces for the two critical state conditions ($D = 0, \partial D/\partial \varepsilon^p_{dev} = 0$). Unlike Cam-Clay, Nor-Sand includes the void ratio e, through the state parameter Ψ [28], as a model variable making the model parameters density independent. Nor-Sand was developed from Nova's stress-dilatancy flow rule (Eq. 8.39) [216] by following the same procedure as Original Cam-Clay [249].

$$\eta' = M + (N - 1)D \tag{8.39}$$

where N is the dilatancy parameter. When $N = 0$, the stress-dilatancy flow rule yields to Original Cam-Clay's stress-dilatancy flow rule.

8.4.1   Yield function

Nor-Sand assumes normality between the stress and the strain increments, which offers simplicity in the modelling. The integration of the stress-dilatancy rule of Eq. 8.39 gives two different yield functions depending on N (Eq. 8.40

when $N > 0$ and Eq. 8.41 when $N = 0$) and shown in Fig. 8.5 (a).

$$F = \eta' - \frac{M}{N}\left[1 + (N-1)\left(\frac{p'}{p_i}\right)^{\frac{N}{1-N}}\right] \qquad \text{for} \quad N > 0 \qquad (8.40)$$

$$F = \eta' - M\left[1 + \ln\left(\frac{p_i}{p'}\right)\right] \qquad \text{for} \quad N = 0 \qquad (8.41)$$

where p_i is the image pressure and p_{cs} is the critical state pressure of the yield surface in Fig. 8.1. It is an equivalent expression of the consolidation pressure p_c in Cam-Clay but offers modelling advantages. It corresponds to the pressure at the summit of the yield surface ($p'_{cs} = p_i$) rather than at the tip. It relates better to the critical state condition, and can be used to characterise the different phases (i.e. hardening, softening, dilation, contraction). The summit point is called the image condition, which is characterised by $q_i = Mp_i$.

It is possible to make Nor-Sand non-associative by making the potential function different by using a different dilatancy parameter N_p as depicted in Fig. 8.5 (b). Although Nor-Sand has a bounding surface called the maximum yield surface used in the hardening rule, it follows the same principles as Cam-Clay. However, it has a different yield surface for $N > 0$ and thus different derivatives given in Eqs. 8.42 to 8.45; the derivatives for $N = 0$ are the same as for Original Cam-Clay (Eqs. 8.21 and 8.23, 8.28).

$$\frac{\partial F}{\partial p'} = -\frac{M}{N}\left[1 + \frac{N-1}{1-N}\left(\frac{p'}{p_i}\right)^{\frac{N}{1-N}}\right] \qquad \text{for} \quad N > 0 \qquad (8.42)$$

$$\frac{\partial F}{\partial q} = 1 \qquad (8.43)$$

$$\frac{\partial F}{\partial \theta} = -\frac{p'}{N}\left[1 + (N-1)\left(\frac{p'}{p_i}\right)^{\frac{N}{1-N}}\right] \qquad \text{for} \quad N > 0 \qquad (8.44)$$

$$\frac{\partial F}{\partial p_i} = \frac{N-1}{1-N}M\left(\frac{p'}{p_i}\right)^{\frac{1}{1-N}} \qquad \text{for} \quad N > 0 \qquad (8.45)$$

8.4.2   Hardening rule

Nor-Sand differs from Original Cam-Clay by its hardening rule. It assumes that the hardening and softening of sand is driven by shearing rather than by compaction, which implies that the yield surface changes size with the deviatoric plastic strain ε^p_{dev} rather than the volumetric plastic strain ε^p_{vol}. It then assumes that the hardening and softening rates are proportional to the

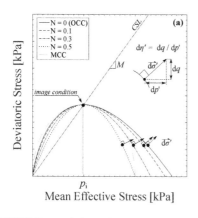
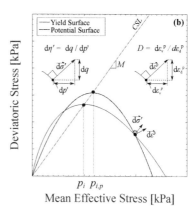

FIGURE 8.5: (a) Yield surface of Nor-Sand for different values of the dilatancy parameter N, and (b) yield and potential function of Nor-Sand (after [101]).

distance between its current state, characterised by the image pressure p_i, and a predicted peak state, characterised by the maximum image pressure $p_{i,max}$. This gives a new hardening rule (Eq. 8.46).

$$\frac{\partial p_i}{\partial \varepsilon_{dev}^P} = H(p_{i,max} - p_i) \tag{8.46}$$

where H is the hardening modulus.

More complex hardening rules have been proposed in the literature for better fits to experimental data (i.e. Eq 8.47 [151]).

$$\frac{\partial p_i}{\partial \varepsilon_{dev}^P} = H \frac{M}{M_{tc}} \exp\left(1 - \frac{\eta'}{M_{tc}}\right)(p_{i,max} - p_i) \tag{8.47}$$

Fig. 8.6 illustrates the hardening/softening concept in Nor-Sand. The model predicts a hardening when $p_i < p_{i,max}$ and a softening when $p_i > p_{i,max}$. The hardening rule is nil at peak state as $p_i = p_{i,max}$ and $\partial D / \partial \varepsilon_{dev}^P = 0$. It is also nil at critical state Therefore, the critical state is also nil and remains nil as $p' = p_i = p_{i,max}$ and $D = \partial D / \partial \varepsilon_{dev}^P = 0$.

The maximum yield surface is a bounding surface sized by the maximum image pressure $p_{i,max}$, and is a prediction of the peak state based on dilatancy consideration (Eq. 8.48).

$$\frac{p_{i,max}}{p'} = \left(1 + D_{min}\frac{N}{M_{tc}}\right)^{N-1/N} \qquad \text{for} \quad N > 0 \tag{8.48a}$$

$$\frac{p_{i,max}}{p'} = \exp\left(-\frac{D_{min}}{M_{tc}}\right) \qquad \text{for} \quad N = 0 \tag{8.48b}$$

where D_{min} is the minimum dilatancy rate and M_{tc} the critical state effective stress ratio in triaxial compression conditions.

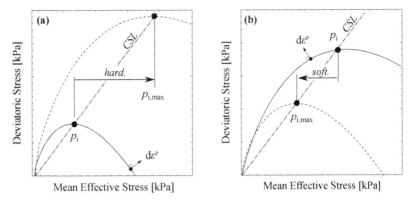

FIGURE 8.6: Hardening and softening concept in Nor-Sand (after [101]).

The dilatancy rate at peak state has to be predicted and this can be done by using state indices as the state parameter [28] (see Section 7.3.2). However, this is done for the current state, which is characterised by the current yield surface (p_i) and, thus, the image state parameter Ψ_i is used (Eq. 8.49), and then converted to a minimum dilatancy rate $D_{\min}$ (Eq. 8.50) [148, 152].

$$\Psi_i = e - e_{cs}(p_i) \tag{8.49}$$

$$D_{\min} = \chi \frac{M}{M_{tc}} \Psi_i \tag{8.50}$$

where $e_{cs}(p_i)$ is the critical state void ratio at image pressure p_i and χ is the dilatancy coefficient

When the image state parameter is positive $\Psi_i > 0$, the maximum yield surface is larger than the yield surface and hardening is expected ($p_i < p_{i,\max} \to \partial p_i / \partial \varepsilon^p_{dev} > 0$), and when the state parameter is negative $\Psi_i < 0$, the maximum yield surface is smaller than the yield surface and softening is expected ($p_i > p_{i,\max} \to \partial p_i / \partial \varepsilon^p_{dev} < 0$).

A CSL must be defined in order to calculate the image state parameter. In this chapter, the relative dilatancy CSL presented in Chapter 7 is used.

8.4.3   Stability for large deformation simulations

Although the inclusion of the void ratio as a model variable is a strong advantage for modelling the peak state, it can cause some numerical instabilities [101]. During dynamic flows of granular material, the MPs undergo significant changes in density caused by the stretching and compression of the flow. This causes the void ratio to change abruptly, which in turn causes sudden changes in the maximum image pressure and, thus, in dilatancy rates. The result is an explosion of MPs during the granular flows. In order to prevent this from happening, additional conditions are implemented in order to guarantee a nil state parameter at very large deformation such that $\Psi = 0$ at large deformation, which ensures a smooth granular flow at large deformation.

8.4.4 Elasticity for sand

The elastic behaviour can be modelled with an isotropic elastic behaviour, with a pressure-dependent shear modulus G (Eq. 8.51) and a constant Poisson ratio ν.

$$G = A^{\mathrm{e}} \left(\frac{p'}{p_{\mathrm{ref}}} \right)^{n^{\mathrm{e}}} \tag{8.51}$$

The bulk modulus K can then be calculated from the shear modulus and the Poisson ratio ν (Eq. 8.52).

$$K = \frac{2(1+\nu)}{3(1-2\nu)} G \tag{8.52}$$

8.4.5 Simulations of drained triaxial compression tests

Fig. 8.7 (a) and (b) illustrate the simulation results of two drained triaxial compression tests, in which the specimens are consolidated isotropically to $p' = 20$ kPa prior to shearing. The simulations are carried out at two different densities. Due to the low initial mean effective stress, the looser specimen ($e_0 = 0.950$) contracts at the end and the denser ($e_0 = 0.850$) dilates at the end. Therefore, these two different cases will be referred to as loose and dense specimens. The model parameters and initial states are given in Table 8.2.

Fig. 8.7 (a) shows the results of the loose specimen in drained conditions. The results show that the specimen hardens in a non-linear way from point A to B and contracts because the stress state is on the contractive side of the

TABLE 8.2
Nor-Sand model parameters for simulations.

Parameter	Symbol	Unit	Value
Shear modulus constant	A^{e}	kPa	2,500
Reference unit pressure	p_{ref}	kPa	1.0
Shear modulus exponent	n^{e}	-	0.5
Poisson ratio	ν	-	0.2
CS stress ratio	M_{tc}	-	1.33
Min. void ratio	e_{min}	-	0.500
Max. void ratio	e_{max}	-	0.946
Crushing pressure	Q	kPa	10,000
Dilatancy parameter	N	-	0.3
Hardening modulus	H	-	200
Dilatancy coefficient	χ	-	3.5

yield surface ($p' > p_i$). As shearing continues, p_i increases due to hardening, and at the same time $p_{i,max}$ reduces due to plastic deformation. Once at point B, the mean effective stress is equal to the image pressure and the maximum image pressure ($p' = p_i = p_{i,max}$), and hence the specimen is at critical state. The dilatancy is nil and no changes in dilatancy are permitted ($D = 0, \partial D/\partial \varepsilon_{dev}^P = 0$).

Fig. 8.7 (b) shows the results of the dense specimen in drained conditions. The results show that the specimen initially contracts from point A to point B because it is on the contractive side of the yield surface ($p' > p_i$). When the stress ratio reaches the critical state stress state ($\eta' \approx M$), there is a nil dilatancy state ($p' = p_i$) because of the normality condition but still allows hardening ($D = 0, \partial D/\partial \varepsilon_{dev}^P \neq 0$) because $p_i < p_{i,max}$. Therefore, the specimen is not at critical state. The shear resistance increases by p_i increasing and $p_{i,max}$ decreasing and reaches the peak state at point C. The stress ratio is maximum and the dilatancy rate becomes D_{min} with $p_i = p_{i,max}$. $p_{i,max}$ continues to decrease by dilation. The specimen starts softening from point C to point D because now $p_i > p_{i,max}$. Eventually, further shearing brings to $p' = p_i = p_{i,max}$ and the effective stress ratio and the void ratio are at critical state ($\eta' = M, e = e_{cs}$). The condition of $D = 0$ and $\partial D/\partial \varepsilon_{dev}^P = 0$ is met and the void ratio remains the same even with continuous shearing.

8.4.6   Simulations of undrained triaxial compression tests

Fig. 8.7 (c) shows the simulated behaviour of the loose specimen during undrained triaxial compression because ($p > p_i$). The specimen undergoes a short hardening phase from point A to B. The stress state is on the contractive side of the yield surface and the excess pore pressure increases. The mean effective stress decreases as the shear resistance mobilises. The peak stress state at point B in undrained conditions is significantly lower than the stress state at point B in drained conditions. The mean effective stress p' at point B is close to the critical state $\eta' \approx M$) but not at critical state because the condition of $p_i < p_{i,max}$ still holds. As the stress ratio is slightly smaller than the critical state slope ($\eta' < M$), the specimen continues to have a contractive tendency, which further increases the excess pore pressure. The reduction in mean effective stress is then accompanied by the reduction in deviator stress and hence the specimen softens from point B to point C with the condition of $p_i > p_{i,max}$. At the same time, the difference between p_i and $p_{i,max}$ reduces with increased shearing. In undrained conditions, Nor-Sand tends asymptotically towards the critical state ($p_i \rightarrow p_{i,max}$), but mathematically it cannot reach $p_i = p_{i,max}$. Therefore, this undrained softening carries on until a nil effective stress is reached. This can be the limitation of the model.

Fig. 8.7 (d) shows the simulated behaviour of the dense specimen during undrained triaxial compression. The stress state is initially on the contractive side ($p' > p_i$) and hence positive excess pore pressure develops from point A to B. At point B ($p' = p_i$), the dilation rate becomes zero, but it continues to

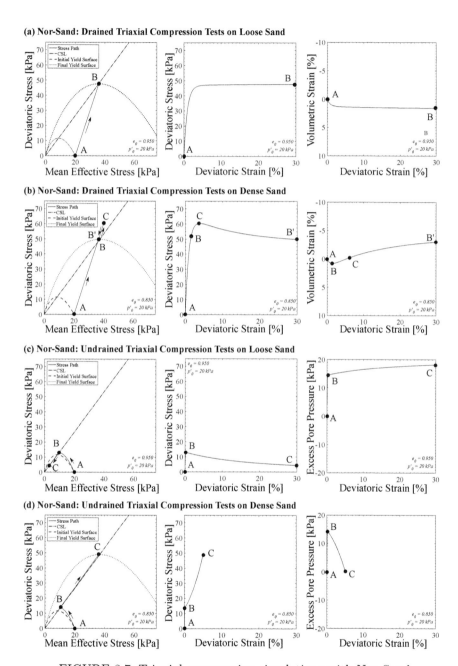

FIGURE 8.7: Triaxial compression simulations with Nor-Sand.

increase its shear resistance because the condition of $p_i < p_{i,max}$. The stress state becomes on the dilative side of the yield surface and hence negative excess pore pressure generates. The excess pore pressure decreases and the mean effective stress increases. The deviator stress continues to rise as the difference between p_i and $p_{i,max}$ reduces with increased shearing. Again the model tends asymptotically towards the critical state $p_i \approx p_{i,max}$ but mathematically it cannot reach $p_i = p_{i,max}$. Therefore, the stress path follows slightly above the CSL. This can be the limitation of the model.

8.5 Mohr-Coulomb

large deformation simulation implies that parts of the material will sustain, at some point, some large deformation, whilst other parts of the material may only experience small strain. The large deformation can also occur at different times for different portions of the soil. This means that the constitutive model must be able to predict both the small and large deformation behaviours (i.e. peak state and critical state).

The CCSM [262] is an ideal framework for large deformation constitutive models and permits modelling soil behaviour as a function of density, pressure and pore pressure, although only few models include the density as a model variable. The advantage of constitutive models developed within the CCSM framework is that they necessarily reach the critical state in terms of stresses and density, and that the critical state model parameters can be directly used for large deformation MPM simulations. However, substantial differences exist between the different models when modelling the peak state. Fig. 8.8 illustrates the difference between the Cam-Clay and Nor-Sand models. Cam-Clay was developed for normally consolidated and lightly overconsolidated clays and the extension to heavily overconsolidated clay is done by modelling the peak state as a yielding point (Fig. 8.8 a), which is not consistent with the stress-dilatancy theory. This means that the elastic parameters contain some plasticity and do not correspond to the actual reversible behaviour. However, the final stress and density states of Cam-Clay are at critical state. Nor-Sand models the peak state as a consequence of the dilatancy and, therefore, the hardening phase is elasto-plastic and the elastic parameters correspond to reversible behaviours.

Mohr-Coulomb (MC) is a failure criterion developed by Terzaghi [296] with two model parameters – the cohesion c' and the friction angle φ'. It aims to predict the shear strength of soil solely from its stress state by assuming elasticity until failure making it stress-path independent. This means that no history variable is required. It was subsequently transformed into a stress-strain relationship and a third parameter was included – the dilatancy angle ψ. None of these parameters are coupled and the user must make sure that these are mechanically consistent. The yield function F and the potential function P are given in Eqs. 8.53 and 8.54, respectively.

$$F = q - \left(\frac{c'}{\tan \varphi'} + p'\right) \frac{\sin \varphi'}{\cos \theta + \dfrac{\sin \theta \sin \varphi'}{\sqrt{3}}} \qquad (8.53)$$

$$P = q - (a_{\text{pp}} + p') \frac{\sin \psi}{\cos \theta + \dfrac{\sin \theta \sin \psi}{\sqrt{3}}} \qquad (8.54)$$

where F is the yield function, P the potential function and a_{pp} the distance to the apex.

The concept of strength in MC is poorly defined as it can equally refer to the peak state or the critical state. Bishop [34] argued that it refers to the peak state and that the cohesion of granular material is a proxy for dilatancy-induced strength ($c' = \sigma'_N \tan \psi$, $\varphi' = \varphi'_{\text{cs}}$, ψ). Bolton [40] suggests that the the cohesion is nil and that the dilatancy-induced strength is captured by the friction angle ($c' = 0$, $\varphi' = \varphi'_{\text{cs}} + 0.8\psi$, ψ). However, the dilatancy angle ψ is a constant value, which leads to unrealistic and forever changing volumes of soil making it unsuitable for large deformation MPM simulations. As the critical state conditions have to be fulfilled at large deformation simulations, the dilatancy angle must be nil and the friction angle has to be the critical state – $c' = 0, \varphi' = \varphi'_{\text{cs}}, \psi = 0$. Fig. 8.9 (a) illustrates this case.

The conflict between the peak state and the critical state is overcome by the inclusion of the strain-induced softening function based on accumulated plastic shear strains $E^{\text{P}}_{\text{dev}}$ allowing the model to go from peak values to critical state values. This model is referred to as Mohr-Coulomb Strain Softening (MCSS). However, the peak state parameters ($c'_{\text{max}} = 0$, φ'_{max}, ψ_{max}) and the critical state parameters ($c'_{\text{max}} = 0$, φ'_{max}, ψ_{max}) have to be explicitly given and the user has to make sure that these are mechanically consistent. Chapter 7 discusses the determination of the peak state parameters. The reduction of the model parameters with an exponential function are given in Eq. 8.55.

$$\frac{\partial \varphi'}{\partial E^{\text{P}}_{\text{dev}}} = -\zeta(\varphi'_{\text{max}} - \varphi'_{\text{res}}) \exp(-\zeta E^{\text{P}}_{\text{dev}}) \qquad (8.55a)$$

$$\frac{\partial c'}{\partial E^{\text{P}}_{\text{dev}}} = -\zeta(c'_{\text{max}} - c'_{\text{res}}) \exp(-\zeta E^{\text{P}}_{\text{dev}}) \qquad (8.55b)$$

$$\frac{\partial \psi}{\partial E^{\text{P}}_{\text{dev}}} = -\zeta(\psi_{\text{max}} - \psi_{\text{res}}) \exp(-\zeta E^{\text{P}}_{\text{dev}}) \qquad (8.55c)$$

where ζ is the shape coefficient which controls the brittleness.

Fig. 8.9 (b) illustrates the predicted behaviour with MCSS and the progressive change between the peak state and the residual state. Although MCSS is better suited for large deformation simulations than MC, it is still not compliant with the critical state theory as the void ratio does not necessarily reach its critical state value ($e \neq e_{\text{cs}}$).

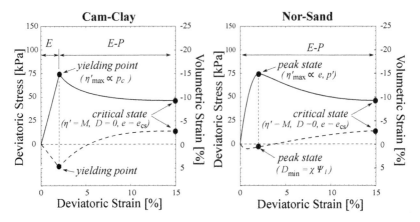

FIGURE 8.8: Schematic description of triaxial compression tests with Cam-Clay and Nor-Sand.

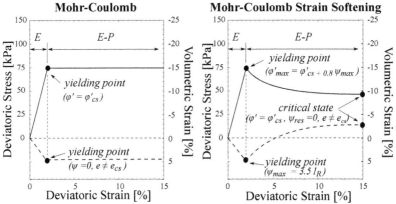

FIGURE 8.9: Schematic description of triaxial compression tests with Mohr-Coulomb.

8.6   Closure

The critical state theory presented in Chapter 7.2 can be extended to develop constitutive models for granular materials. The critical state models give realistic stress-strain relationships as well as the ultimate state when sheared to large strains. These models differentiate material parameters from the state parameters (such as void ratio or consolidation pressures). Hence, the same material parameters can be used at different depths, different densities or different stress histories as long as they are soil of the same origin. This is why these models are powerful compared to classical Mohr-Coulomb models, in which the model parameters have to be assigned carefully depending on the state of the soil.

Cam-Clay was the first constitutive model to relate the stresses with strains based on the critical state framework. It can predict the behaviour of a wide variety of soils, from normally consolidated to overconsolidated, in both drained and undrained conditions with only 5 plastic parameters which have physical meanings and are quantifiable from element tests. However, Cam-Clay assumes that the two conditions of the critical state theory ($D = 0$, $\partial D/\partial \varepsilon_{\text{dev}}^{\text{p}} = 0$) are fulfilled simultaneously and hence cannot model the peak strength as a consequence of dilatancy. The peak strength can only be modelled by sudden yielding from elastic state to plastic state, which prevents energy from being dissipated during the hardening phase.

Nor-Sand decouples the two conditions of the critical state theory ($D = 0$, $\partial D/\partial \varepsilon_{\text{dev}}^{\text{p}} = 0$) by introducing a maximum yield surface. Nor-Sand uses the density as a model variable and a single set of model parameters is required to model the behaviour of a wide range of conditions. It can simulate the mechanical behaviour of sand rather well in both drained and undrained conditions.

The models presented in this chapter were developed by considering soil as an ideal homogeneous continuum. Real soil consists of an aggregation of solid grains with local variations in voids. This particulate nature can lead to local failures such as strain localisation, which may not be easily captured by these models. Nevertheless, these models provide a solid foundation to simulate the development of stiffness and strength with strain and they have contributed in improving the understanding of the mechanical behaviour of clay and sand in the field of soil mechanics and geotechnical engineering.

Part II

Application

9

The Granular Column Collapse

E.J. Fern and K. Soga

9.1 Introduction

Many natural disasters consist of a sudden release of material, which flows and can cause substantial damage to property and economical losses. For instance, the 2013 Bingham copper mine landslide in Utah (USA) consisted of a series of two catastrophic landslides with a total volume of 55 million m^3, making it the largest non-volcanic landslide in North America [221], and seriously compromising the economical sustainability of the mine. However, case studies are often complex with many uncertainties and limited site information. It is hence difficult to use them to explore, benchmark and validate MPM for large deformation simulations of mass movements. For this reason, simple experiments with well-defined conditions are favoured. The granular column collapse is a good example of a well-established experiment, which permits exploring the failure of granular materials, its post-failure fluid-like behaviour, and its deposition mechanism.

This chapter presents MPM simulations of granular column collapses and demonstrates the influence of some numerical features, including the the constitutive model, on the failure mechanism and run-out distance.

9.2 Experiment description

The granular column collapse is a well-documented experiment [159, 160, 177–179], which consists in releasing the lateral supports of a column filled with sand on a flat surface, and looking at the failure mechanism, post-failure flow mechanism, and deposition profile. This is achieved in the experiment by using a high-speed camera. The duration of a collapse is typically less than 3 s.

The experimental layout can be axisymmetric or plane strain as shown in Fig. 9.1. These give similar run-out distances, although some researchers argue that the plane strain experiment gives slightly smaller run-out distances due to friction with the side walls [298]. Fig. 9.2 shows an experimental trial of a granular column collapse of dry sand in axisymmetric conditions. The cylinder is lifted in less than 0.5 s removing the lateral support of the sand, which fails and starts flowing. The frictional contact between the cylinder and the sand is assumed to be negligible. However, this is not the case for the frictional contact between the column and the base. The sand then flows until it reaches a final deposition profile. The experiment compares the final deposition configuration with the initial configuration with three dimensionless variables (Eq. 9.1).

$$a = \frac{h_0}{r_0} \quad , \qquad r^* = \frac{r_f - r_0}{r_0} \quad , \qquad h^* = \frac{h_0}{h_f} \tag{9.1}$$

where a, r^* and h^* are the aspect ratio, the normalised run-out distance and normalised height, respectively. h_0 and h_f are the initial and final height, respectively. r_0 and r_f are the initial and final radius or distance, respectively.

Fig. 9.3 shows the experimental relationship between the initial aspect ratio and the final deposition configuration. It was found that a relationship exists between the initial aspect ratio a and the final run-out distance r^* and height h^* available in the literature [159, 160, 177–179]

Two groups of flow regimes exist [159, 179]. Regime 1 is for small aspect ratio columns. A small volume of mass is mobilised and slides in a single flow motion. Two sub-categories exist depending on whether the deposition is truncated (Regime 1a) or not (Regime 1b) [179]. Regime 2 is for large aspect ratio columns for which the final deposition profile has the shape of a "Mexican hat" and for which the relationship between the initial aspect ratio and the final run-out distance differs from Regime 1. The transition is gradual going from the "slow avalanches of shallow columns" (Regime 1) to "violent cascading collapses of tall columns" (Regime 2) [19] and that this transition is material and state dependent [76, 103, 104].

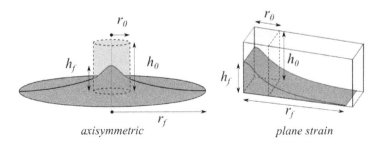

FIGURE 9.1: Layout of granular column collapse experiment (after [103]).

FIGURE 9.2: Experiment of a granular column collapse with an axisymmetric layout.

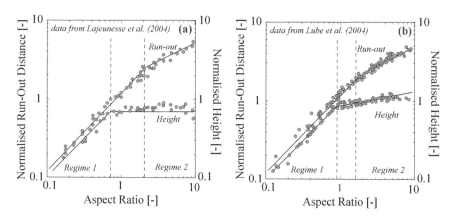

FIGURE 9.3: Experimental results for the normalised run-out distance and height (after [103]).

9.3   Model description

A model is built for a granular column with an initial aspect ratio of $a = 2.0$. This model is used to carry out parametric analyses illustrating the influence of the different modelling choices on the collapse behaviour and run-out distance. The MPM simulations are carried out using *Anura3D* [14] and the sand is modelled as a dry material with the one-phase single-point MPM formulation.

9.3.1   Model layout

The granular column collapse is modelled with a two-dimensional (2D) plane strain model for a column with an initial aspect ratio of $a = 2$ (Fig. 9.4). That is an aspect ratio between Regime 1 and Regime 2. The size of the model has to be sufficiently big to allow MPs to flow and stop without hitting the end boundary. Boundary conditions are placed around the box to prevent the MPs from leaving the defined volume. Different models are generated with different mesh but all are regular and 4 MPs per cell are initialised as default. The column is modelled as a solid continuum for which the initial state and constitutive model have to be defined. The column is maintained in place by additional boundary conditions, which will be removed once the stresses are initialised. A base layer is defined in order to provide some surface roughness and it allows the column to slow down. It is modelled as an elastic continuum. A contact condition is placed on the base layer. This permits controlling the amount of friction and adhesion between the collapsing column and the base layer. Note that the contact condition is not compulsory in these simulations but it does improve the results.

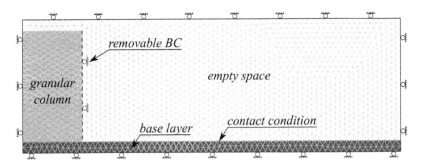

FIGURE 9.4: MPM model for a granular column with an aspect ratio of $a = 2$.

9.3.2 Constitutive models

All materials require a constitutive model to convert the strains into stresses and to give the material a certain stiffness. The granular column is modelled with three different constitutive models - an elastic-perfectly-plastic Mohr-Coulomb model (MC), an elastic-plastic Mohr-Coulomb model with strain softening indexMohr-Coulomb!strain softening(MCSS), and an elasto-plastic non-associative Nor-Sand model (NS). These models are described in Chapter 8. However, the material used in the experiments are not characterised in a geotechnical way, which makes the calibration process very difficult. For this reason, the model parameters used in the simulations are those of Chiba sand [101] and they are calibrated using single element triaxial compression tests such that the peak state, softening phase and critical state are identical. The parameter values are chosen such that they are mechanically consistent. That means that the peak strength is the consequence of dilatancy and the critical state strength. A more extensive discussion on the calibration process is given by Fern and Soga [103]. Tables 9.1, 9.2 and 9.3 give the model parameters for MC, MCSS and NS, respectively. The base layer is modelled as linear elastic with $E = 20$ MPa and $\nu = 0.2$.

9.3.3 Stress initialisation

The stresses are initialised at the beginning of the simulations by gravity loading and during which the system is damped in order to avoid straining the soil. A quasi-static convergence criterion can be defined in order to guarantee equilibrium of the system. Stresses can also be imposed, if known. In case of a flat surface, the stresses can be defined with the earth pressure coefficient K_0 (see Chapter 6). This technique assumes that the ground surface coefficient is constant within a given material and that ground surface is flat. However, the surface of the base is located at a lower elevation, which implies that the initial stresses in it are higher than they should be. As the base is modelled as an elastic continuum, this results in a sudden drop of stress at the beginning of the simulation making the base layer vibrate. As no damping is used, this vibration persists until the end of the simulation and it can cause MPs to be ejected from the surface and fly in the empty space. This is more significant for MPs with small masses. However, gravity loading can be significantly more computationally costly than imposing the initial stresses.

9.4 Mesh sensitivity

A series of simulations are carried out using different meshes and number of MPs as shown in Fig. 9.5. The coarse mesh has 2,700 cells and 2,184 MPs

TABLE 9.1
Default model parameters for MC.

Parameter	Symbol	Unit	Value
Young modulus	E	MPa	20
Poisson ratio	ν	-	0.2
Cohesion	c'	kPa	0.0
Friction angle	φ'	°	33
Dilatancy angle	ψ	°	0.0
Initial porosity	n_0	-	0.444
Unit weight solid	ρ_s	kN/m^3	2,700

TABLE 9.2
Default model parameters for MCSS.

Parameter	Symbol	Unit	Value	
			Loose	Dense
Young modulus	E	MPa	20	20
Poisson ratio	ν	-	0.2	0.2
Peak cohesion	c'_{max}	kPa	0.0	0.0
Residual cohesion	c'_{res}	kPa	0.0	0.0
Peak friction angle	φ'_{max}	deg	39	50
Residual friction angle	φ'_{res}	deg	33	33
Peak dilatancy angle	ψ_{max}	deg	6	15
Residual dilatancy angle	ψ_{res}	deg	0	0
Shape function	ζ	-	4	5
Initial porosity	n_0	-	0.444	0.394
Unit weight solid	ρ_s	kN/m^3	2,700	

TABLE 9.3
Default model parameters for NS.

Parameter	Symbol	Unit	Value
Shear modulus constant	A^e	MPa	2.5
Shear modulus exponent	n^e	-	0.5
Poisson ratio	ν	-	0.2
CS stress ratio	M_{tc}	-	1.33
Min. void ratio	e_{min}	-	0.500
Max. void ratio	e_{max}	-	0.946
Crushing pressure	Q	MPa	10
Dilatancy parameter yield	N_f	-	0.6
Dilatancy parameter potential	N_p	-	0.3
Hardening modulus	H	-	50 - 1850Ψ
Dilatancy coefficient	χ	-	2.6
Initial porosity	n_0	-	0.444 or 0.394

(4 MPs/cell). The medium-fine mesh has 11,058 cells and 8,472 MPs (4 MPs/cell), and 11,058 cells and 16,176 MPs (10 MPs/cell). The fine mesh has 20,379 cells and 16,128 MPs (4 MPs/cell). Fig. 9.5 shows the contour plot of the deviatoric strains in which black zones are extensively sheared ($\varepsilon_{\text{dev}} > 100\%$), grey zones are sheared ($0\% < \varepsilon_{\text{dev}} < 100\%$), and white zones are not sheared ($\varepsilon_{\text{dev}} = 0\%$). This colour convention will be kept throughout the chapter.

The comparison of the results (Fig. 9.5) illustrates the mesh dependency for the failure surface, although the run-out distances are similar. This mesh dependency disappears as the mesh is refined with better-defined surfaces for the medium-fine and fine meshes than for the coarse mesh. Increasing the number of MPs for a given mesh does not improve significantly the results but it does increase the computational cost. The medium-fine mesh with 4 MPs/cell shows the same results as the fine mesh but is computationally cheaper and, hence, will be used for the other series of simulations.

9.5 Surface friction

The experimental results show that the granular flow is immobilised by frictional contact with the base layer, building static layers from the base upwards, and permits the existence of the static cone. It is hence necessary to model this surface condition in order to obtain the correct mechanics. This can be achieved by placing the base layer, which models a perfect contact between bodies. Fig. 9.6 (a) shows the reference case in which a static cone is formed and the MPs located at the top of column, which have the highest potential energy, end up at the distal end of the final deposition profile. This is referred to as a top-down failure mechanism [101].

It is possible to apply a contact condition at the surface of the base layer, which controls the forces transmitted to the base layer and, hence, the amount of energy dissipated. The contact formulation in *Anura3D* requires defining a friction coefficient and an adhesion term, which controls the amount of shear and tensile normal forces transmitted to the base layer, respectively. Granular material typically transfers forces by frictional contact without adhering to the surface of the base. Therefore, the adhesion is nil and only the frictional coefficient μ is defined.

Fig. 9.6 (b) shows the simulation with a surface friction of $\mu = 0.3$, which corresponds to half the critical state friction angle of the sand ($\varphi' = \varphi'_{cs}/2 = 16.5° \rightarrow \mu \approx 0.3$), and no adhesion. The results of the simulation show that run-out distance is shorter with a slightly larger static in comparison with Fig. 9.6 (a). The inclusion of a nil-adhesion contact condition prevents the ejection of MPs by the vibrating base layer and improves the simulation. This is because the normal tensile stresses are numerically removed from the system at the location of the contact condition.

Fig. 9.6 (c) shows the simulation with a surface friction of $\mu = 0.15$, which is a quarter of the critical state friction angle, and dissipates less energy than the previous case. The results show a longer run-out distance with a more fluid-like flow behaviour and a smaller static.

Fig. 9.6 (d) shows the simulation with no friction ($\mu = 0$) between the base and the granular column. This is equivalent to suppressing the base from the simulation. The results show that the entire column fails, flows in a fluid-like manner and no static cone is formed. The granular material flows in a liquid-like manner until reaching the model boundary. It then accumulates in the corner and layers of granular material slip one on top of each other dissipating energy. Although the granular material has a liquid-like flow behaviour, it remains a granular material because of the existence of inter-granular friction. This is a fundamental difference between granular materials and liquids. The final deposition profile looks like a pile of granular material in the far right corner. This mechanism is referred to as a bottom-out failure mechanism as the MPs located at the base of the column end up at the distal end of the final deposition profile [101].

There is experimental evidence that the granular flow forms static layers from the base upwards, forming successive layers of static material [177]; the flow is immobilised from the base upwards. Fig. 9.7 shows the incremental deviatoric strain, highlighting the flowing mass, for different time steps for the simulation with a contact coefficient of $\mu = 0.3$. It can be seen that a static cone is formed and the mobilised mass slips on it (Fig. 9.7 a, $t = 0.3$ s). Static layers are rapidly formed at the toe of the static cone (Fig. 9.7 b, $t = 0.7$ s) and progressively build upwards towards the free surface (Fig. 9.7 c-d, $t = 1.0$-1.3 s). This demonstrates the role of the base friction in the top-down flow mechanism, which does not exist in the bottom-out mechanism (i.e. $\mu = 0.0$).

9.6   Damping

There are numerous examples in the literature of numerical simulations of the granular column collapse in which the bottom-out failure mechanism is simulated instead of a top-down failure mechanism. Many simulations control the amount of dissipated energy with contact conditions and/or numerical local damping, and the results are validated by only comparing the run-out distances and not the failure mechanism. As explained in Chapter 6, local damping is used to gain stability in the computation but it also leads to additional dissipation of energy in the system. It is therefore possible to curve-fit the run-out distance of a granular column collapse by increasing the damping.

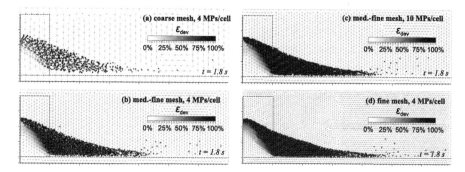

FIGURE 9.5: Influence of mesh for simulations with MC (no strain smoothing, no damping, no contact condition).

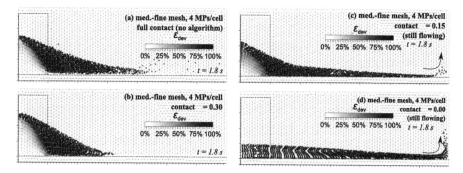

FIGURE 9.6: Influence of contact conditions for simulations with MC (no strain smoothing, no damping).

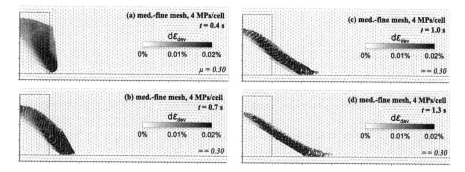

FIGURE 9.7: Moving mass (incremental deviatoric strain) for simulation with MC and a contact condition ($\mu = 0.3$) (no strain smoothing, no damping).

Fig. 9.8 shows simulation results for different damping coefficients – (a) $\alpha = 0$, (b) $\alpha = 0.15$, (c) $\alpha = 0.3$ and (d) $\alpha = 0.6$. The results show that the first consequence of damping is a reduction of the run-out distance from $r^* = 3.5$ ($\alpha = 0.0$) to 1.5 ($\alpha = 0.6$). The increase in damping steepens the failure surface resulting in a larger static cone and a smaller mobilised mass. Damping changes the dynamics of the problems and it should be used with caution.

9.7 Strain smoothing

Kinematic locking is an important issue for both FEM and MPM. It is the build-up of fictitious stiffness due to the inability of an element to reproduce the correct deformation (i.e. [196]). One technique to overcome this issue is to smooth the volumetric strains over neighbouring cells (see Chapter 6). However, this has an effect on the softening behaviour of some constitutive models.

Fig. 9.9 (a-b) shows the simulations with MC ($\varphi'_{cs} = 33°$, $\psi = 0°$) with and without strain smoothing. The inclusion of strain smoothing gives a better-defined failure surface and final deposition profile as well as reducing the mesh dependency. However, this is not always the case. Fig. 9.9 (c-d) shows the same simulations but with MCSS (dense sand: $\varphi' = 50° \rightarrow \varphi'_{res} = 33°$, $\psi_{\max} = 15° \rightarrow \psi_{res} = 0°$). The simulation without strain smoothing shows a steep failure surface due to the higher peak friction angle and a short run-out distance in comparison with the cases with MC. However, the simulation with strain smoothing gives a different final deposition profile, which resembles the one with MC. This is caused by a progressive softening of the cone induced by accumulated plastic strain in the cone and, subsequently, softening until reaching the critical state.

9.8 Integration scheme

Fig. 9.10 shows the same simulation as Fig. 9.9 but with a Gauss integration scheme (see Section 3.6). The results show similar results than previously with slightly better-defined deposition profiles. However, the case with MCSS with strain softening also shows a poorly-defined and diffused failure surface, which is made worse by the mixed integration. Although the use of strain smoothing is often recommended to mitigate locking, it also transpires to have a strong effect on strain-driven softening behaviours.

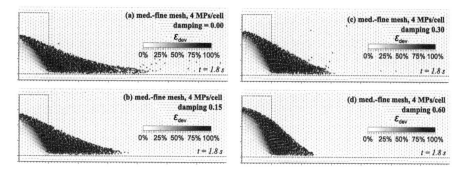

FIGURE 9.8: Influence of damping for simulations with MC (no contact condition, no strain smoothing).

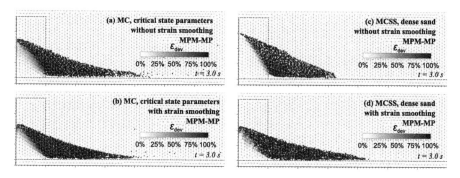

FIGURE 9.9: Influence of strain smoothing for simulations with MC and MCSS (dense sand) (no damping, no contact condition).

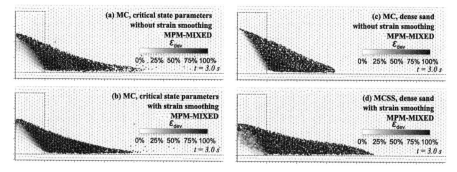

FIGURE 9.10: Influence of strain smoothing for simulations with MC and MCSS (dense sand) and a Gauss integration (no damping, no contact condition).

9.9 Constitutive models

Fern and Soga [103] showed that the constitutive model plays a twofold role in the granular column collapse. It controls (1) the failure surface at small deformation defining the size of the static cone and (2) the post-failure liquid-like flow defining the run-out distance. This can be demonstrated by using a fundamentally different constitutive model than MC such as Nor-Sand model (NS).

A series of simulations is carried out with a non-associative NS ($N_f \neq N_p$) [101], which differs from the associative NS model ($N_f = N_p$) presented in Chapter 8 by the inclusion of a potential function. Fig. 9.11 shows the simulation for loose sand with the non-associative NS. The results show that the failure surface is slightly concave and steeper at $t = 0.3$ s than for the simulations with MC, although these are small differences. This is caused by the way the model predicts the peak strength with the state parameter Ψ. The mobilised mass forms a wedge, which slides on a shear band. The wedge then crumbles upon impact with the base layer. The mobilised mass then flows in a liquid-like manner but remains a granular material as inter-granular friction persists. It slows down progressively by frictional contact with the base and static layers build up from the base to the surface ($t = 0.8$ to 1.3 s) until reaching the final deposition profile at $t = 3.0$ s. However, a slow progressive avalanching persists but has no influence on the run-out distance. This avalanching process is more significant with the non-associative flow rule than for the associative one [101], which is described in the experiment [177, 179], and it is caused by localisation effects.

Fig. 9.12 shows the final run-out distance for different column sizes and in which the experimental run-out distance is marked with an arrow. The mesh size is kept identical for all simulations for comparison reasons. No damping or frictional contacts are used in these simulations. The small column with an initial aspect ratio of $a = 0.5$ (Fig. 9.12 a) has a run-out distance of $r^* = 0.4$, which is slightly shorter than the experiment, but it can be argued that the mesh is too coarse to fully capture the correct run-out distance. The simulation of a column with an initial aspect ratio of $a = 1.0$ (Fig. 9.12 b) has a run-out distance of $r^* = 1.0$, which corresponds to the experimental results. Although the normalised run-out distance increased linearly from columns with an initial aspect ratio of $a = 0.5$ to 1.0, the failure surfaces are identical and, hence, the mobilised mass is also identical with the same potential energy. The simulation of a column with an initial aspect ratio of $a = 1.5$ (Fig. 9.12 c) differs from the two previous cases. The failure surface defining the static cone appears to be concave but this is due to a certain mesh dependency. As for the two previous cases, the final deposition profile is still the same height as the initial column and, hence, these first three column collapses are Regime 1 collapses. The simulation of a column with an initial aspect ratio of $a = 2.0$

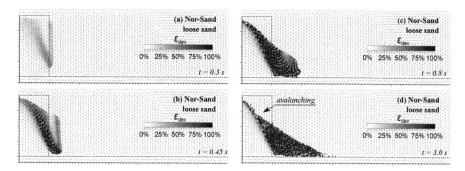

FIGURE 9.11: Granular column simulation for loose sand ($I_D = 33\%$) with a non-associative NS model.

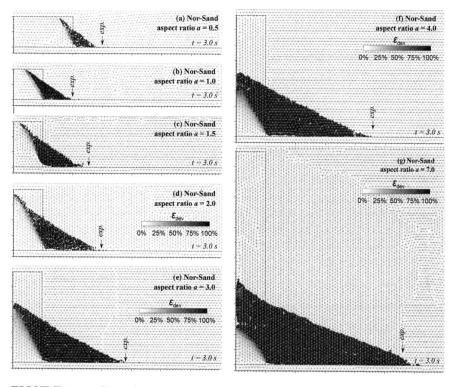

FIGURE 9.12: Granular column simulation for different initial aspect ratios of loose sand with a non-associative NS model.

(Fig. 9.12 d) has a well-defined failure surface as it coincides with the mesh. The run-out distance is $r^* = 1.8$ and it is a little shorter than the experimental results. However, the experimental data shows a progressive transition from Regime 1 to 2 between $a = 1.0$ and 2.0 and this mismatch varies from one set of data to another. The prediction with NS is in the range of the experimental results which was not the case for the MC simulations. The simulations of columns with an initial aspect ratio of $a = 3.0$ and 4.0 (Fig. 9.12 e and f) have a well-defined failure surface which consumed the summit of the column. The run-out distances are $r^* = 2.8$ and 3.6, respectively, and correspond to the experimental data. The simulation of a large column with an initial aspect ratio of $a = 7.0$ (Fig. 9.12 g) has a run-out distance of $r^* = 4.8$.

Fig. 9.13 shows simulations with both the associative and non-associative NS models. The results show that the run-out distance is not significantly affected by the associativity but the development of the failure surface is. The avalanching process taking place at the end of the simulations is very apparent for the non-associative simulations. This is caused by the ability of the model to localise strains.

Although NS performs better than MC for MPM simulations of granular column collapses, this cannot be generalised to all large deformation simulations as the failure mechanism may differ. MC is a failure criterion developed by Terzaghi [296] which was developed from the work of Mohr [202] on brittle steel and Coulomb's work [73] on friction in which a pressure dependency is included. It aims to empirically predict the (peak) stress state at which failure occurs rather than describing the stress-strain paths which lead to failure. It was only later that it was converted to a stress-strain relationship by assuming an elastic hardening phase and including the dilatancy angle to describe the changes in volume. Although MC is very practical and computationally efficient, it was not developed from an energy consideration and does not model the correct stress-strain behaviours. In contrast, NS follows the same principles as Original Cam-Clay and is based on energy considerations [249,262] (see Chapter 8). This allows the model to dissipate the correct amount of energy. It is therefore not surprising that NS and MC give very different predictions.

9.10   Initial state and material type

The comparison between simulations with MCSS and NS show the final deposition profile of a column with an initial aspect ratio of $a = 2.0$ for loose and dense sands. NS includes the density (void ratio) as a model variable. This means that the model parameters do not need to be changed as the model predicts the peak strength automatically (see Chapter 8). However, MCSS does not include the density as a model variable and, hence, its model parameters are density specific. This means that the peak friction angle φ'_{max}

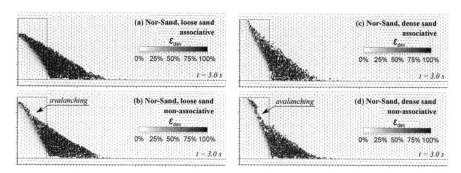

FIGURE 9.13: Influence of the associativity in NS on the granular column collapse.

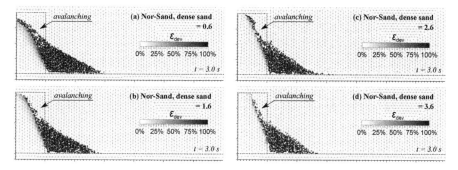

FIGURE 9.14: Influence of the dilatancy coefficient on the granular column collapse.

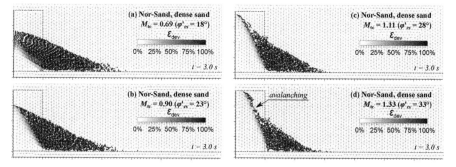

FIGURE 9.15: Influence of the critical state strength on the granular column collapse.

and maximum dilatancy angle ψ_{max} are different for the loose and dense sand case. Simulations for loose and dense sand with MCSS show a steeper failure surface for the dense sand than for the loose sand. These simulations suggest that the run-out distance for loose sand is significantly longer than for dense sand. Conversely, results with NS suggest that this difference in run-out distance is smaller. It is shown later in this chapter that the experimental results and the theoretical prediction suggest that there is a certain density dependency for the run-out distance but it is mild. The different run-out distances for the MCSS simulations can be explained by the way the model dissipates energy. MCSS models the hardening phase as elastic, which is more significant for dense sands than for loose sands. Thus, no energy is dissipated during the early stage of the failure. Once the material has yielded and the failure surface is defined, MCSS can dissipate energy. In contrast, NS is an elasto-plastic model which dissipates energy during the hardening phase and therefore can dissipate more energy for the dense sand case than for the loose sand case, albeit limited.

Tatsuoka [291], followed by Jefferies and Been [149, 150], pointed out that the conversion of state indices to dilatancy angles is fabric dependent. This means that the conversion factor used in stress-dilatancy theories can change depending on the orientation of the soil grains (see Chapter 7). Fern [101] showed that this dilatancy coefficient could change substantially for sands with elongated grains but remained somewhat constant for spherical-grained sands. Fig. 9.14 shows the final deposition profile of a column with an initial aspect ratio of $a = 2.0$ for different dilatancy coefficients for dense sand; the effect of the dilatancy coefficient is density dependent and is better illustrated for the dense sand case. The results show that an increase in the dilatancy coefficient, which increases the peak strength and the stiffness, steepens the failure surface, changes the final height of the static cone, and shortens a little the run-out distance. The avalanching process at the end of the simulations is more significant for $\chi = 4.6$ than for $\chi = 1.6$.

The critical state strength is a material property, which affects both the small deformation peak strength and the large deformation residual strength. Fig. 9.15 shows the final deposition profile for a column with an initial aspect ratio of $a = 2.0$ for different critical state strengths. The results show that the critical state strength strongly influences the peak strength and, hence, the failure surface. It also controls the amount of energy dissipated during the run-out in turn influencing the run-out distance.

9.11   Energy consideration

The granular column collapse is an energy problem. The total amount of potential energy is defined by its initial size and mass but the amount of energy involved in the collapse is defined by the mobilised mass, which is the total column minus the static cone. Fig. 9.16 shows a schematic description of the potential energy of the mobilised mass for the column. Two different groups appear. The small aspect ratio column has a failure surface which reaches the summit of the column and the mobilised mass is triangular. The large aspect ratio column has a failure surface which reaches the side of the column and the mobilised mass takes the shape of a trapeze. Therefore, only a small part of the potential energy of the total column is mobilised for the small aspect ratio column but a large part is mobilised for the large aspect ratio columns. The energy equations can be expressed mathematically as Eq. 9.2.

$$E_{\text{mob}}^{\text{pot}}(t = 0) = E_{\text{mob}}^{\text{pot}}(t) + E_{\text{mob}}^{\text{kin}}(t) + C(t) \tag{9.2}$$

where E_{mob}^{pot} and E_{mob}^{kin} are the potential and kinetic energies of the mobilised mass, respectively, C the energy dissipation and t the time.

The constitutive model has a twofold influence on the simulations [103]. (1) It defines the failure surface and, hence, the mobilised mass and its potential energy ($E_{\text{mob}}^{\text{pot}}$ at $t = 0$) in which the initial state and the material type play an important role. (2) It controls the dissipation of energy $C(t)$ and, hence, influences the run-out distance. It is possible to estimate the potential energy of the mobilised mass as Eqs. 9.3 and 9.4.

Small aspect ratio:

$$E_{\text{pot}}^{\text{mob}} = m_{\text{mob}} g h_{\text{mob}}^{\text{CG}}$$

$$\rightarrow E_{\text{pot}}^{\text{mob}} = \frac{1}{3} h_0{}^3 \cot \varphi_{\text{max}}'(1 - n)\rho_{\text{s}} g \tag{9.3}$$

Large aspect ratio:

$$E_{\text{pot}}^{\text{mob}} = m_{\text{tot}} g h_{\text{tot}}^{\text{CG}} - m_{\text{stat}} g h_{\text{stat}}^{\text{CG}}$$

$$\rightarrow E_{\text{pot}}^{\text{mob}} = \left(\frac{1}{2} h_0{}^2 - \frac{1}{6} r_0{}^3 \tan^2 \varphi_{\text{max}}' \right) (1 - n)\rho_{\text{s}} g \tag{9.4}$$

where h is the height and r with the subscripts/superscripts 0 refers to the initial configuration and CG to the centre of gravity. n is the porosity, ρ_{s} the density of the solid grains and g the gravity.

The potential energy equations (Eqs. 9.3-9.4) show a twofold dependency to the type of material through the critical state friction angle φ_{cs}', which is included in the peak friction angle φ_{max}' and the solid density ρ_{s}. It also shows a twofold dependency on the state through the maximum dilatancy angle ψ_{max}, which is included in the peak friction angle φ_{max}', and the porosity n. This state dependency has been observed experimentally [76].

The potential energy of the mobilised mass can be estimated by predicting the maximum dilatancy angle ψ_{max} with the relative dilatancy index I_{R} [40] (see Chapter 7), and subsequently predicting the peak strength. Fig. 9.17 (a) shows the predictions . A bi–linear relationship between the energy of the mobilised mass and the size of the column in a log-log scale with a stronger

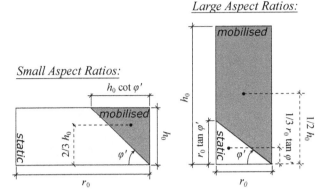

FIGURE 9.16: Schematic description of the centre of gravity of the mobilised mass of a granular column (after [103]).

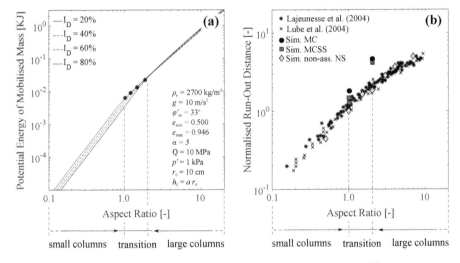

FIGURE 9.17: (a) Potential energy of the mobilised mass and (b) comparison with experimental and simulated normalised run-out distance (modified from [103]).

dependency on the initial density for small aspect ratio columns than for large ones. This relationship is similar to the experimental relationship exists between the normalised run-out distance and the initial aspect ratio (Fig. 9.17 b). The transition from small columns (Regime 1) to large columns (Regime 2) is density dependent. Fig. 9.17 (b) also compares the results of the simulations with the experimental data and it shows that the NS simulations predict the correct run-out distance in absence of any damping or contact condition but the MC simulations overestimate the run-out distance.

9.12 Closure

The granular column collapse is a well-established experiment, which aims to investigate how granular materials fail and flow on a flat surface. Despite its simplicity in execution, it offers insight in the mechanics of granular materials transitioning from solid-like to liquid-like behaviours and back to solid-like. The column fails at small deformation as two rigid blocks sliding on a failure surface and then flows at large deformation in a liquid-like manner. However, the mechanics of dry granular flows differ from those of liquids by the existence of inter-granular friction and dilation, which allows the formation of a static cone and a top-down failure mechanism.

Series of parametric analyses are shown in order to illustrate the influence of these different choices. It is shown that the failure surface and, to some extent, the run-out distance are mesh-dependent, and that the surface friction is indispensable in order to capture the correct failure mechanism. It is also shown that some numerical features, such as numerical damping, strain smoothing or the integration scheme, can influence the results and that these features should be used with caution.

The role of the constitutive model is twofold. It governs the small strain failure by positioning the failure surface and it controls the energy dissipation, which controls the run-out distance. It is illustrated by a series of simulations and by theoretical considerations. It is shown that a simple constitutive model cannot dissipate sufficient energy and, hence, over-predict the run-out distance. This insufficient energy dissipation is often artificially compensated by numerical damping but excess damping is cheating as it changes the dynamics of the problem. The choice of the constitutive models plays a pivotal role in collapse predictions.

The material type and initial state also play a central role in the collapse behaviour by controlling the critical state strength and dilatancy characteristics (fabric). It is illustrated by a series of simulations and by theoretical considerations.

10

Inverse Analysis for Modelling Reduced-Scale Laboratory Slopes

S. Cuomo, M. Calvello and P. Ghasemi

10.1 Introduction

Slope instability may include the formation of multiple and successive shear bands, progressive or retrogressive slides, soil static liquefaction and large propagation run-out. The role played by static liquefaction for the evolution of slope instabilities in flow-like landslides and the modelling of landslide propagation have been extensively investigated in the literature (e.g. [51, 223]). MPM offers a comprehensive tool to have a global understanding of complex phenomena, including multiple processes. Yet, detailed datasets from the field are still difficult to obtain and only few examples of instrumented sites are available in the literature to properly calibrate MPM models. On the other hand, reduced-scale tests are becoming popular as they provide new opportunities to deepen the knowledge on slope instability.

Such laboratory slope experiments are useful to investigate the features of landslide propagation, which is a challenging issue from a scientific point of view and deserves attention for landslide risk mitigation. Parameters of constitutive models and rheological laws, such as those considered in the relationship between shear stress and shear strain rate, are usually estimated through the back-analysis of case histories, adopting heuristic assessments and/or trial-and-error procedures to identify values of the input parameters best-fitting the observed response.

In this chapter, automated inverse analysis procedures are used to calibrate the MPM models of two well-instrumented laboratory experiments on reduced-scale slopes, respectively dealing with (1) large slope deformation and (2) long run-out soil propagation. The first laboratory test refers to a retrogressive slope instability combined with soil liquefaction. The second test reproduces a soil mass rapidly propagating along an inclined plane and depositing over a flat area.

Depending on the specific tasks of the MPM analysis, different approaches can be used to model the soil mechanical behaviour. The Hypoplasticity model is used for the slope deformation test, while a Mohr-Coulomb model is employed to capture the dynamic behaviour of a fast-moving soil mass propagating over a non-deformable frictional surface. In both cases, the model parameters are assessed by means of inverse analysis. Details on the adopted inverse analysis procedures are reported in [50] and [74]. In both examples, the MPM model accurately reproduces the temporal and spatial evolution of soil geometry and topography. The results of the analyses indicate that proper calibration of the constitutive models adopted in MPM models of reduced-scale slopes improves the prediction of the main observed features of large slope deformations and long run-out soil propagations.

10.2   Simulating a flow-slide test

Eckersley [92] provided one of the very first contributions about a comprehensive observation (in a reduced-scale slope) of slope deformations, retrogressive failure, build-up of pore water pressure after failure, and propagation of the failed material up to rest along a gentle profile. The slope is brought to failure by water seepage from a lateral boundary towards the toe of the slope. The slope is 1 m tall, composed of Northwich coking coal and constructed in a glass-sided tank. Inclination of the ground surface is 36° and the floor extends to 1.5 m beyond the slope toe. The material ranges from fine sand and silt to gravel characterised by a specific gravity of $G_{\mathrm{s}} \approx 1.34$ and a grain size $D_{10} = 0.06 - 0.3$ mm which is close to typical values of organic soils [199] and other materials involved in flow-slides [93]. The minimum and maximum void ratios were $e_{\min} = 0.21$ and $e_{\max} = 1.23$, respectively. Thus, the material was characterised by high compressibility and hydraulic conductivity.

In the experiment named No.7, the slope failure started by shallow sliding in the zone of the saturated coal. The 1st Stage comprises two fairly distinct shallow slides over a period of 4 s, each extending rapidly uphill by slipping of the overstepped dry coal face. In the 2nd Stage, a 0.2-m thick slab, comprising the whole face, failed moving the previous debris ahead. In the 3rd Stage, a deep compound slide initiated, pushing the previously failed material horizontally, with the whole mass decelerating and coming to rest. Previous studies performed using the limit equilibrium method (LEM) and finite element method (FEM) were able to explain how the rising water table led to the triggering of the slide [75].

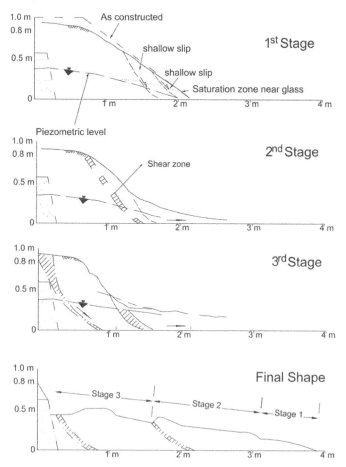

FIGURE 10.1: Experiment results: failure and post-failure stages (after [92]).

10.2.1 Calibration of the constitutive model by inverse analysis

A Hypoplastic constitutive model is selected to properly consider the influence of the pressure level, soil relative density and possible rotation of stress paths, the latter being related to the large displacements in the failure and post-failure stages. The fundamentals of hypoplasticity were developed by Kolymbas [158], and the later model proposed by Gudehus [120] is used in this chapter. The model has 8 main material parameters and 5 secondary parameters. The 8 main parameters are the critical state friction angle φ'_{cs}, two parameters h_s and n controlling the normal consolidation line (NCL) and the critical state line (CSL), respectively, the critical void ratio at zero stress e_{c0}, the maximum void ratio at zero stress e_{i0}, the minimum void ratio e_{d0}, a model parameter α controlling the dependency of the peak friction angle on relative density, and a parameter β for the dependency of the soil stiffness on relative density.

There are 5 other parameters. m_r is controlling the initial shear module, at very small strain, upon 180° strain path reversal and in the initial loading, m_t controls the initial shear module upon 90° strain path reversal, R_{max} relates to the size of the elastic strain range, and β_r and χ are used to control the rate of degradation of the stiffness with strain.

The void ratio e_p at a given mean isotropic stress p' is related to both the values of the void ratios at that stress level (critical state void ratio e_{cs}, maximum void ratio e_i and minimum void ratio e_d) and the void ratio at zero stress e_{p0} [26] as given in Eq. 10.1.

$$\frac{e_p}{e_{p0}} = \frac{e_{cs}}{e_{c0}} = \frac{e_d}{e_{d0}} = \frac{e_i}{e_{i0}} = \exp\left[-\left(\frac{3p}{h_s}\right)^n\right] \qquad (10.1)$$

The model calibration is performed through the minimisation of an objective function, which is a measure of the error between the experimental data and the model predictions. The Species-based Quantum Particle Swarm optimisation (SQPSO) algorithm is used. This technique was designed to calibrate simultaneously a large number of parameters and to increase the likelihood of finding the global optimum [140]. It has also been used recently for estimating the Hypoplastic model parameters of a reference sand [74]. The fit between the observed and the simulated values is quantified by defining an error of the numerical simulation that is the objective function to minimise by inverse analysis. The error function EF is defined as the sum of the error functions of each considered experimental curve $EF(k)$ as given in Eqs. 10.2-10.4.

$$EF = \sum_{k=1}^{N} EF(k) \qquad (10.2)$$

$$EF(k) = \sum_{i=1}^{m_k} e_k^2(i) \qquad (10.3)$$

$$e_k(i) = [(y_k(i) - y_k'(i)]w_k(i) \qquad (10.4)$$

where N is the number of experimental curves considered, m_k is the number of observations adopted to define the k-th experimental curve, $e_k(i)$ the weighted residual related to the i-th observation of the k-th experimental curve, $y_k(i)$ the value of the i-th observation of the k-th experimental curve, $y_k'(i)$ the value computed by the model which corresponds to the i-th observation of the k-th experimental curve, $w_k(i)$ the weight assigned to the i-th observation of the k-th experimental curve.

The weights are assigned to produce dimensionless weighed residuals $e_k(i)$ such that the error functions of the considered experimental curves $EF(k)$ can be summed to produce a global dimensionless error function EF. The weight assigned to the i-th observation of the k-th experimental curve is thus defined as given in Eq. 10.5.

$$w_k(i) = \frac{1}{s_k(i)} \qquad (10.5)$$

where $s_k(i)$ is the acceptable error related to the i-th observation of the k-th experimental curve.

The acceptable error $s_k(i)$ has the units of measure of the observation to which it refers. It defines an acceptable range for the difference between $y_k(i)$ and $y'_k(i)$, and produces dimensionless weighted residuals lower than 1.00 when the difference falls within that range. In this chapter, the acceptable errors used to compute the weights of the i-th observation of a given experimental curve are set to $s_k(i) = 5\%$ of that observation.

Six of the eight main model parameters (φ'_{cs}, h_s, n, e_{c0}, e_{i0} and e_{d0}) are derived directly from the experimental results, from prior information, or from literature. The same is done for two of the five secondary parameters (χ and R) [327]. That leaves 2 main model parameters (α and β) and 3 secondary parameters (m_r, m_t, R_{max}) to be determined by the inverse algorithm with the assumptions that the small strain stiffness is controlled by parameters m_r, m_t and R_{max}, and that m_t is equal to the maximum value between 1.0 and $0.5m_r$.

Three undrained triaxial tests with different initial void ratios, representing the contractive behaviour upon shearing, are taken into account. The match between the soil response and the prediction of the Hypoplastic model is analysed with reference to the two curves ($q - \varepsilon_a$) and ($p - \varepsilon_a$) for each test. The observations for each curve varies from 14 points (test NP-05) to 18 points (test NP-02, test NP-13). Globally, 100 observations are used, each weighted considering a coefficient of variation equal to 0.05. The optimisation is carried out through 80 iterations, not specifying a starting value of the parameters. Fig. 10.2 shows the satisfactory matching between the experimental data and the results of the constitutive model achieved with the optimised estimates of the soil parameters. Not only the ($q - \varepsilon_a$) and ($p - \varepsilon_a$) curves are used in the objective function (Fig. 10.2) but also the effective stress paths are accurately reproduced (Fig. 10.2 c). The final estimated values of the parameters are reported in Table 10.1. Details of the calibration procedure of the Hypoplastic model are given in [74].

10.2.2 Description of MPM model

The initial condition of the MPM modelling is the piezometric line observed at the first failure of the slope during the experiment. The slope is modelled as an upper dry part lying over a saturated one. The upper dry part of the slope is simulated by a one-phase material whereas the saturated one is simulated by a two-phase material using the two-phase single-point formulation presented in Chapter 2. The plywood floor is simulated by a linear elastic material. A frictional contact algorithm is adopted above the floor using a Mohr-Coulomb criterion with a friction coefficient of $\mu = 0.5$ as reported in [92]. The wire cage

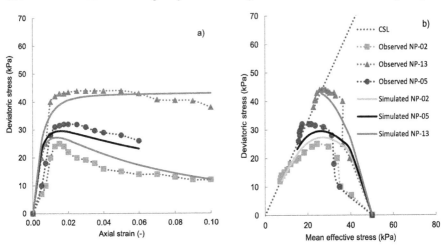

FIGURE 10.2: Experimental data and simulation of undrained triaxial tests.

TABLE 10.1

Estimated values of the parameters for the Hypoplastic constitutive model.

Parameter	Symbol	Unit	Value
Critical state friction angle	φ_{cs}	$\circ$	40
Granular stiffness	h_s	kPa	93.27
Exponent	n	-	0.08
Critical void ratio at zero pressure	e_{c0}	-	0.93
Maximum void ratio at zero pressure	e_{i0}	-	1.23
Minimum void ratio at zero pressure	e_{d0}	-	0.21
Exponent	α	-	0.38
Exponent	β	-	1.05
Initial shear module controller	m_r	-	2.23
Shear module upon 90° rotation stress	m_t	-	1.11
Range of small strain tensor	R_{max}	-	$4 \cdot 10^{-4}$
Stiffness degradation controller	β_r	-	0.08
Stiffness degradation controller	χ	-	1

at the rear of the slope is simulated by a linear elastic material. A static water pressure is applied as boundary condition for the saturated layer to simulate the water pressure imposed from the reservoir. The presence of suction is not considered in the MPM simulations for the sake of simplicity.

The computational mesh is composed of 13,062 4-node unstructured tetrahedral elements, and 4 MPs are initialised in each element. The typical length of the element is 0.05 cm. Fig. 10.3 shows the MPM model.

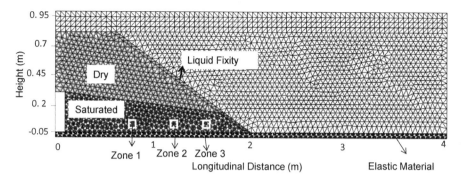

FIGURE 10.3: Geometry and spatial discretisation of the computational domain.

The initial effective stress and pore water pressures are computed from a preliminary static simulation considering the water table of the 1st Stage, and setting the velocity of the liquid phase to zero over the slope surface. These fixities are removed once the initial stresses are initialised. The initial static simulation is performed using a non-associative Mohr-Coulomb model with friction angle of $\varphi'_{cs} = 40°$, a nil cohesion $c' = 0$, a Poisson ratio of $\nu = 0.33$, a Young modulus of $E = 1,000$ kPa, and a dilatancy angle of $\psi = 0°$.

The stresses are initialised by gravity loading and assuming an elastic behaviour. The constitutive model is then changed to the Hypoplastic model.

10.2.3 MPM results

The results of the MPM simulation are able to capture the retrogressive slides induced by static liquefaction. The evolution of the effective stress is related to the onset and development of static liquefaction. First, liquefaction starts at the toe of slope, where stress levels are the lowest. Then, the mean effective stress reduces in the middle part of the slope and, finally, at the rear. The contour plots of the mean effective stress are shown in Fig. 10.4 for different times indicating how much and where the localised shear bands with low value of mean effective stress evolved inside the slope during the large deformations process. s

Fig. 10.5 shows the evolution of the pore water pressure over time for the three control zones Z1, Z2 and Z3. At the onset of each slope instability, the pore water pressure reaches its maximum value and the mean effective stress reaches a nil value, allowing liquefaction to occur. Afterwards, the excess pore water pressure dissipates and the mean effective stress increases inferring a higher shear strength to the material.

Fig. 10.6 shows the evolution of the void ratio, the critical state void ratio and the mean effective stress for MP P1 during the first stage. The void ratio remains during this stage, whilst the the mean effective stress decreases. At $t = 1.0$ s, the void ratio starts decreasing and the declining rate of effective stress

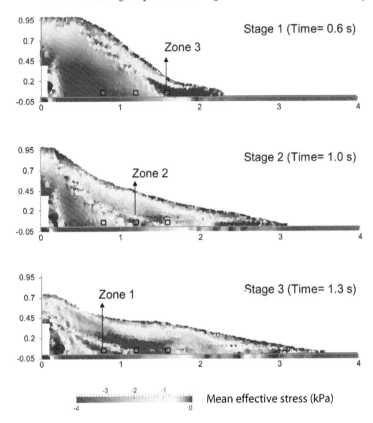

FIGURE 10.4: Mean effective stresses during the three slope instability stages.

changes. The point at which the void ratio starts to change and the dramatic drop in stress initiated can be considered an instability point. Afterwards, the mean effective stress decreases rapidly and the material liquefies when the effective stress is nil. The slight reduction of mean effective stress before the instability point is due to the changes in the geometry of the slope that happened in the previous failure stages. The MP does not reach the critical state, i.e. the situation in which void ratio and critical void ratio converge, until $t = 1.3$ s, which is far from the initiation of liquefaction and instability.

10.3   Modelling the propagation of a debris flow

The small scale experiment herein considered for soil propagation was carried out by Denlinger and Iverson [80] using loosely packed and well-sorted dry

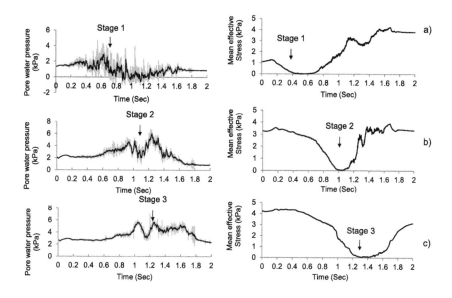

FIGURE 10.5: Evolution of the pore water pressure and mean effective stress over time at the control zones.

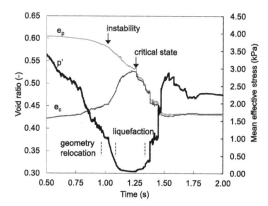

FIGURE 10.6: Point P1 tracked over time of the current and critical void ratios, and the mean effective stress.

sand within a rectangular flume with a Formica bed surface inclined at 31.4° (Fig. 10.7). The bulk density of the soil is about $\gamma \approx 16$ kN/m³, the reported friction angle of the sand is $\varphi' = 40°$, the reported contact friction angle between the soil and the Formica surface is 29°.

A vertical gate in the uppermost part of the slope is used to confine and then suddenly release about 290 cm^3 of dry sand. A non-invasive optical shadowing technique is used to measure the soil thickness during the flow. The flow accelerates, elongates, and thins rapidly after the gate opens. The sand deposition is complete 1.5 s after the flow release. The sand mainly deposits around the break in slope, which is located at a horizontal distance of 42.5 cm from the gate. Sand thicknesses are reported at 0.32 s, 0.53 s, 0.93 s and 1.5 s after the gate release [80]. The longitudinal cross-sections of the propagating soil are drawn at different experimental times using the four reported graphs of contour lines.

10.3.1 Description of MPM model

The MPM model is shown in Fig. 10.8 and has 12,555 elements. The soil is modelled with 1,760 MPs for 440 elements. The base of the flume is modelled with linear elastic material adopting 1,000 kPa for the Young modulus. The experimental gate is simulated by applying horizontal fixities at the right boundary of these elements. Stresses are initialised by gravity loading with a quasi-static calculation. Subsequently, the horizontal fixities are removed and the soil propagates downwards along the slope. Artificial bulk viscosity damping is employed, with values of C_0 and C_1 equal to 0.4 and 1.2, respectively.

Fig. 10.8 (b) shows the experimental position of the sand at the end of the test, corresponding to an experimental time of $t = 1.5$ s. The soil surface is discretised by considering 18 locations, almost equally spaced along the horizontal axis. The observations used in the inverse analysis of the MPM model (see next section) are the values of the soil elevation at these locations. When, at any given experimental time, the soil is not present at these locations, the elevation of the base of the flume is used.

The soil is modelled with a Mohr-Coulomb model, which has 5 model parameters – a Young modulus E, a Poisson ratio ν, a cohesion c, a friction angle φ, and a dilatancy angle ψ. The base of the flume is simulated by a linear elastic material. The contact between the soil and the base of the apparatus is simulated with a frictional contact condition defined by a friction coefficient μ.

The initial values of the six input parameters of the constitutive models are determined considering the values of the sand properties reported by [80], and the MPM results of numerical simulations performed by Bandara [20] and Ceccato [53]. They are equal to $E = 1,000$ kPa, $c = 0$, $\nu = 0.3$, $\varphi = 40°$, $\psi = 0$, and $\mu = 0.55$. The time needed to run one model simulation is approximatively equal to 60 s.

10.3.2 Parameter calibration by inverse analysis

Differently from the procedure adopted in the previous case study, the calibration by inverse analysis of the constitutive models is here performed by

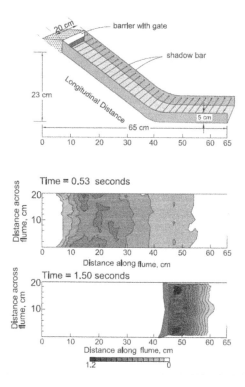

FIGURE 10.7: Schematic of the flume apparatus and overview of some experimental results (modified from [80]).

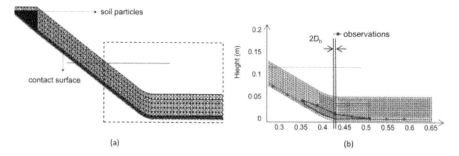

FIGURE 10.8: (a) MPM model and (b) experimental observations at the end of the test ($t = 1.5$ s).

directly comparing the experimental observations of the flume test with the results of the MPM model. To this aim, the soil elevations at different locations along the flume, at different experimental times, are used as observations. A numerical algorithm is defined to extract the values of the elevations of the MPs corresponding to the adopted observations. That is 18 locations fixed in space for any given experimental time (Fig. 10.8 b). The algorithm uses buffer zones, which have a width equal to $2 \cdot D_b$ at the location of each observation, i.e. longitudinal distance X_i:

$$X_{i,min} = X_i - D_b \quad , \qquad X_{i,max} = X_i + D_b \qquad (10.6)$$

where $X_{i,min}$ and $X_{i,max}$ are the initial and final longitudinal distance, respectively, of the buffer zone for the i-th observation.

The numerical value to compare to the elevation of the i-th observation at any given experimental time is equal to the maximum elevation of all the MPs falling within the corresponding buffer zone at that time.

Fig. 10.9 shows the comparison between the experimental observations and the results of the MPM simulation, which is based on the input parameter values defined as initial estimates of the parameters (Table 10.2). The comparison is shown for two experimental times – $t = 0.53$ s and $t = 1.5$ s. The model, coherent with results already reported in the literature [20, 61], approximately reproduces the final run-out distance of the soil, but it does not correctly model the location of the soil during propagation ($t = 0.53$ s) nor does it accurately identify the shape of the deposited soil ($t = 1.5$ s).

The MPM model has been then optimised by simultaneously calibrating two of the six input parameters. The inverse analysis is run by minimising a weighted least-square objective function (Eq. 10.7), $S(\vec{b})$, defined as Eq. 10.7.

$$S(\vec{b}) = [\vec{y} - \vec{y}'(\vec{b})]^t W [\vec{y} - \vec{y}'(\vec{b})] = \vec{e}^t W \vec{e} \qquad (10.7)$$

where $\vec{b}$ is a vector containing values of the number of parameters to be estimated, $\vec{y}$ the vector of the observations being matched by the regression, $\vec{y}'(\vec{b})$ the vector of the computed values corresponding to observations, W the weight matrix, and $\vec{e}$ the vector of residuals.

TABLE 10.2

Initial and estimated optimal values of the input parameters of constitutive models.

Parameter	Symbol	Unit	Initial Value	Optimal value
Friction angle	φ	°	40	43.43
Young module	E	kPa	1,000	1,000
Poisson ratio	ν	-	0.3	0.3
Dilatation angle	ψ	°	0.00	0.00
Cohesion	c	kPa	0.00	0.00
Basal friction coefficient.	μ	-	0.55	0.48

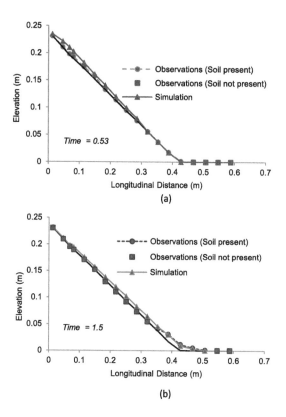

FIGURE 10.9: Comparison between experimental observations and results of initial MPM simulation at (a) $t = 0.53$ s and (b) $t = 1.5$ s.

The vector of residuals is defined considering the elevation data of the propagating soil at the experimental stages for which the observations are reported, and corresponding to the times $t = 0.32$ s, $t = 0.53$ s, $t = 0.93$ s and $t = 1.5$ s. Considering that the soil surface profile along a longitudinal cross section is discretised with 18 points, a maximum of 72 observations is thus used to compute the objective function. Model calibration by inverse analysis is conducted using UCODE [228], which is a computer code adopting a modified Gauss-Newton method to perform inverse modelling as a parameter estimation problem. At each iteration step, to compute the updated estimates of the parameters being calibrated, $\vec{b}_{r+1}$, the following two equations are solved.

$$(\boldsymbol{C}^{\mathrm{T}}\boldsymbol{X}_{\mathrm{r}}^{\mathrm{T}}\boldsymbol{W}\boldsymbol{X}_{\mathrm{r}}\boldsymbol{C} + \bar{\boldsymbol{I}}m_{\mathrm{r}})\boldsymbol{C}^{-1}\vec{d}_r = \boldsymbol{C}^T\boldsymbol{X}_{\mathrm{r}}^T\boldsymbol{W}\left(\vec{y} - \vec{y'}(\vec{b}_r)\right) \qquad (10.8)$$

$$\vec{b}_{\mathrm{r}+1} = \rho_{\mathrm{r}}\vec{d}_{\mathrm{r}} + \vec{b}_{\mathrm{r}} \qquad (10.9)$$

where $\vec{d_r}$ is the vector used to update the parameter estimates $\vec{b}$, r the iteration index of the parameter estimation, $\mathbf{X_r}$ is the sensitivity matrix with $(X_{ij} = \partial y_i/\partial b_j)$ evaluated at parameter estimate $\mathbf{X_r}$, $\mathbf{C}$ is a diagonal scaling matrix with elements $C_{jj} = 1/\sqrt{(\mathbf{X^T W X})_{jj}}$, $\mathbf{I}$ is the identity matrix, m_r is a parameter used to improve regression performance, and ρ_r is a damping parameter.

The relative importance of the input parameters to be simultaneously estimated by inverse analysis, given the assumed observations, can be defined by performing a sensitivity analysis. To this aim, the adopted optimisation algorithm allows the computation of statistics representative of the sensitivity of the predictions to changes in parameters values derived from a variance-covariance matrix $\mathbf{V}(\vec{b})$ (Eqs. 10.10 and 10.11).

$$\mathbf{V}(\vec{b}) = s^2(\mathbf{X^T W X})^{-1} \tag{10.10}$$

$$s^2 = \frac{S(\vec{b})}{ND - NP} \tag{10.11}$$

where s^2 is the model error variance, $\mathbf{X}$ the sensitivity matrix, $\mathbf{W}$ the weight matrix; ND is the number of observations, NP is the number of estimated parameters.

The results of the sensitivity analysis, performed for this case study, indicate that the model results are highly sensitive only to the following two input parameters – the contact coefficient between the soil and the base of the apparatus μ and the friction angle of the soil φ.

For this reason, as already mentioned, these are the only two parameters optimised by inverse analysis, while the values of the other 4 input parameters are kept constant (Table 10.2). The other 6 input parameters are taken from the experimental description. Many different sets of observations are used to run the inverse modelling algorithm. In other words, different objective functions are minimised to find different sets of optimal values of the two considered input parameters. In particular, for each experimental time, the points adopted to describe the soil surface are subdivided into two classes in relation to whether the observations referred to areas with soil or without soil. This distinction allows differentiating between experimental data carrying information of the absolute value of the soil depth at a given location and data only reporting the "absence" of soil at that location. Six inverse analyses are thus performed, considering the following observation sets: all the available elevation data (72 observations), elevation data carrying information of the absolute value of the soil depth (36 observations), and single stage elevation data from each one of the four experimental stages considered (18 observations at each stage).

10.3.3   Inverse analysis results

The best results are obtained when the inverse analysis is conducted using only the 18 soil elevation values at the end of the propagation, that is for

TABLE 10.3

Observations used in the performed inverse analyses (obs. = observations).

Analysis	Type of obs.	Exp. time	Nb. of obs.
No.1	all available observations	all 4 exp. stages	72
No.2	observations (soil present)	all 4 exp. stages	36
No.3	all available observations	$t = 0.32$ s	18
No.4	all available observations	$t = 0.53$ s	18
No.5	all available observations	$t = 0.93$ s	18
No.6	all available observations	$t = 1.5$ s	18

inverse analysis No.6 (Fig. 10.10). The best-fit between the computed and experimental data is obtained by increasing the values of both the parameters being calibrated – from 40° to 43.43° for parameter φ, and from 0.55 to 0.48 for parameter μ. The success of the inverse analysis is verified by the fact that the calibrated model is almost perfectly reproducing the position of the deposited soil mass and only slightly over-predicting the soil deposition heights. It is important to note, however, that the calibrated model is not able to adequately reproduce the evolution of the soil during propagation (i.e. observation times 0.32 s, 0.53 s, 0.93 s). In other words, the adopted MPM model can be calibrated to adequately predict the final condition of the soil mass but it is not able to adequately reproduce, at the same time, both the evolution and the final deposition characteristics of the debris flow. This is due to the inherent limitations of the constitutive elastic-perfectly plastic law adopted to model the propagating soil mass using a continuum modelling approach, as is already reported in the literature (e.g. [103]). The latter is further confirmed by the fact that the optimal inverse analysis does not use all the observations, but only the ones related to the final deposition stage of the soil mass.

Despite these limitations, the considered example shows that simple frictional constitutive models, if well calibrated, can be used to reproduce the most relevant outcome of dry debris flows, the position of the soil mass at its final deposition stage. This finding is comforting for a series of reasons, mainly related to the potential use of MPM models to simulate real debris flows: (1) the computational time needed to run MPM models employing complex elasto-plastic constitutive laws can be too demanding, (2) soil height data related to intermediate stages of the soil propagation are not easy to retrieve outside the laboratory, and (3) the final position of the soil mass is the "observation" most typically adopted by analysts evaluating the performance of a propagation model simulating real granular flows.

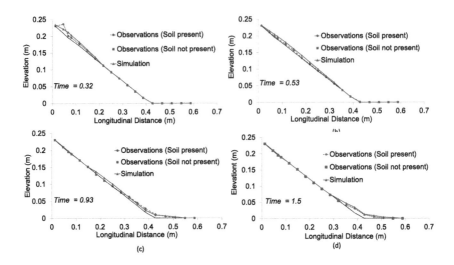

FIGURE 10.10: Comparison between experimental results and optimal MPM simulations calibrated by inverse analysis.

10.4   Closure

Two modelling examples of laboratory slope tests were provided to show how inverse analysis approaches can be effectively used for the calibration of MPM models that are able to reproduce complex slope evolution scenarios. Particularly, a progressive-retrogressive failure including soil liquefaction and a fast large run-out propagation problem were satisfactorily simulated. In both cases, inverse analysis was used but with different scopes. For the slope case the used constitutive model included several parameters to be estimated. Therefore, inverse analysis was used to calibrate the mechanical soil parameters from triaxial tests, and those parameters were then used for the MPM model of the boundary value problem. On the other hand, for the flume test, inverse analysis was directly used to calibrate a relatively simple MPM model of the slope, adopting an elastic perfectly plastic constitutive law to simulate the propagating soil mass and a purely frictional law to model the contact between the soil and the base of the experimental apparatus. The two case studies show that inverse analysis can contribute to enhance the potential of MPM modelling. More generally, MPM models calibrated using reduced-scale laboratory slopes can be used to analyse the effect played by some soil parameters on landslide triggering and propagation.

11

Dyke Embankment Analysis

B. Zuada Coelho and J.D. Nuttall

11.1 Introduction

Slope stability of dykes is an important issue in geotechnical engineering as dykes often form the first line of defence against flooding. Traditionally, slope stability analyses are carried out using limit equilibrium methods (LEM) or finite element methods (FEM). While LEMs are computationally efficient, they suffer from drawbacks in comparison with other numerical methods. The main drawback of LEM is the definition of safety, which strongly depends on the assumptions of the shape of the slip surface [118, 167]. Duncan [89] provides a detailed review of LEMs.

FEM is a more advanced numerical method that, for the analysis of slope stability, has significant advantages. The slip surface is not predefined, with its shape and location being determined by solving governing equations (see Chapter 2). FEM also provides insight into the initiation of failure and the failure mechanism. Advanced constitutive models can be used allowing for the modelling of complex mechanical behaviour of the soil [98, 118, 167].

LEM defines failure as the moment when the shear strength of the soil is equal to the shear strength required to guarantee the slope equilibrium, whereas FEM traditionally defines failure as the moment when there is non-convergence of the solution [118]. Numerical non-convergence and slope failure simultaneously occur and are accompanied by a significant increase in the nodal displacements within the mesh. For smaller deformations, FEM is an ideal tool for analysis up to the onset of failure. However, its use for large deformation post-failure analysis is hampered by problems related to mesh distortion.

MPM can be used to overcome several FEM limitations. The problems related to mesh distortion under large deformations are circumvented, as well as the diffusion associated with the convective terms of the Eulerian approach [285, 287] (Chapter 2). The use of Lagrangian MPs conserves mass and allows the use of complex history-dependent constitutive models (Chapter 8). Similar to FEM, the discrete equations defining the momentum balance are obtained on the background grid with an updated Lagrangian formulation. Geomechanical problems involving slope stability have been successfully analysed using MPM by e.g. [102, 278, 341].

In this chapter, MPM is used to study the stability of a dyke. After validation of the method against literature results, MPM is applied to a typical Dutch dyke profile, and the differences with standard methodologies are discussed.

11.2   Verification

Anura3D [14] has been used to perform the MPM analyses. Hicks and Wong [131] presented the slope stability analysis for which the validation is performed. Fig. 11.1 illustrates the cross section of the slope geometry and mesh for the MPM analysis. A fully saturated slope is assumed under undrained conditions (effective stress analysis). The system of equations is solved explicitly in the time domain. Low-order tetrahedral elements are used to discretise the domain with 1,086 active elements with initially 4 MPs in each element. The displacements of the bottom boundary are constrained in both the vertical and horizontal directions, and are constrained in the horizontal direction along the vertical boundaries. A Mohr-Coulomb constitutive model is used to model the slope material and the parameters are given in Table 11.1.

The factor of safety of the slope is assessed by increasing the gravity multiplier, while keeping all material parameters constant. The definition of the factor of safety is the gravity multiplier at failure [167, 289, 346]. The LEM analysis is calculated using the software *D-GeoStability* [79] using Bishop's method [35]. Fig. 11.2 shows the vertical displacement of Point A, located at the edge of the slope crest (Fig. 11.1), obtained using LEM, FEM [131] and MPM.

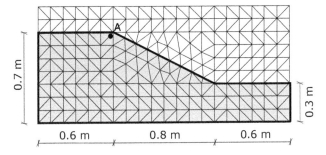

FIGURE 11.1: Cross section of the slope geometry and mesh (after [351]).

TABLE 11.1

Material parameters for the slope (Mohr-Coulomb).

Parameter	Symbol	Unit	Value
Volumetric weight	γ	kN/m^3	20
Young modulus	E	kPa	1,000
Poisson ratio	ν	-	0.3075
Undrained Poisson ratio	ν_u	-	0.499
Cohesion	c'	kPa	0.6
Friction angle	φ'	°	30
Dilatancy angle	ψ	°	0

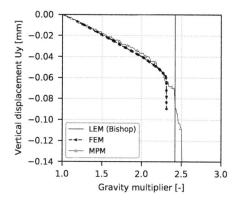

FIGURE 11.2: Comparison between LEM, FEM [131] and MPM results for the vertical displacement of the slope crest (after [351]).

The FEM and MPM simulations are in agreement up to the onset of failure, i.e. a gravity multiplier of approximately 2.3, which demonstrates the correctness of the MPM solution for small strain analysis. While the FEM computation is unable to provide any insight into the dyke behaviour after the initial failure, MPM is able to simulate beyond this point. The LEM result is roughly in agreement with the FEM analysis, demonstrating that the methodology provides a good solution when estimating the initial factor of safety, and using less computational resources.

Following the traditional design methodology, it is assumed that the factor of safety of the slope is 2.3. However, at this factor of safety, the vertical displacement of point A, located on the crest, is only 0.08 mm (0.002% of the slope height), meaning that the slope still fulfils its function of retaining high water levels. This is further discussed in the next section.

11.3   Embankment analysis and safety

A generic dyke section is studied to illustrate the benefits of using MPM for slope stability analysis. The dyke geometry is representative of a typical Dutch dyke section with asymmetric slope inclination on either side of the dyke (1:4 outer slope, 1:5 inner slope). The dyke is considered fully saturated with the water level placed at the height of the crest. Fig. 11.3 presents the cross section of the three-dimensional (3D) dyke geometry and mesh for the MPM analysis.

The domain is discretised using low-order tetrahedral elements (in total 7,017 active elements) with initially 4 MPs in each element, while the system of equations is solved explicitly in the time domain. At the vertical boundaries the displacements are constrained horizontally, while the displacements are constrained in both the vertical and horizontal directions along the base. The dyke material is considered to be uniform and homogeneous, and is modelled using a Mohr-Coulomb constitutive model. The model parameters are given in Table 11.2. Mesh refinement is applied where higher shear strains are expected.

The application of distributed loads in MPM is not straightforward as MPs can move through the mesh. To overcome this issue, an elastic material is used

TABLE 11.2
Material parameters for the dyke (Mohr-Coulomb) and water (linear elastic).

Material	Parameter	Symbol	Unit	Value
Dyke	Volumetric weight	γ	kN/m^3	20
	Young modulus	E	kPa	7,500
	Poisson ratio	ν	-	0.33
	Undrained Poisson ratio	ν_u	-	0.499
	Cohesion	c'	kPa	1
	Friction angle	φ'	$^\circ$	35
	Dilatancy angle	ψ	$^\circ$	0
Water	Volumetric weight	γ	kN/m^3	10
	Young modulus	E	kPa	150,000
	Poisson ratio	ν	-	0
	Undrained Poisson ratio	ν_u	-	0.499

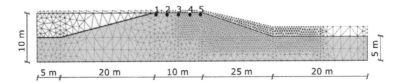

FIGURE 11.3: Cross section of dyke geometry and mesh (after [351]).

to represent the water loading by the water reservoir with the parameters presented in Table 11.2. Note that the water is not infiltrating the dyke. The vertical effective stress and pore water pressure, after the gravity initialisation (gravity multiplier 1g), are shown in Fig. 11.4, and it shows that the proposed methodology is appropriate to model the water load.

As in the previous section, dyke safety is assessed using a gravity multiplier, which is problematic when assessing the deformation with the model; the deformation due to shearing and the deformation due to the elastic deformation caused by the gravity increase are combined. To distinguish between these two components, an additional representative one-dimensional (1D) soil column subjected to the gravity increase is computed with MPM, and the resulting strain is subtracted from the dyke results. The deformation of the 1D soil column is entirely caused by the gravity increase as no shearing occurs. This procedure is illustrated in Fig. 11.5. The results presented are corrected in this way, and only concern the shear deformation.

11.3.1 Results

The vertical displacement at several locations along the dyke crest, indicated in Fig 11.3, are shown in Fig. 11.6 (a), together with the factor of safety and the result obtained by means of LEM analysis. Larger displacements are exhibited at points towards the inner slope, while Point 1, which is located close to the outer slope and next to the water loading, has the smallest vertical deformation.

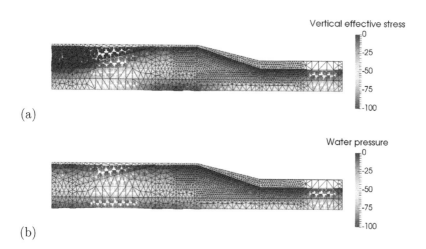

FIGURE 11.4: Initialisation stage: (a) vertical effective stress and (b) water pressure at gravity multiplier 1g (after [351]).

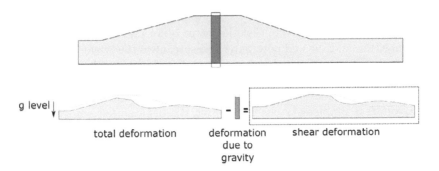

FIGURE 11.5: Procedure for the correction of the dyke deformation.

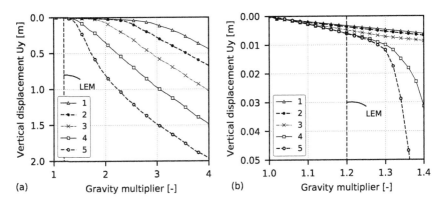

FIGURE 11.6: Vertical displacement for several points along the dyke crest: (a) full range of gravity multipliers and (b) low gravity multipliers (after [351]).

For the factor of safety calculated by LEM analysis, i.e. 1.2g, the crest deformation is very small, as illustrated in Fig. 11.6 (b), which presents the vertical crest displacement for lower values of the gravity multiplier. At this gravity multiplier level the maximum dyke crest deformation is smaller than 7 mm and occurs at Point 5. Therefore, the dyke would be considered as failed with a maximum displacement of 7 mm, if analysed with the conventional LEM approaches, although under such conditions the dyke maintains its primary functionality to retain water.

The dyke crest profile for a range of gravity multipliers is presented in Fig. 11.7. As previously observed under a gravity multiplier of 1.2, the dyke crest exhibits negligible displacement. At a gravity multiplier of 1.5, the dyke crest begins to visibly deform towards the inner slope. At gravity multipliers of 3 and above a clear deformation is observed along the entire dyke crest. It should also be noted that the dyke crest deforms horizontally, and is not limited to settlement.

The ability of MPM to analyse the dyke response beyond the initial failure is advantageous in dyke safety assessment. With an increasing gravity

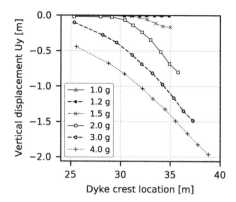

FIGURE 11.7: Dyke crest profile for different values of gravity multiplier (after [351]).

multiplier, the dyke continuously deforms and new equilibrium positions are reached. This is because the failure mechanism is progressive and not abrupt as it is assumed with LEM.

By defining a maximum allowable displacement for the dyke crest, a new safety assessment and factor of safety can be established based on this displacement criterion. For instance, in this example, if a maximum crest displacement of 25 cm is acceptable, the factor of safety would be larger than 1.5 compared with the 1.2 predicted with LEM.

Fig. 11.8 presents the displacement and shear strain fields of the dyke for different gravity multipliers. It is observed that the dyke exhibits a classic macro-stability failure mechanism with the movement of the inner slope. No significant displacement or shear bands occur, for gravity multipliers of 2.0 or lower, while for higher values a well-defined shear band develops and its effects on the dyke displacement are clearly visible.

11.3.2 Material behaviour

The standard dyke safety assessment using LEM does not take into account the dyke deformation. Therefore, the use of a constitutive model has no influence on the calculated safety factor. However, any safety assessment based on displacement criteria would be influenced by the constitutive model and its parameters, which are fundamental to correctly model the displacement field. A parametric study of the Young modulus is performed by varying its value from 2,500 to 15,000 kPa, in order to illustrate the importance of the parameter determination.

The results of the dyke crest displacement for several cases are presented in Fig. 11.9. The results show the effect of the Young modulus on the displacement of the dyke crest. The deformation pattern at low gravity multipliers

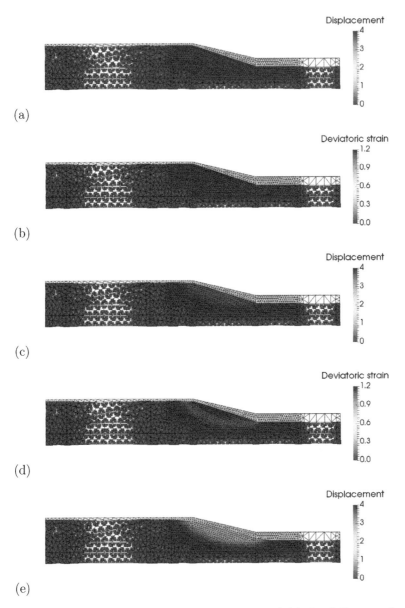

FIGURE 11.8: Displacement and deviatoric strain fields for different values of gravity: (a-b) 1.2g, (c-d) 1.5g, (e-f) 2g, (g-h) 3g, and (i-j) 4g (after [351]).

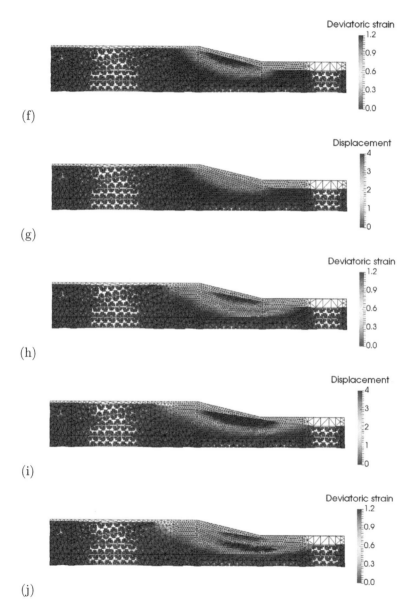

FIGURE 11.8: Displacement and deviatoric strain fields for different values of gravity (continued): (a-b) 1.2g, (c-d) 1.5g, (e-f) 2g, (g-h) 3g, and (i-j) 4g (after [351]).

(Fig. 11.9 a-b), is the same for all Young moduli; the lower the Young moduli result in larger crest deformations.

The deformation pattern changes as the gravity multiplier increases. At a gravity multiplier of 2.0 (Fig. 11.9 c), the displacement at the outer slope is larger for the lower Young modulus, while at the inner slope, the inverse is observed with larger displacement for the larger Young moduli. At the centre of the dyke crest an inflection point can be found where the influence of the Young modulus on the displacement is inverted. This can be explained by the stress redistribution caused by the large deformations, and is one illustration of the complexities of large deformation analysis. Under higher gravity multipliers (as shown in Fig. 11.9 d-e) this effect is enhanced, with higher crest displacements occurring for higher Young moduli.

The importance of the correct selection of the constitutive model is highlighted in Fig. 11.10. For the sake of comparison, additional analyses are performed for which the dyke body is modelled with the Mohr-Coulomb strain softening model [335] (see Chapter 8). The parameters for the dyke material are presented in Table 11.3. The parameters for the water are the same as previously presented in Table 11.2. The residual friction angle is assumed to be 15° and 25°.

Fig. 11.10 shows the comparison of the vertical crest displacement for the analysis with Mohr-Coulomb with and without strain softening. It follows that the choice of material for the dyke body has a significant impact on the results and, hence, on the achieved safety level. Analyses with the Mohr-Coulomb strain softening model result in higher displacements of the dyke crest at a given gravity level. Moreover, lower residual friction angles provide higher displacements. This follows the same findings as for the granular column collapse (see Chapter 8) for which the different predictions are explained as a consequence of different energy dissipating mechanisms [102]. Although these observations are expected, it illustrates that it is important to choose a constitutive model that correctly describes soil behaviour, but also to correctly evaluate the model parameters, when performing more advanced

TABLE 11.3

Mohr-Coulomb strain softening material parameters for the dyke.

Material	Parameter	Symbol	Unit	Value
Dyke	Volumetric weight	γ	kN/m^3	20
	Young modulus	E	kPa	7 500
	Poisson ratio	ν	-	0.33
	Undrained Poisson ratio	ν_u	-	0.499
	Cohesion	c'	kPa	1
	Peak friction angle	φ'_{max}	°	35
	Residual friction angle	φ'_{res}	°	varying
	Dilatancy angle	ψ	°	0
	Shape factor	β	-	100

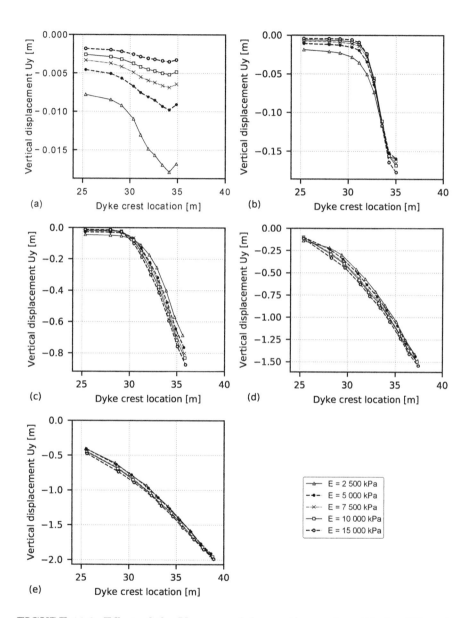

FIGURE 11.9: Effect of the Young modulus on the crest profile for different values of gravity: (a) 1.2g, (b) 1.5g, (c) 2g, (d) 3g, and (e) 4g (after [351]).

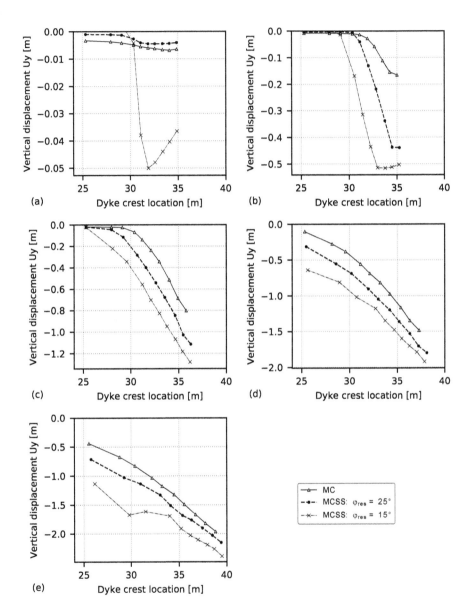

FIGURE 11.10: Effect of the constitutive model and constitutive model parameters on the crest deformation for different values of gravity: (a) 1.2g, (b) 1.5g, (c) 2g, (d) 3g and (e) 4g.

computations by means of MPM. This means that, in order to shift towards a displacement criterion during design, more soil investigation is needed in order to be able to correctly describe the constitutive model and determine its model parameters.

11.4 Closure

This chapter presents the validation of MPM for dyke safety analysis based on displacement criteria. The standard methodology for the safety assessment of dykes is based on LEMs, which are unable to take into account deformation as a criterion. Based on the MPM calculations, it is shown that, for the same factor of safety given by LEM, dyke crest displacements are very small. This means that the traditional factor of safety does not imply that the dyke does not fulfil its requirement to retain water despite an initial failure taking place. After this initial failure, the dyke is stable with a new equilibrium position.

The development of MPM permits large deformations analyses beyond this initial failure. From the analysed dyke section, it is shown that the dyke failure is progressive, with the dyke continuing to deform with increasing gravity, beyond the initial failure. Dyke deformation profiles for the crest are computed, which, in combination with criteria for the maximum allowable crest displacement, could be used to assess the dyke safety. The development of a displacement based safety assessment of dykes could lead to more economical dyke design, while having a positive impact for existing dykes by providing a better prediction of their safety level. The presented analysis shows that more attention is required in the correct use of constitutive models, and their parameters.

This chapter and its analyses can be considered as a first step in a transition towards a safety assessment based on displacement criteria. This has to be accompanied by establishing a definition of the maximum allowable crest displacement, and by placing the methodology within the context of a probabilistic framework for dyke safety. This transition is likely to require more in-depth soil investigations in order to determine reliable material parameters.

12

Landslides in Unsaturated Soil

A. Yerro and E.E. Alonso

12.1 Introduction

Different applications in geotechnical engineering involve large deformation of soils under unsaturated conditions. In particular, one of the most important problems is the instability of slopes and landslides triggered by heavy rainfall. A global survey of landslide occurrence [190] revealed that in 2003 there were 210 damaging landslide events worldwide and that over 90% were triggered by heavy rainfall. The main goal of this chapter is to examine the behaviour of unsaturated slopes subjected to rainfall and water infiltration by means of the three-phase single-point MPM (see Chapter 2). The conditions leading to the onset of instability, the dynamic behaviour of the unstable mass, and the final run-out are investigated. The study and prediction of these variables is essential to assess and evaluate the risk and destructive power of landslides. Initially, a suction-dependent elastic-plastic Mohr-Coulomb model is presented. It is expressed in terms of net stress and suction variables. Then, the instability of a slope subjected to rain infiltration, inspired from a real case, is solved and discussed. The model shows the development of the initial failure surface in a region of deviatoric strain localisation, the evolution of stress and suction states in some characteristic locations, and the dynamics of the motion characterised.

12.2 Constitutive modelling of unsaturated soils

In the past two decades, the advances in constitutive modelling of unsaturated soils have been numerous and it is still an area of very active research. Many constitutive models require two independent stress measures to describe unsaturated soil behaviour. The choice of appropriate stress variables is subjected to continuous debate, but in general the main stresses include suction and degree of saturation. A number of constitutive models are available in the literature based on continuum mechanics and, hence, are compatible with the MPM framework.

In this chapter, it is assumed that the instability of a slope is essentially governed by the evolution of soil shear strength variables (cohesion and friction angle), which evolve with suction. The most commonly used shear strength criterion is proposed by Fredlund et al. [107]. It considers a linear increase of the cohesion with suction. However, further investigations show that such increase is not constant but decreases for high suction values, and friction angles can change with suction too [96, 101]. More recently, it has been found that shear strength also depends on the current degree of saturation [314], but this variation has not been considered here for simplicity reasons. Here, the effect of suction in the soil shear strength is written in terms of net stress and suction by means of a Mohr-Coulomb yield criteria with suction dependence (Eq. 12.1).

$$q = c \cos \varphi + \bar{p} \sin \varphi \tag{12.1}$$

where q is the deviatoric stress, $\bar{p}$ the volumetric net stress, c the cohesion, and φ the friction angle.

The suction dependence is accounted in the definition of shear strength parameters (c and φ) (Eqs. 12.2 and 12.3).

$$c = c' + \Delta c_{\max} \left[1 - \exp\left(-B \frac{s}{p_{\mathrm{atm}}} \right) \right] \tag{12.2}$$

$$\varphi = \varphi' + A \left(\frac{s}{p_{\mathrm{atm}}} \right) \tag{12.3}$$

where c' and φ' are the cohesion and friction angle under saturated conditions. The last terms of Eqs. 12.2 and 12.3 represent the effect of suction, which increases the soil apparent strength. It is accepted that cohesion increases with suction exponentially up to a maximum value $\Delta c_{\max}$. p_{atm} is the atmospheric pressure and B controls the rate of variation of apparent cohesion with suction. On the other hand, the friction angle is assumed to have a linear dependence with suction depending on the parameter A. Obviously, other expressions could be introduced.

In order to reduce the singularities of the Mohr-Coulomb yield surface, a hyperbolic approximation is considered. In addition, an explicit sub-stepping algorithm with error control and a correction of the yield surface drift are taken into account to ensure an efficient integration of the model.

12.3   Case description

The problem presented in this chapter is inspired from a real case described by Alonso et al. [9], and was previously simulated with MPM by Yerro [334]. Several road embankments of medium height (6-8 m) are subjected to heavy rainfall immediately after the end of construction. Shallow failures are observed in some slopes, and the mobilised material moves downwards an estimated distance of 2-4 m, damaging the road side shoulders. The embankments are built with low to medium plasticity sandy clay, that is compacted dry of optimum. The loss of strength upon soil saturation is investigated by means of a set of suction controlled direct shear tests.

The simulated slope has a height of 7 m and an angle of 32.5° (Fig. 12.1). The flat upper and lower surfaces reproduce the actual embankment geometry. The calculation is performed in three dimensions and plane strain conditions are assumed. The thickness of the model is 0.4 m.

Fig. 12.1 shows the computational mesh formed by tetrahedrons, in which 4 material points (MPs) are initially distributed within each filled element. The computational mesh covers a larger volume to allow for the expected large displacements associated with the slope instability. It has been refined and made homogeneous in the region where the failure is expected to minimise mesh size effects in order to get accurate results.

Other numerical parameters are presented in Table 12.1. In this calculation, a small value of damping is adopted ($\alpha = 0.05$), which may represent the friction that can occur between grains. A low value allows capturing the acceleration of the mass motion and reduces spurious numerical instabilities.

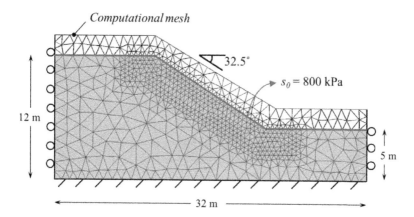

FIGURE 12.1: Geometry of the embankment slope, computational mesh and initial distribution of the MPs (after [334]).

TABLE 12.1

Numerical parameters.

Parameter	Symbol	Unit	Value
Element type		-	Tetrahedron
Number of elements	N_e	-	3,654
Number of MPs	N_p	-	7,593
Damping factor	α	-	0.05
Time step	Δt	s	$2 \cdot 10^{-4}$

A high value of damping would increase the energy dissipation and any movement of the soil would slow down in a non-realistic manner.

Because this problem is solved considering the three-phase single-point formulation, it is important to highlight that boundary conditions must be applied for the three phases involved in the computation (solid, liquid, and gas). Taking into account the mechanics of the solid phase, the lower boundary is fixed and horizontal displacements along vertical contours are prevented. Lateral and bottom contours are impervious for the liquid phase, and a constant zero gas pressure in excess of atmospheric pressure is prescribed at all the boundaries ($p_G = 0$ kPa), which implies that the gas can move easily through the boundaries.

The simulation of atmospheric conditions is considered in the model by prescribing the suction value ($s = p_G - p_L$) along the ground surface free boundary. Initially, when the conditions are relatively dry, the prescribed suction is $s_0 = 800$ kPa. This value is estimated from the known values of the compaction water content and the water retention curve determined experimentally. Initial stresses and pore pressures in the slope are in equilibrium with the gravity force and the prescribed suction s_0 along the free surface. Then, the wetting induced by the rainfall is modelled by applying a decrease of suction along the top free boundary surface from 800 kPa to 0 during 10 seconds. Afterwards, the saturated boundary condition of $s = 0$ kPa is maintained constant along the ground surface during the entire simulation period. As a result, an essentially downward flow is generated in the embankment due to suction gradients.

The embankment is assumed to be homogeneous material and the properties of the different phases forming the soil (solid, liquid, and gas) must be defined (Table 12.2). Note that the liquid phase is pure water and the gas phase is considered to be dry air. Hence, neither water vapour nor dissolved gas is taken into account. The water saturated permeability of the embankment is increased to accelerate the wetting process and to reduce the computational cost.

The elastic-plastic suction-dependent Mohr-Coulomb model (Section 12.2) is considered, and strength parameters are summarised in Table 12.3. A small cohesion of 1 kPa is assumed under saturated conditions to avoid numerical difficulties in zones of very low effective confinement. The friction angle at

TABLE 12.2

General characteristics of the soil.

Parameter	Symbol	Unit	Value
Soil density	ρ	kg/m^3	2,700
Porosity	n	-	0.35
Poisson ratio	ν	-	0.33
Young modulus	E	MPa	10
Liquid density	ρ_{L}	kg/m^3	1,000
Gas density	ρ_{G}	kg/m^3	1
Liquid bulk modulus	K_{L}	MPa	100
Gas bulk modulus	K_{G}	MPa	0.01
Liquid viscosity	μ_{L}	kg/ms	0.001
Gas viscosity	μ_{G}	kg/ms	10^{-6}
Intrinsic permeability liquid	κ_{L}	m^2	10^{-10}
Intrinsic permeability gas	κ_{G}	m^2	10^{-11}

TABLE 12.3

Mohr-Coulomb parameters.

Parameter	Symbol	Unit	Value
Cohesion	c	kPa	1
Effective friction angle	φ'	°	20
Maximum cohesion increment	$\triangle c_{\mathrm{max}}$	kPa	15
Model parameter	A	-	10^{-4}
Model parameter	B	-	$7{\cdot}10^{-4}$

saturated conditions is found to be close to $20°$ in direct shear tests performed on recovered samples. These parameters lead to unstable conditions, in a limit equilibrium analysis, in a situation of full saturation of the slope ($s = 0$ kPa). The slope remains initially stable due to the additional strength induced by the effect of suction, which depends on parameters A, B and $\triangle c_{\mathrm{max}}$ (Eqs. 12.2 and 12.3). The estimated A value leads to a very small variation in friction with suction: less than $1°$ for the maximum range of change on suction of 800 kPa. The selected B and $\triangle c_{\mathrm{max}}$ values lead to a progressive reduction of cohesion with suction from a value $c = 67$ kPa at the initial state ($s = 800$ kPa) to $c = 1$ kPa for saturated conditions.

The water retention curve, which is the relationship between suction s and degree of saturation S_r, is also required to perform the three-phase MPM analysis. In this case, the Van Genuchten water retention curve is considered and the parameters are listed in Table 12.4.

TABLE 12.4

Retention curve parameters.

Parameter	Symbol	Unit	Value
Minimum degree of saturation	S_{min}	-	0
Maximum degree of saturation	S_{max}	-	1
Reference pressure	p_0	kPa	50
Model parameter	λ_w	-	0.09

12.4   Embankment response

In order to analyse the embankment behaviour, attention will focus on seven control MPs distributed along the slope at depths of 0.5 m, 1.7 m, and 3.5 m as indicated in Fig. 12.2 (a) at $t = 0$ s. These depths are selected since a relatively shallow failure is expected in this case because the loss of suction is faster the closer to the boundary subjected to rainfall.

The evolution of suction in the embankment may be followed in Fig. 12.2 (a). Contour plots of equal suction, at three different times, are selected: $t = 0$ s, $t = 15$ s, and $t = 200$ s. The first 15 seconds result in a major change in suction if compared with the initial state characterised by an essentially constant value ($s = 700\text{-}800$ kPa). The initial ($t = 0$ s) vertical suction gradient reflects flow equilibrium conditions in view of the imposed boundary conditions. It is also important to note that as the wetting process progresses, the saturation is faster in the toe side of the slope. This is a consequence of the position of the impervious bottom boundary, which is closer to the ground surface (5 m below the surface) than on the left side (12 m below the surface).

Fig. 12.2 (b) shows the contours of deviatoric strain. High shear strains begin to develop at the slope toe soon after the beginning of wetting. A shear band defining a potential shallow failure surface at an average depth of 1.6 m is already defined at $t = 15$ s. As wetting continues to increase, the apparent shear strength along the failure surface decreases and the slope becomes unstable. The final run-out, defined as the distance between the initial toe of the slope and the toe of the final geometry, can be quantified to be 2.5 m.

12.4.1   Dynamics of the motion

The model provides information on the overall dynamic behaviour of the slide. This is a significant improvement over static formulations. The calculated total displacement, velocity and acceleration are plotted in Figs. 12.3 and 12.4. Velocities and accelerations represented in Fig. 12.4 (a) and (b) have been calculated by applying a smoothing on the total displacements. Analysing these plots it can be seen that the embankment remains essentially stable during

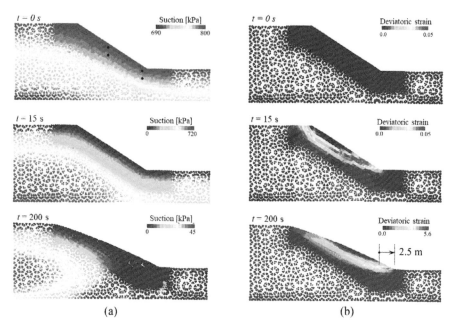

(a)   (b)

FIGURE 12.2: Calculated suction and deviatoric strain contours at three different times ($t = 0$ s, $t = 15$ s, and $t = 200$ s). Initial position of control MPs (A1, A2, B1, B2, B3, C1, and C2) is indicated.

the first 20 seconds after the initiation of wetting. At time $t = 20$ s, the failure mechanism is fully developed and control points clearly located in the mobilised volume (A1, B1, B2, C1, and C2) start moving. They accelerate fast during fifteen additional seconds. Peak velocity is attained around $t = 35$ s. After a peak value, the velocity decreases rapidly followed by a progressive reduction towards a new state of equilibrium. This reaction cannot be generalised and it will depend strongly on the slope geometry. In this example, the lower horizontal platform contributes to arresting the motion after a relatively small displacement. The control point A2 moves 4 m, reaches a maximum velocity of 0.1 mm/s and achieves an acceleration of 9 mm/s^2. Points located below the mobilised material remain essentially motionless (A2 and B3). Note that the sliding motion is far from a classic solid rigid motion and dynamic variables completely depend on the position of the point within the slope. Finally, it is important to highlight that the final distance travelled by some MPs in the slope can be larger than the final run-out of the toe of the slope (2.5 m, Fig. 12.2).

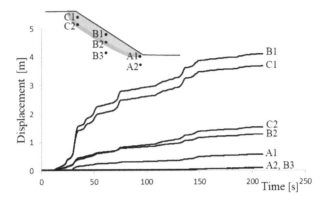

FIGURE 12.3: Evolution of total displacement of control MPs (after [334]).

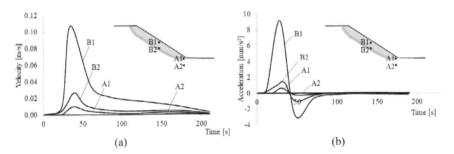

FIGURE 12.4: Evolution of (a) velocity and (b) acceleration of control MPs (after [334]).

12.4.2   Suction, degree of saturation and time

Figs. 12.5 and 12.6 show the evolution of suction and the degree of saturation of different control points (A2, B1, B2, B3, and C2). Both figures are directly related between them by the water retention relationship. According to the initial suction distribution, the degree of saturation at $t = 0$ s is approximately 0.758. Comparing points located along a vertical middle section (B1, B2, and B2) (Fig. 12.5 a) it is clear that the reduction of suction is faster in shallow points than in deeper points. On the other hand, points located at the same depth (A2, B2, and C2) (Fig. 12.5 b) have similar wetting evolution. As the calculation progresses the degree of saturation increases and approaches almost fully saturated conditions at $t = 200$ s in all controlled MPs.

12.4.3   Stress, suction and time

The stress evolutions have been analysed for five control points (A1, A2, B1, B2, and B3) and are presented in Figs. 12.7 and 12.8. Fig. 12.7 shows the

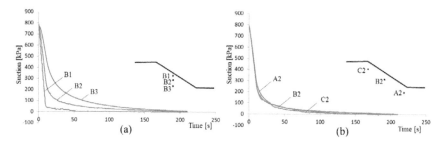

FIGURE 12.5: Evolution of suction at control MPs (a) at the middle section (B1, B2, and B3), and (b) at depth 1.7 m (A2, B2, C3). Initial location of points is sketched (after [334]).

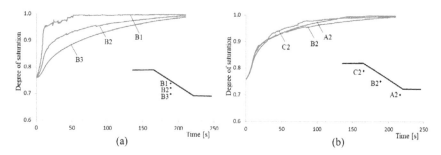

FIGURE 12.6: Degree of saturation evolution at control MPs (a) at the middle section (B1, B2, and B3), and (b) at depth 1.7 m (A2, B2, C3). Initial location of points is sketched (after [334]).

evolution of net mean stress while Fig. 12.8 shows the evolutions of the mobilised shear stress and the corresponding yield Mohr-Coulomb envelope. The plot of the evolution of the available strength provides significant information, because it allows easy evaluation whether a point is under plastic conditions.

In these two figures, some oscillations are observed. These can be the consequence of the dynamic formulation, but also can be related to some remaining cell crossing noise during the motion of the slide. Another reason that can explain the onset of these instabilities is a sudden change of the stiffness of the soil when it reaches full saturation.

The MP A1 at the slope toe is essentially yielding at the start of the simulation (Fig. 12.8 b) and it maintains plastic conditions throughout the sliding process. Point B1, a shallow point at mid slope, also yields very quickly after the beginning of rainfall and both remain in a plastic state. B2, which is located within the shear band, behaves essentially as B1. Points A2 and B3, both located below the slip surface, are in an elastic state throughout most of the sliding process (Fig. 12.8).

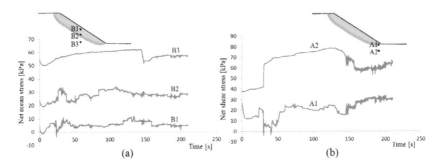

FIGURE 12.7: Calculated volumetric net stress evolution of control points (a) A2, B2, and C2, and (b) B1, B2, and B3. Initial location of points is sketched (after [334]).

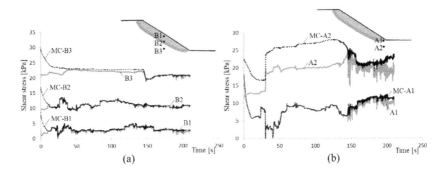

FIGURE 12.8: Calculated shear stress evolution of control points (a) A2, B2, and C2, and (b) B1, B2, and B3. Evolution of Mohr-Coulomb (MC) yield shear stress is also indicated for each point with a dotted line. Initial location of points is sketched (after [334]).

Shallow MPs are subjected to a rapid wetting and a loss of shear strength is associated with suction softening. As a result, those points yield faster than deeper MPs, which are more confined and are capable of offering more shear strength.

12.5 Closure

The general three-phase single-point MPM formulation is applied to unsaturated soils described by an elastic-plastic suction dependent Mohr-Coulomb model formulated in terms of net stress and suction.

The instability of unsaturated slopes induced by rainfall wetting is a relevant practical problem in virtually all climate and soil conditions. The method handles in a natural way the kinematics of sliding and it provides information on velocities, accelerations and run-outs, which help to estimate the expected damage in case of sliding.

A simple embankment slope, for which the characteristics are taken from a real case, is analysed. It involves surface instability induced by heavy rains. The model provides an insight into the coupled flow-stress-strain mechanisms developing in the slope. Suction decrease results in a marked strength softening. Deviatoric strain localisation starts at the slope toe and eventually materialises into a full sliding zone. The slope motion starts when a shallow band of soil reaches a low, but non-zero, suction value and accelerates in a few seconds. The slide does not displace as a rigid body, however. Points close to the surface experience a faster and more intense suction reduction and their strength soon reaches the minimum value (saturated conditions). They are capable of "flowing" over the more resistant zones at depth. The end result is a complex motion which makes it difficult to define run-out, velocity and acceleration in a clear and simple way. In fact, these variables depend on the MP position within the sliding volume. This is believed to be the case in practise when observing rain-induced instabilities.

Other large deformation problems, such as wetting induced collapse or swelling, may be analysed by the same method but they will require the consideration of a different constitutive model. However, the general formulation of the three phase approach described will remain unchanged.

13

Preliminary Analysis of a Landslide in the North-West Pacific

E.J. Fern

13.1 Introduction

A large and catastrophic landslide took place on Whitman Bench on 22nd March 2014 near the town of Oso (WA, USA). The landslide was outstanding by its size [153], catastrophic failure and long run-out distance. The landslide destroyed an entire neighbourhood killing 43 people and making it the deadliest landslide in continental US. Fig. 13.1 shows two photographs illustrating the size of the 2014 landslide with (a) showing the head scarp from the middle of the landslide debris, and (b) the graben, which is a ditch formed by a large back-rotated block (left) and the head scarp (right).

A series of landslides have taken place on Whitman Bench since the 1950s (e.g. [267]), including the 2006 Hazel Landslide [290] which is believed to have been remobilised in 2014 [2, 235, 246, 280]. These landslides were significantly smaller than the 2014 landslide and did not claim lives or cause significant damage to property in comparison with the 2014 landslide. However, the post-2014 investigations revealed the existence of large ancient landslides nearby (e.g. Rowan Landslide) and they were larger and more mobile than the 2014 landslides [125]. This is suggests that the 2014 landslide was not unique in the geological time scale.

A number of expert teams carried out surface and subsurface explorations as well as extensive laboratory testing (e.g. [2, 16, 147, 153, 235, 243, 246, 280, 326]). They showed that Whitman Bench is entirely made of soil and that the landslide was not triggered by an earthquake or rainfall infiltration [235], although the rainfall of the 2013-14 winter was significant, but not exceptional, with a return period of 88 years [127]. Therefore, the exact cause of the 2014 landslide and its long run-out distance remain scientifically unexplained.

Large deformation simulations were used to explore the post-failure behaviour of the 2014 Oso Landslide in an attempt to explain its long run-out distance. Iverson and George [146] carried out two-dimensional (2D) depth-

FIGURE 13.1: Photographs of the 2014 Oso Landslide (a) around the location of the neighbourhood and (b) in the graben at borehole EB4 (Photo credit: E.J. Fern, June 2017).

averaged finite volume simulations within the first few days of the event (class-A prediction, see Chapter 1). The landslide was modelled as a single debris flow caused by liquefaction and the surface of rupture was imposed. The simulations were able to match the observed run-out distance and these simulations were crucial in directing the emergency services. Aaron et al. [2] carried out 2D depth-averaged SPH simulations (class-C/C1 predictions) to assess the run-out distance considering a 2-slide scenario, and Yerro et al. [235, 337] carried out 2D plane strain MPM simulations (class-B1 prediction). Although all these simulations are of excellent quality, they all relied on limited geotechnical information and imposed the surface rupture despite its exact position was still unknown at the time of the numerical predictions.

A detailed geotechnical site investigation was carried out at the University of California at Berkeley (UCB) [235, 246] in collaboration with the US Geological Survey (USGS) [243] and the Department of Transportation of Washington State (WSDOT) [16, 17]. The detailed analysis of these documents (e.g. [2, 147, 153, 280, 326]) permitted a better understanding of the site conditions and geotechnical characteristics of the soils as well as a better estimation of the position of the surface of rupture. It also permitted the identification of weak layers within Whitman Bench. The full discussion of the mechanical characteristics of the site is extensive and falls out of the scope of this chapter. It is unclear at this stage if these weak layers played an important role in the failure mechanism.

This chapter presents a preliminary numerical analysis of the onset of the 2014 Oso Landslide in an attempt to demonstrate the ability of MPM to model small strain behaviours and the transition towards large deformation behaviours.

13.2 Field situation

The first step of any simulation is to build an understanding of the problem to be modelled and identify key mechanical features. This analysis is the sole work of the author and it is based on previous expert investigations (e.g. [2, 16, 17, 125, 153, 235, 243, 246, 280, 326]).

13.2.1 Physiography

The LiDAR scan data was obtained and plotted. Fig. 13.2 (a) shows the topography prior to the landslide in 2013 in which the contour of the 2006 Hazel Landslide deposit and the highway SR530 are marked. Fig. 13.2 (b) shows the topography shortly after the landslide in 2014 in which the contour of the 2014 Oso Landslide is marked. It also shows the different surface appearances of the 2014 Oso Landslide deposit with large blocks and mounds in the centre portion, and a thinner and flatter deposition towards the distal end of the landslide. The difference in elevation between different scans gives an estimation of the run-out distances (Fig. 13.3) and shows that the 2014 Oso Landslide is significantly larger than the 2006 Hazel Landslide. The diffused red zone on the right of the 2014 Oso Landslide run-out is due to upstream flooding caused by the damming of the river; Aaron et al. [2] obtained independently similar results using the same data.

Fig. 13.4 shows a fish-eye photographic montage of the landslide with, from left to right, the head scarp followed by a back-rotated block on which all the trees have toppled over in the same direction. These trees were never overridden by a landslide (Fig. 13.1 b) and are also visible in Fig. 13.1 (b)

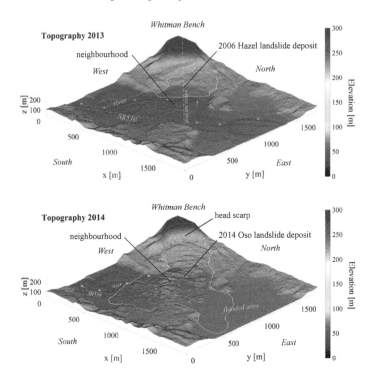

FIGURE 13.2: Topography before and after the 2014 Oso Landslide.

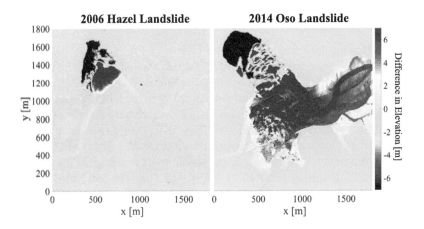

FIGURE 13.3: Run-out distances of the 2006 Hazel and 2014 Oso Landslide obtained by computing the difference in elevation between the different LiDAR scans.

13.2.2 Geology context and stratigraphy

Pyles et al. [235, 246] carried out an in-depth and site-specific geological investigation, which is used as the basis of the geological understanding and interpretation by the author. The geology is the result of complex interactions between different lithologies and several glaciers with ice flowing both up and down the valley [235]. The pushing-and-pulling of these glaciers caused disruptions in the stratigraphy. The presence of a lake allowed the glaciers to be buoyant.

The sequence of the soil units follows a general pattern for the northwestern Pacific [219, 258], although there are a few differences related to the presence of an ancient lake called Bear Lake by Pyles et al. [235]. The soil units exhibit significant spatial variation over the site and the soil units were disrupted by glacial activity, which gives a complex stratigraphy. The thickness of each soil unit varies significantly over the area of the site. The soil units are described in the following sub-section

The stratigraphy and the position of the surface were determined by analysis of the borehole log, non-standard penetration tests and cone penetration tests (CPT) provided by [16, 17]. The obtained stratigraphy was cross-checked with those of Pyles et al. [235] and Stark et al. [2, 280]. Fig. 13.4 shows a 2D idealisation of the stratigraphy of the site interpreted by the authors.

13.2.3 Soil units and geotechnical characteristics

The geotechnical description of each soil unit is based on the work of Pyles et al. [235, 246] and Stark et al. [280] as well as on interpretation of data obtained from [16, 17, 235, 243, 280]. Samples were collected from the borehole cores and sent for grain size, index and strength testing. Mean geotechnical parameters were calculated by the author for each soil unit and these are given in Table 13.1. A summarised description of the soil units, from top down, is as follows.

The recessional outwash (RO) forms the top layer of Whitman Bench. It is composed of well-graded sands (SW-SM), silts (ML), gravels and cobbles (GW-GM) with occasional boulders. It is densely compacted and lightly cemented with thin inter-layered to laminated beds. A thin layer of silt is present at the bottom of this layer and forms an acquitard [235]. The laboratory tests show that the median grain size is typically $D_{50} > 0.1$ mm, excluding the silty layer. The water content varies from $w = 5\%$ in the upper section (partially saturated) to $w = 30\%$ in the bottom silty layer (saturated). The hydraulic conductivity was calculated by the author from laboratory tests and is estimated between $k = 10^{-6}$ to 10^{-4} m/s.

The till complex (TC) regroups all soils that originate or were modified by ice. It is made of a wide range of lithologies and grain sizes. A matrix of fine-grained soil binds coarser material together giving it concrete-like

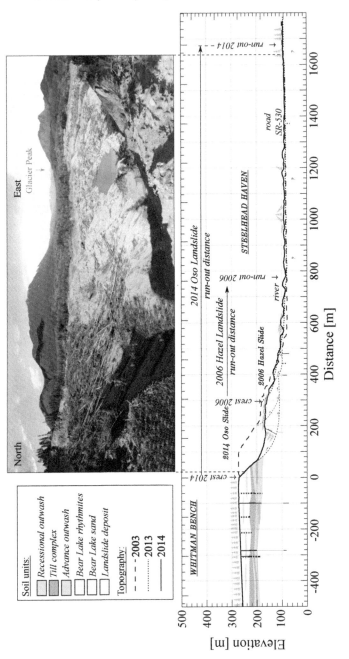

FIGURE 13.4: Schematic interpretation of the stratigraphy along the reference line (Fig. 13.2) with interpreted failure surfaces and panoramic fish-eye photographic montage of the 2014 Oso Landslide.

TABLE 13.1

Mean geotechnical characteristics ($\pm$ standard deviation) obtained from the geotechnical characterisation.

Soil	γ kN/m^3	w %	G_s -	e -	w_L %	w_P %
RO[1]	18.9±1.6	16.9±1.6	2.74±0.07	0.625±0.174	-	-
TC[2]	22.0*	9.7 ±3.1	2.75±0.03	0.373*	26.8±7	16.6±1.1
AO	-	25.6*	2.76*	-	-	-
BLR[3]	19.0±0.5	28.5±7.9	2.79±0.07	0.892 ±366	46.6±11	25.0 ±3.8
BLS	22.0*	21.0±7.0	2.77*	0.524*	-	-

[1] varies with depth, [2] few tests on matrix, [3] mostly clayey samples.
* less than 5 tests (not representative of soil unit).
where γ is the unit weight, G_s the specific gravity, e the void ratio, w_L the liquid limit and w_P the plastic limit.

characteristics. Its thickness varies from thin discontinuous layers to thick blocks of tens of metres as shown in Fig. 13.1 (b). The water content ranges from $w = 7\%$ to 13%. The mean liquid and plastic limits are $w_L = 26.8\%$ and $w_P = 16.6\%$, respectively, qualifying the matrix as a low-plasticity clay (CL) ($I_P = w_L - w_P = 10.2\%$). However, only a limited number of tests were carried out on the matrix [243] and, hence, the results are not representative of the entire soil unit. No strength tests were carried out on the TC.

The advance outwash (AO) is composed of laminated lacustrine silts and silty sands in the east and coarse sands and pebbles in the west. The thickness of the upper silty sand increases westward [235]. Few index tests and no strength tests were carried out on this unit.

The Bear Lake rhythmites (BLR) is the most important soil unit as most of the surface of rupture was located in this unit. It is composed of fine-grained glacio-lacustrine rhythmites, possibly varves, of clay (CL, CH) and silt (ML) with local inclusions of sandy lenses [235]. It is generally light to dark grey with occasional pale yellowish stains due to weathering [235]. Its thickness is around 100 m making it the thickest soil unit. The upper contact is generally located at elevations ranging from approximately 190 to 210 m and the bottom contact at an elevation around 98 m. The median grain size is $D_{50} = 0.001$ to 0.01 mm. The water content ranges from 20% to 40% with a mean of $w = 28.5\%$ and the degrees of saturation are between 90% and 100%, suggesting saturated conditions. The Atterberg limits show medium-low plasticity clays ($w_L = 46\%$, $I_P = 21\%$) and low-plasticity silts ($10\% < I_P < 18\%$), which have transitional behaviours between clays and sands [43]. The relationship between the liquid limit w_L and the plasticity index I_P follows the North American trend [281]. Clays are also characterised by their activity [271] and the mean activity is $A = 0.66$, which is low. X-ray diffraction analyses revealed the presence of smectite and montmorillonite [235], which are usually highly active. It is possible that the fresh water lacustrine environment deactivated the clay and/or that the clay contains additional low-plasticity mineral giving a lower activity [271]. The liquidity indices (see Chapter 7) were computed giving a mean value of $I_L = 0.16 \pm 0.56$, which is slightly lower than the values obtained by Stark et al. [280]. This indicates that the water content is close to the plastic limit. Stark et al. [280] suggested a sensitivity of 1.4 to 3.3, which is considered as a normal sensitivity [280]. Similar values were obtained when computing the sensitivity with CPT data [16, 17].

Direct shear tests on intact specimens were carried out [16, 235] and the results show a well-defined bi-linear peak failure surface caused by the microstructure of the soil. The failure surface is fitted with a Mohr-Coulomb failure envelop with $\varphi'_{max} = 36°$ for $\sigma'_N < 350$ kPa, and $\varphi'_{max} = 24°$ and $c' = 100$ kPa for $\sigma'_N > 350$ kPa. However, the results also reveal the existence of weak specimens ($\varphi'_{max} = 12°$) originating from an elevation around 103 m, which is the elevation of the bottom part of the surface of rupture.

Fully-softened ring shear tests on remoulded specimens were also carried out [16, 235] and showed residual strength between $\varphi' = 10°$ and 24°. The results confirmed the presence of the weak soil with a strength around $\varphi'_{res} = 5°$ to 6°. Skempton [272, 273] showed that the residual strength is controlled mainly by sliding friction of the clay minerals and that the residual strengths can be as low as $\varphi'_{res} = 15°$ for kaolinite, $\varphi'_{res} = 10°$ for illite, and $\varphi'_{res} = 5°$ for montmorillonite.

The mineralogy analysis of the clay showed that both illite and montmorillonite are present in the BLR [235, 246]. The origin of these weak layers is still unknown but they could be (1) due to the migration of pore water into silty layers causing local liquefaction, (2) due to the presence of bentonite of volcanic or lacustrine origin, or (3) due to the inclusion of an external lubricant such as graphite.

Weak bentonite can be formed by the devitrification and chemical alteration (weathering) of glassy volcanic ash or tuff, and this weak bentonite is typically composed of montmorillonite [251]. These layers are common throughout Washington and British Columbia [66, 277] and they are laterally continuous [88]. They are often pre-sheared to residual state by valley rebound deformation and give progressive failures [145]. In the case of Oso, the volcanic ash could come from nearby volcanoes, such as Glacier Peak located a few miles up the valley (Fig. 13.4). Volcanic soils have been found in the Stillaguamish valley, both upstream and downstream of Oso [29].

The Bear Lake sand (BLS) is a horizontally-bedded poorly-graded sand. The thickness of this unit is unknown but the upper contact of this unit is located at an elevation of around 97 m. Experts (e.g. [235, 280]) showed that the BLS was not involved in the landslide and acts as a stiff base.

Only one dry density test was carried on the BLS and it had a unit weight of $\gamma_{dry} = 18.3$ kN/m^3 [243].

The landslide debris (LD) is not a soil unit in the geological sense as it is composed of the other soil units. However, it differs from the other units in a geotechnical sense as its state has been altered by the landslide. The inclusion of the LD as a soil unit facilitates the interpretation of the site as shown in Fig. 13.4. The 2006 Hazel Landslide debris (Fig. 13.2 a), is of interest as some experts (i.e. [235, 280]) believe that this landslide debris deposit was remobilised in 2014, unbuttressing Whitman Bench and allowing it fail. This landslide deposit was made of disturbed BLR and its mechanical strength would relate to the residual state of BLR.

13.3 Scenario and objectives

The field data revealed the existence of a weak layer at an elevation of 103 m, which coincides with the elevation of the surface of rupture. The aim of this numerical analysis is to understand the role that this weak layer played in the failure mechanism and the post-failure behaviour. Other failure mechanisms remain possible. The simulations are carried out with *Anura3D* [14]. MPM is an ideal numerical method as it permits modelling both the small and large deformation behaviour. Unlike other simulations of the 2014 Oso Landslides [2, 146, 235, 337], the failure surface is free to develop anywhere in the bench and the analysis focusses on the early stages rather than on the long run-out distance. The run-out behaviour of the landslide is discussed by Yerro et al. [235, 337].

Four scenarios are generated. (1) The first scenario assumes the remobilisation of the 2006 Hazel landslide and the presence of a weak layer at its base. (2) The second simulation assumes the remobilisation of the 2006 Hazel landslide without the weak layer. (3) The third simulation assumes the presence of the weak layer without the mobilisation of the 2006 Hazel landslide. (4) The fourth simulation assumes no remobilisation of the 2006 Hazel landslide and the absence of the weak layer.

13.4 Model idealisation

The Oso Landslide should ideally be a three-dimensional (3D) model due to its complex topography and run-out trajectory. However, this is computationally challenging and 3D models are more difficult to generate and interpret. For this reason, many researchers favoured 2D models (e.g. [235, 337]). For the present example, the 2013 LiDAR data is discretised with a 10 m × 10 m grid and a cross section is extracted from the middle section of the bench in order to obtain the exact topography.

The model idealises the stratigraphy as horizontally-bedded stratigraphy and neglects the local variation of the soil (Fig. 13.5). The contacts between the RC and the TC are at an elevation of 232 m, between the TC and the AO at 212 m, between the AO and the BLR at 180 m, and between the BLR and the BLS at 98 m. The LD of the 2006 Hazel landslide is modelled as a single soil unit and its position is deducted from Fig. 13.2 and is consistent with Pyles et al. [235] and Stark et al. [280]; minor differences do exist due to interpretation. A weak layer of BLR is placed at the base of the BLR close to the LD. The positions, lengths and thickness are model assumptions. The BLS constitutes an elastic base of the model and stretches over the valley floor for modelling convenience.

The size of the model and its mesh are defined by carrying out a series of preliminary simulations seeking computational efficiency without compromising the development of the failure surface. The computational power is dedicated to the onset of the failure and, hence, the mesh is refined within the bench and coarsened in the valley. The model is intentionally not rectangular and the amount of empty space is reduced in order to reduce the required memory for the simulation. All soil units have 4 MPs/cell with the exception of the BLS, which has 1 MP/cell. The model has a total of 87,309 MPs and 39,630 cells. The boundary conditions are "on box", which prevents horizontal displacements on the sides of the model and vertical displacements at the base. No other boundary conditions are used.

13.4.1 Type of analysis

It is well known that effective stress analyses cannot model correctly the undrained strength of soil with first generation models (i.e. Mohr-Coulomb, Tresca); the undrained shear strength of normally and lightly overconsolidated

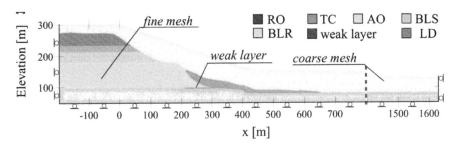

FIGURE 13.5: 2D plane strain model with stratigraphy, mesh and boundary conditions.

soil can be overestimated and the one of overconsolidated soil underestimated. The groundwater regime and pore water pressure profiles of Whitman Bench are largely unknown and this limits the ability to carry out an effective stress analysis. The simulations are carried out as total stress analyses for which the stress paths to failure are known and can be modelled. A slope that is at the point of collapse should appear critical whether the stability analysis is carried out in terms of total stresses or effective stresses [207]. However, these simulations assume the absence of seepage in the bench and this is a crude assumption, which qualifies these simulations as preliminary.

13.4.2 Material parameters

The coarse-grained soils are modelled as drained material with a Mohr-Coulomb strain softening model and the fine-grained soils as undrained with a Tresca strain softening model. The base layer is modelled as linear elastic as it was not involved in the failure of Whitman Bench. The material parameters for each soil unit are as follows.

RO: The model parameters are derived from CPT data [16] and give a peak friction angle of $\varphi_{max} = 40°$ and a Young modulus of $E = 150$ MPa at a depth of 10 m. The residual (or critical state) friction angle is assumed to be $\varphi_{cs} = 33°$ and the Poisson ratio to be $\nu = 0.3$, which ara typical values for silica sand. A small peak cohesion of $c = 100$ kPa is assumed in order to take into account the cementation of the natural soil. It is assumed to be destroyed when sheared giving $c_{res} = 0$. The shape factor is set to $\zeta = 10$ in order to model a smooth and progressive evolution towards the residual conditions.

TC: Few tests were carried out on this soil unit and its mechanical characteristics are largely unknown. However, Milligan [200] reviewed the geotechnical characteristics of till in North America and suggested a typical undrained strength of $c_u \approx 1$ MPa with $E_u \approx 500$ MPa. These values are consistent with the Vashon till [258] and similar to values given by the ICE manual of geotechnical engineering [47]. However, the TC is a soil mass with discontinuities and for which the geotechnical characteristics differ from those obtained for intact soil in the laboratory in the same way that rock mass differs from intact rock [47]. For this reason, the stiffness is reduced to $E = 300$ MPa. This reduction in stiffness does not influence the onset of the failure. The TC is assumed to be in undrained conditions and, thus, the Poisson ratio is $\nu = 0.49$.

AO: Due to the absence of data, the AO is assumed to have the same mechanical characteristics as the RO and to be in drained conditions. However, the AO is located at a greater depth and the stiffness is assumed to be $E = 450$ MPa.

BLR: The BLR has a low hydraulic conductivity and is assumed to be in undrained conditions. Despite extensive testing in drained conditions, The undrained strength of the BLR is largely unknown. However, it is estimated to be between 750 kPa and 1,500 kPa. Three series of simulations were carried out with $c_u = 750$ kPa, 1,000 kPa and 1,250 kPa. A residual strength of $c_{res} = 100$ kPa is assumed with a shape factor $\zeta = 10$, which models a slow and progressive transition from the peak strength to the residual strength. The shear strength is 219 kPa at $\varepsilon_d = 20\%$ and 103 kPa at $\varepsilon_d = 50\%$ for $c_u = 1,000$ kPa. The stiffness is assumed to be $E = 450$ MPa with $\nu = 0.49$.

BLR WL: The weak layer (WL) is assumed to have the same characteristics as the BLR but with smaller strengths – $c_u = 100$ kPa. The other parameters are identical to the BLR.

BLS: No surface of rupture was found in the BLS and this layer acts as a base. It can hence be modelled as an elastic medium. The Young modulus is estimated around $E = 650$ MPa and The Poisson ratio $\nu = 0.3$.

LD: Little information on the mechanical characteristics of the landslide debris (LD) are available and the model parameters are derived theoretically. Vardanega and Haigh [310] reviewed the use of the liquidity index I_L (Eq. 7.11) to predict the undrained strength of disturbed clay (Eq. 13.1).

$$c_u = 1.7 \times 35^{(1-I_L)} \quad \text{for} \quad 0.2 < I_L < 1.1 \tag{13.1}$$

It is assumed that the soil is in a looser state than the BLR in Whitman Bench and that its water content would be $w = 31.6\%$. Assuming that the mean Atterberg limit values of the BLR hold true for the disturbed landslide mass (Table 13.1), the liquidity index is $I_L = (0.316-0.25)/(0.466-0.25) = 0.31$, which gives $c_u = 20$ kPa. Bjerurm [36] found that the undrained strength is often over-estimated by 10%. This gives $c_u = 18$ kPa. The Young modulus is estimated at $E_u = 43$ MPa. The Poisson ratio is set at $\nu = 0.49$.

13.4.3   Loading sequence

All simulations presented in this chapter follow the same loading scheme. (1) The stresses are initialised by gravity loading assuming an elastic behaviour. The K_0-procedure is computationally faster but it cannot be used in this case due to the non-horizontal ground surface (see Chapter 6). A damping factor of $\alpha = 0.20$ is applied in order to remove some of the dynamic effects related to the loading. This avoids generating large strains during the stress initialisation reducing the shear strength of the soils. A convergence criterion is applied in order to ensure that the stress state is the final state. (2) The material parameters are reset and yielding is permitted. The failure surface is thus free to develop in the bench and defines the mobilised mass, which is free to accelerate and slide. No damping or strain smoothing are applied during this stage of the analysis.

13.5 Back-rotated blocks and progressive failure

A distinctive morphological feature of the 2014 Oso Landslide is the appearance of a concave head scarp and a concentration of strain forming a thin surface of rupture. This allowed the upper part of the bench to rotate backwards and slide forming the graben (Fig. 13.1 b). It is necessary to check if *Anura3D* [14] can reproduce such a failure. A simple model of a small embankment is created in order to verify these numerical features. It is composed of a 2-m high embankment with a slope of 2:3 and an elastic base. A contact condition (see Chapter 6) is imposed between the embankment and the base in order to facilitate the slippage of the embankment ($\mu = 0.05$). The embankment is modelled as a dry soil with a Mohr-Coulomb strain softening model (see Chapter 8) with $c' = 5 \rightarrow 0$ kPa, $\varphi' = 19° \rightarrow 5°$, and $E = 50 \rightarrow 25$ MPa. The shape factor is set at $\zeta = 1,000$ in order to model a brittle material. The Poisson ratio is kept constant at $\nu = 0.3$. A no-tension and a stiffness-reduction criteria are implemented in the code and verified with these simulations. The model has a fine mesh composed of 58,800 elements and 80,496 MPs (4 MPs/active cell). The loading sequence is the same as for the Oso model.

Fig. 13.6 shows the results of the simulation. The failure starts at the toe of the embankment where the deviatoric stresses are the highest and the strain localises along the base before rotating and reaching the top of the embankment. A first surface of rupture is formed and a mass of soil slides down as a rigid block. The low friction between the embankment and the base layer allows the strains to propagate further back in the embankment and a second block slides, followed by a third, fourth and fifth block.

13.6 Preliminary simulations

The first phase of the simulations is the stress initialisation, which is achieved by increasing the gravity and applying a convergence criterion in order to ensure that the stresses are in equilibrium. The soil layers are modelled as elastic by setting their shear strength to a very high value and will be reduced in the second phase of the simulation. Fig. 13.7 shows the initial deviatoric stresses after initialisation. The maximum stresses are reached at the bottom of the BLR and within Whitman Bench. The 2006 Hazel Landslide debris has

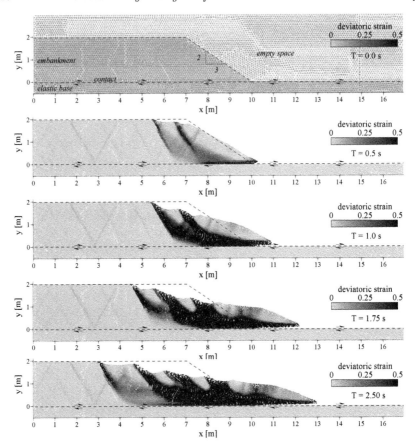

FIGURE 13.6: Model validation for strain localisation and back-rotated failure in an embankment.

a lower stress as it is modelled with high initial void ratios. Fig. 13.8 shows the strain state after the stress initialisation and shows that no plastic strains are accumulated during the gravity loading. This is an important condition as the shear strength is reduced with plastic strains. The accumulation of plastic strains is controlled by the damping coefficient, which is set at $\alpha = 0.20$ for this case.

The second phase of the simulations allows the soil to yield. The first simulation has a weak layer at the toe of the BLR layer and permits the mobilisation of the landslide debris. Fig. 13.8 shows the development of the deviatoric strains. The failure of the 2006 Hazel landslide debris (black mass at the toe of the slope) causes an unloading of the BLR in Whitman Bench and induces some plastic deformation. This causes in turn a reduction of the strength of Whitman Bench and facilitates its failure. The strains are concentrated in a thin layer at the bottom of the BLR soil unit and where the

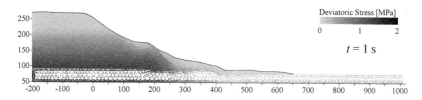

FIGURE 13.7: Initial stress state after gravity loading: deviatoric stress.

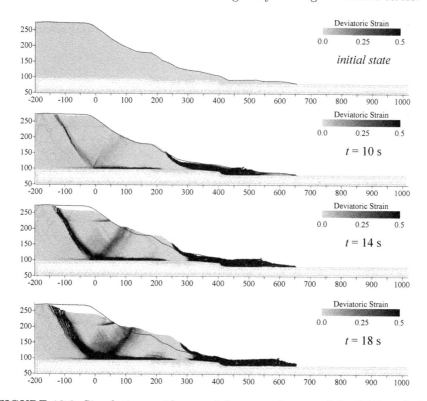

FIGURE 13.8: Simulations with a weak layer at the toe of the BLR and the failure of the LD.

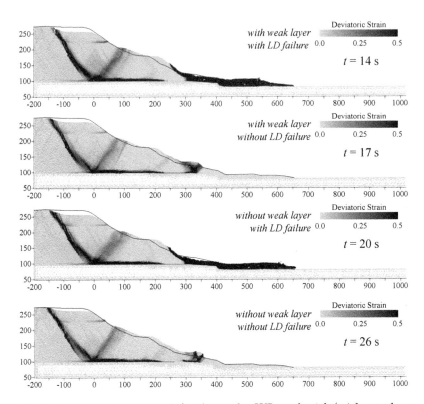

FIGURE 13.9: Simulations with/without the WL and with/without the mobilisation of the LD.

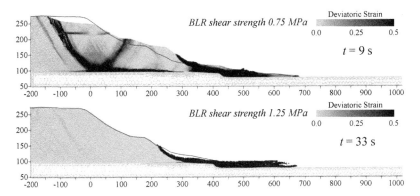

FIGURE 13.10: Simulations for different values of maximum shear strength of the BLR.

stresses are the largest. It is only after 10 s of simulated time, during which the LD is flowing, that the surface of rupture propagates throughout the bench up to the top surface ($t = 10$ s). The surface of rupture outcrops the bench at position $x = -100$ m, which corresponds to the field case (Fig. 13.4). A mobilised mass is thus defined and starts sliding down ($t = 14$ s) pushing the LD further out. The mobilised mass is then broken in two distinct parts due to the shape of the surface of rupture ($t = 18$ s). The simulations are stopped at this stage as the interest of the numerical analysis is the development of the surface of rupture and the early post-failure behaviour, and because of the computational cost of running these MPM simulations.

Fig. 13.9 shows the surface of rupture for cases with and without the weak layer at the toe of the BLR, and with and without the failure of the 2006 Hazel Landslide debris (LD). The results show that Whitman Bench fails in all cases suggesting that it is the cause of the landslide. However, the time at which the failure occurs differs from one case to another. This suggests that the weak layer and the failure of the LD are aggravating factors. The absence of failure of the LD postpones the failure by 3 s of simulated time ($t = 17$ s) and by 11 s in absence of the weak layer ($t = 26$ s). The absence of the weak layer with the failure of LD postpones the failure by 6 s ($t = 20$ s). This is because additional plastic strains are required to reduce the strength of the BLR and this is achieved by further unloading the toe of the landslide. A parametric analysis on the shear strength of the BLR is undertaken. Fig. 13.10 shows the surface of rupture for the cases with $c_u = 1.25$ and 0.75 MPa. The simulation with $c_u = 1.25$ kPa shows that insufficient plastic strain propagates into the bench for it to yield and fail. However, the present models only assume a small layer of weak material and it is not impossible that this layer is large in reality. The simulation with $c_u = 0.75$ kPa shows an acceleration of the development of the surface of rupture, which reassembles the default case for $c_u = 1,000$ kPa but it is rapidly followed by a second surface of rupture.

13.7 Closure

The 2014 Oso Landslide (WA, USA) is an exceptional landslide by its size and run-out distance. Although extensive investigation work has already been carried out, the causes of the landslide remain unknown. The geotechnical site characterisation revealed the existence of weak layers of soil, including one at the base of the Bear Lake Rhythmite (BLR). The numerical investigation presented in this chapter aims to understand if this weak layer is a cause of the landslide or an aggravating factor. Experts believe that the unbutressing of Whitman Bench by the remobilisation of 2006 Hazel Landslide debris is the cause. This was also investigated by the MPM simulations. The numerical results suggest that the failure of Whitman Bench depends on the shear strength of the BLR and that both the weak layer and the remobilisation of the 2006 Hazel Landslide debris are aggravating factors. Although the precipitation could not have seeped through the entire bench to the base of the BLR [235], it is possible that the water infiltrated the bench through the Bear Lake sand (BLS) only providing water to the base of Whitman Bench. This would be sufficient to decrease the shear strength and trigger strain localisation and the failure of the slope.

14

Thermal Interaction in Shear Bands: the Vajont Landslide

M. Alvarado, N.M. Pinyol and E.E. Alonso

14.1 Introduction

The localisation of strains in landslide shearing surfaces can induce relevant weakening processes due to the thermal pressurisation of the pore water. The physical phenomenon, which is also invoked in the field of earthquakes in the study of fault slips, consists in the dissipation of frictional-work heat. The heating produces dilation of soil skeleton and water, which can result in an increase in pore water pressure. This reduces the effective stress and leads to an acceleration of the landslide due to the subsequent loss of strength. This can explain the high velocity and sudden acceleration observed in some cases. This acceleration and the subsequent increase in velocity will induce additional increments of temperature and, therefore, the thermo-hydro-mechanical (THM) problem outline may become a self-feeding process that results in a catastrophic accelerated landslide.

The Vajont landslide (Italy, 1963) [111] generated a great impact in the scientific and engineering communities because of the large velocity reached by an ancient slide reactivated by the construction of a dam and its reservoir that submerged the landslide toe. In October 1963, the reservoir level was close to its maximum elevation and an estimated total volume of 280 million m^3 of rock invaded the reservoir with an estimated speed of 20-30 m/s. The mobilised mass displaced the water generating a 220-m high wave, which over-topped the dam. Whilst the dam withstood the wave, several downstream villages were destroyed causing 2,000 casualties. The sliding mass reached the opposite side of the valley 400 m away.

Since 1963, the Vajont landslide became the topic of a large number of publications providing detailed geological and geotechnical descriptions of the site and the materials involved (i.e. [91, 126, 222, 266]). According to several analyses of the Vajont landslide [306, 313, 320], the high velocity of the slide requires a mechanism leading to the total loss of basal shear strength. The most-favoured explanation for the total loss of frictional strength is associated with the development of frictional heat at the sliding surface; a detailed description is given by Alonso et al. [10].

Thermal failure can be demonstrated with a simple heating laboratory test. A piece of saturated Opalinus-clay rock is heated in a microwave. During the test, the specimen cracks and breaks into smaller fragments accompanied by an audible cracking noise, and water comes out of the specimen. Fig. 14.1 shows the specimen (a) before and (b) after the experiment. The specimen has very low permeability coefficients $(0.8 \cdot 10^{-13}$ m/s $< k < 7.3 \cdot 10^{-13}$ m/s) and a uniaxial compressive strength of $1,000$ MPa $< UCS < 7,000$ MPa [38, 299].

An explanation for the failure of the specimen is as follows. When the temperature of a saturated porous material increases, the solid skeleton and its pore water dilate. Local equilibrium of temperature is achieved soon and, therefore, the temperatures of water and solid skeleton will be essentially equal. The thermal dilation of water and solid will result in an internal volumetric expansion, which can induce an increment of pore water pressure and, hence, a reduction of the effective stress. The increase in pore pressure is proportional to the stiffness of the geomaterial. The thermally-induced increase of pore water pressure induces a gradient, which generates an outflow. This ex-

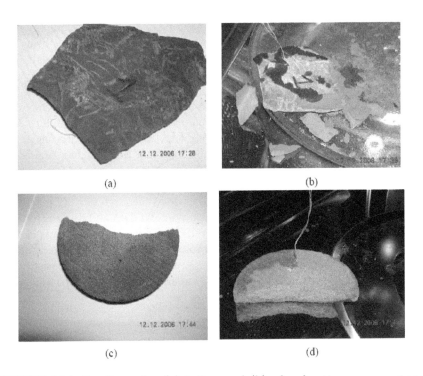

FIGURE 14.1: Opalinus clay (a) before and (b) after heating, porous stone, and (c) before and (d) after heating (after [7]).

plains the presence of water at the surface of the specimen after the test. Due to the reduction of the effective strength in the absence of external stresses, the tensile strength of the rock is reached and the specimen fails (Fig. 14.1 b).

The same experiment was carried out on a saturated porous stone used in the laboratory tests (Fig. 14.1 c). In contrast with the Opalinus-clay rock, the porous stone remained intact, although some water was also seen to escape from the stone and water drops were observed on the microwave dish. The high value of permeability of the porous stone led to the dissipation of the thermally-induced pore water pressure, the effective stress did not decrease, and the specimen remained intact. Therefore, the attained pore water pressure for a given rate of temperature increase is the result of two competing mechanisms – (1) the rate of increase of water volume, directly related to the rate of increase of temperature, and (2) the rate of dissipation governed by the permeability of the porous material, and also by the rock stiffness in a process similar to the more familiar consolidation phenomenon.

In case of landslides, the heat is caused by the landslide motion, which depends on the difference in velocity across sliding surfaces. The result is a non-linear problem where the landslide motion, the changes of temperature and water pressure are combined to determine the post-failure behaviour of the unstable slopes.

Many researchers (i.e. [225, 312, 319]) solved this problem by formulating the mass, energy and momentum balances at and around the shear band of the basal surface, and solving the dynamic equation of the motion of the whole mass. All of these works share two simplifications: (1) the slope kinematics are solved in simple geometries defined by interacting solid blocks – the motion is often simply defined by a rigid body sliding on a planar surface, and (2) the dissipation of the mechanical work is concentrated in shear bands defining the contacts between rigid bodies.

MPM is a powerful tool to simulate thermally-driven landslides and overcomes the restriction of previous works with the purpose of generalising the basic concepts. The ultimate aim of the analysis is to explore the implications of thermally-induced pore pressures on landslides, irrespective of their size, kinematic deformation mechanisms or assumptions concerning the dominant mode of energy dissipation.

The work presented in this chapter is developed in more detail in [226] and most of the figures included here are taken from the mentioned paper.

14.2   Thermo-hydro-mechanical coupled analysis of saturated landslides

The equations are expressed in terms of the following unknown variables: acceleration of solid particles $\vec{a}_s$, liquid pressure p_l and temperature Θ, which is denoted with a capital Θ in order to avoid confusion with the Lode angle θ. Relative acceleration of the liquid with respect to the solid skeleton is assumed negligible (Eq. 14.1), and further simplified.

$$\vec{a}_{l/s} = \vec{a}_s - \vec{a}_l = 0 \rightarrow \vec{a}_s = \vec{a}_l \tag{14.1}$$

The formulation is then simplified to a $\vec{u} - p_l$ formulation commonly used in FEM [175, 348, 349]; $\vec{u}$ is the displacement and p_l the pore liquid pressure.

The governing equations, which provide a theoretical explanation to the coupled phenomena and the acceleration of landslides, are described below. A summary of the equations can be found in Appendix A.1, and include some modifications of the equations according to Section 14.2.3 in an attempt to overcome some limitations of the basic formulation as discussed in Section 14.2.2.

- The momentum balance of the mixture formed of solid particles and water (Cauchy's Equations) controls the dynamic motion of the mass of a continuum medium formed by solid particles and water (Eq. 14.2).

$$\rho \vec{a}_s = \nabla \cdot \vec{\sigma} + \rho \vec{b} \tag{14.2}$$

with $\rho = n\rho_l + (1 - n)\rho_s$

where the acceleration of the solid skeleton $\vec{a}_s$ results from the spatial variation of Cauchy's stress field σ and the mass force $\rho \vec{b}$. ρ is the density of the saturated soil, n is the porosity and ρ_l and ρ_s are the fluid and solid density, respectively.

- The conservation of the fluid momentum is an equivalent expression of the generalised equation for Darcy flow rate $\vec{q_l}$, which is given in Eq. 14.3.

$$\vec{q_l} = -\frac{k}{\gamma_l}\left(\nabla p_l - \rho_l \vec{b} + \rho_l \vec{a}_l\right) \tag{14.3}$$

Eq. 14.3 establishes that the water flows through the porous media due to the gradient of the liquid pressure ∇p_l to the massive forces $\rho_l \vec{b}$, and to the acceleration-induced drag forces $\rho_l \vec{a}_l$. The flow is proportional to the hydraulic conductivity k. γ_l is the specific weight of liquid.

- The mass balance equation includes the mass of solid particles and liquid (Eq. 14.4).

$$n\alpha_l \frac{dp_l}{dt} - \beta \frac{d\Theta}{dt} + \nabla \cdot \vec{v}_s + \nabla \cdot \vec{q_l} = 0 \tag{14.4}$$

The compressibility of the solid particles with respect to changes in stress is assumed negligible in Eq. 14.4. Changes in volume can be caused by different terms as follows.

- Changes of the water density due to liquid pressure variation, which are proportional to the compressibility coefficient of the liquid phase α_1.

- Changes of the density of solid and liquid phases induced by temperature changes, which are proportional to the volumetric thermal expansion coefficient for the mixture (Eq. 14.5).

$$\beta = (1 - n)\,\beta_s + n\beta_l \qquad (14.5)$$

where β_s and β_l are the volumetric thermal expansion coefficients for the solid and liquid phases, respectively.

- Volumetric changes of the solid skeleton expressed as the spatial variation of the solid phase velocity $\nabla \cdot \vec{v}_s$, which are equivalent to the time variation of the volumetric strains.

- Water in- or out-flow expressed in terms of Darcy (Eq. 14.3).

- The energy balance equation of the mixture is given in Eq. 14.6 and assumes that the specific heat of the phases remains constant.

$$(\rho c)_m \frac{d\Theta}{dt} + \nabla \cdot \vec{q}_h + \rho_l c_l \Theta \nabla \cdot \vec{q}_l + (\rho c)_m \Theta \nabla \cdot \vec{v}_s = \dot{H} \qquad (14.6)$$

where the specific heat of the mixture $(\rho c)_m$ is defined as given in Eq. 14.7 and the heat flow conduction $\vec{q}_h$ as given in Eq. 14.8.

$$(\rho c)_m = n\rho_l c_l + (1 - n)\,\rho_s c_s \qquad (14.7)$$

$$\vec{q}_h = -\Gamma_\Theta \nabla \Theta \qquad (14.8)$$

Eq. 14.6 states that the external supply of heat rate $\dot{H}$ is equal to the sum of the following terms.

- The internal energy in the solid and liquid phases depends on their specific heats c_s and c_l, respectively.

- The heat flow conduction, driven by temperature gradients (Fourier's law), depends on the thermal conductivity coefficient Γ_Θ.

- The convective heat transport is due to liquid and solid flows.

- The first law of thermodynamics supports the assumption that the plastic work generated in the continuous media is dissipated as heat (Eq. 14.9).

$$\dot{H} = \vec{\sigma}' \cdot \vec{\dot{\varepsilon}}^p \qquad (14.9)$$

$$\text{with } \vec{\sigma}' = \vec{\sigma} - p_l \vec{m} \qquad (14.10)$$

where $\vec{\sigma}'$ is the effective stress tensor and $\vec{\dot{\varepsilon}}^p$ is the plastic strain rate.

- Constitutive models correlate the increments of the effective stress and strain, as described in Chapter 8.

$$d\vec{\sigma}' = \boldsymbol{D} \cdot d\vec{\varepsilon} \tag{14.11}$$

where $\boldsymbol{D}$ is the stiffness matrix.

In this chapter, the soil behaviour is modelled with an elastic-plastic Mohr-Coulomb strain softening model (see Chapter 8).

The thermally-induced acceleration of a landslide can be explained by following the equations summarised above as follows. Consider, a planar landslide for simplicity reasons. Once unstable, the mobilised mass accelerates (Eq. 14.2). In the basal shear band, where the strains rate is localised and assuming that the rest of the unstable mass moves as a rigid solid, heat is generated (Eq. 14.9). This induces changes in temperature (Eq. 14.6). According to Eq. 14.4 and depending on the material characteristics (compressibility, dilation coefficients and permeability), pore water pressure increases. This reduces the effective stress. The Mohr-Coulomb model gives a reduction of strength because of the reduction in effective stress and, hence, the landslide accelerates.

These governing equations are implemented in MPM with an explicit Euler time integration. The solid acceleration and the increments of temperature, liquid pressure and stress at time $^{k+1}t =^k t + \Delta t$ are calculated as a function of the variables evaluated in the previous time step. The computational algorithm of MPM discretises equations as given in Appendix A.2.

14.2.1   Pathological mesh dependency

The thermal phenomenon controlling the acceleration of landslides is highly dependent on the shear band thickness. This is because the dimension of the shear band affects the accumulation of plastic work per unit of volume of soil inside of the band as well as the pore pressure and heat dissipation toward the surrounding soil. This effect can be understood if one examines the following simple example. Fig. 14.2 shows two reference volumes ("samples") subjected to shear deformation defined by a displacement rate $\dot{\delta}$. It is assumed that the shear band of each sample has a different thickness. For simplicity, a planar deformation is assumed. Accepting a linear distribution of the shear strains inside the sheared zone and assuming that the soils are at failure, the rate of the generated mechanical work per unit volume for each sample, can be calculated by Eq. 14.12 and Eq. 14.13.

$$\dot{W}_1 = \tau_{\mathrm{f}} \dot{\gamma}_1 = \tau_{\mathrm{f}} \frac{\dot{\delta}}{h_1} \tag{14.12}$$

$$\dot{W}_2 = \tau_{\mathrm{f}} \dot{\gamma}_2 = \tau_{\mathrm{f}} \frac{\dot{\delta}}{h_2} \tag{14.13}$$

where τ_f is the shear strength, $\dot{\gamma}$ is the shear strain rate and 1 and 2 refer to the reference volumes represented in Fig. 14.2. It follows that the dissipated work per unit of volume depends on the thickness of the shear band. However, the total work on the band volume is equal for both cases as shown in Eq. 14.14.

$$\dot{W}_1 h_1 l = \dot{W}_2 h_2 l = \tau_f \dot{\delta} l \tag{14.14}$$

Since temperature increments are computed from the heat generated per unit of volume (Eq. 14.9), higher increments of temperature and pore water pressure are expected in the case of the thinner shear band (Fig. 14.2 c).

The thickness of a shear band is related to grain size distribution [11,247, 311], and can be estimated as $e \approx 200D_{50}$ for clays [313]. The size of the shear band in fine soils ranges between a few millimetres to centimetres. Based on a FDM analysis, Alonso et al. [10] demonstrated that there is no effect on the results in terms of landslide motion when the shear band thickness varies in the order of magnitude of a few millimetres. In this work, the thickness of the shear band is introduced as an input parameter, irrespective of the spatial discretisation. This suggests that it is not necessary to specify the shear band thickness, which is always difficult to quantify. However the effect is relevant in case of increasing the shear band thickness to tens of centimetres.

In numerical analysis (e.g. FEM, FDM, MPM), the size of the shear band depends on the element size and on the inclination of the band with respect to the mesh lines. In practice, strains tend to localise in a single or a few elements. Therefore, in view of the previous discussion, the size of the elements of the computational mesh in a MPM should be similar to the thickness of the actual shear band (a few millimetres) in order to obtain realistic results. Otherwise, the computed heat generated in the slip zone per unit volume will be lower than the heat actually generated. However, in real landslide modelling, the size of landslides requires element sizes in the order of tens of centimetres to metres for computational cost reasons. This is increasingly important for coupled THM analysis.

A direct application of Eq. 14.9 results in work rates per unit volume significantly smaller that the rates dissipated in real sliding surfaces. The

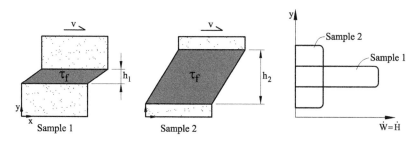

FIGURE 14.2: Schematic representation of two shear bands exhibiting different shear band thickness (after [226]).

TABLE 14.1

Model parameters.

Parameter	Symbol	Unit	Value
Water:			
Density	ρ_l	$\mathrm{kg/m^3}$	1,000
Bulk modulus	α_l	MPa	2,200
Thermal dilatation coef.	β_l	$1/°C$	0.00034
Specific heat	c_l	$\mathrm{N \cdot m/(kg \cdot ° C)}$	4,186
		$\mathrm{cal/(kg \cdot ° C)}$	1
Solid:			
Density	ρ_s	$\mathrm{kg/m^3}$	2,700
Thermal dilatation coef.	β_s	$1/° C$	0.00003
Specific heat	c_s	$\mathrm{N \cdot m/(kg \cdot ° C)}$	837
		0.2	$\mathrm{cal/(kg \cdot ° C)}$
Clay:			
Porosity	n	-	0.4
Permeability	k	$\mathrm{m/s}$	10^{-11}
Young modulus	E	kPa	20,000

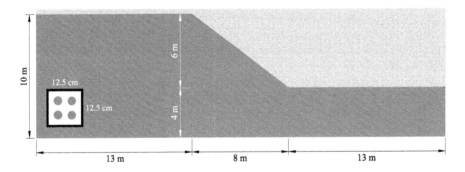

FIGURE 14.3: Slope geometry and discretisation of the reference case (after [226]).

consequence is an underestimation of excess pore pressures generated in shear bands.

This pathological effect is checked for the analysis of a homogeneous saturated slope. For this case, the soil is modelled with a Mohr-Coulomb strain softening model with $\varphi' = 28°$, $c' = 2$ kPa and $\psi = 0°$. The other parameters are given in Table 14.1. A support square mesh (12.5×12.5 cm) defines the computational domain and is shown in Fig. 14.3. Four MPs per element are initially located in each active element. The slope failure is triggered by reducing the cohesion from 2 kPa to 1 kPa.

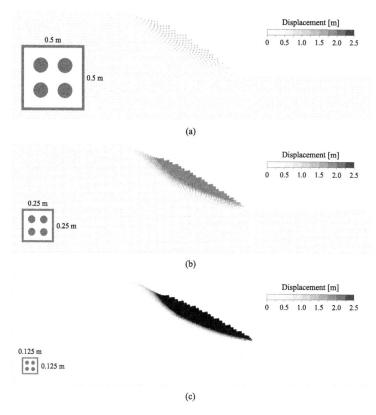

FIGURE 14.4: Maximum displacement calculated for the reference case for a permeability value of $k = 10^{-11}$ m/s and for different mesh sizes (after [226]).

The case is run again using two additional coarser meshes – 0.5×0.5 m and 0.25×0.25 m, respectively. The calculated displacement and excess pore water pressure at the end of the motion are shown in Figs. 14.4 and 14.5, respectively. The results show a certain mesh dependency. A numerical alternative to overcome such dependency is described in the next section.

14.2.2 Embedded shear bands

A numerical procedure proposed by [226] and [12] was developed in order to consider the real thickness of a shear band when performing numerical modelling of THM phenomena in landslides. Note that the proposed procedure does not provide a methodology to solve the mesh dependency for strain localisation problems. The proposed procedure provides a solution to account for the magnitude of the heat and excess pore water pressure generated because the generated frictional work and its dissipation depend on the shear band thickness.

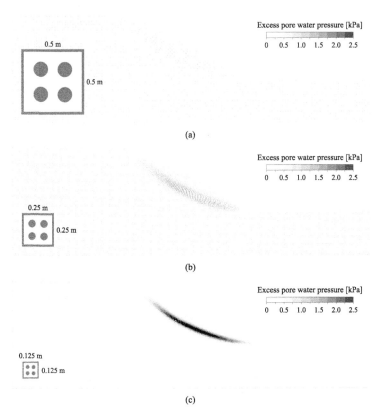

(a)

(b)

(c)

FIGURE 14.5: Excess pore water pressure at the end of the motion calculated for the reference case for a permeability value of $k = 10^{-11}$ m/s and for different mesh sizes (after [226]).

The idea is to include a set of embedded shear bands into the material domain whenever the plastic deviatoric strain exceeds a reference value (Fig. 14.6). A reference volume corresponding to the volume initially assigned to each MP is characterised by a reference length L_{ref}. The strain computed at each MP is assumed to be localised in an embedded shear band for which the thickness is given as an input parameter L_B and depends on the soil type. Heat due to frictional work dissipation and the induced liquid pressure is assumed to be generated in the embedded elements. Different variables for temperature and liquid pressure in the embedded shear bands and the rest of the soil are defined as Θ^B, Θ^M, p_l^B, p_l^M, respectively. Dissipation processes of heat and liquid flow between the embedded shear bands and the rest of the material domain, called matrix, are formulated at the local level. Energy and mass balance equations are now formulated at matrix and embedded shear

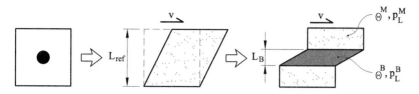

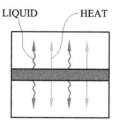

FIGURE 14.6: Schematic representation of embedded shear bands (after [226]).

band level as follows.

$$\frac{\mathrm{d}}{\mathrm{d}t}\left[(\rho c)_{\mathrm{m}}\Theta^{M}\right] + \nabla \cdot \left[-\Gamma_{\Theta}\nabla\Theta^{M}\right] + \nabla \cdot \left[\rho_{\mathrm{l}}c_{\mathrm{l}}\vec{q_{\mathrm{l}}}\Theta^{M} + (\rho c)_{\mathrm{m}}\vec{v_{\mathrm{s}}}\Theta^{M}\right] = f_{\Theta}^{B-M} \tag{14.15}$$

$$\frac{\mathrm{d}}{\mathrm{d}t}\left[(\rho c)_{\mathrm{m}}\Theta^{B}\right] = -f_{\Theta}^{B-M} + \dot{H}^{B} \tag{14.16}$$

$$\frac{\mathrm{d}}{\mathrm{d}t}\left[\beta\Theta^{M}\right] + n\alpha_{\mathrm{l}}\frac{\mathrm{d}p_{\mathrm{l}}^{M}}{\mathrm{d}t} + \nabla \cdot \vec{v_{\mathrm{s}}} + \nabla \cdot \vec{q_{\mathrm{l}}} = \frac{1}{\rho_{\mathrm{l}}}f_{\mathrm{l}}^{B-M} \tag{14.17}$$

$$\frac{\mathrm{d}}{\mathrm{d}t}\left[\beta\Theta^{B}\right] + n\alpha_{\mathrm{l}}\frac{\mathrm{d}p_{\mathrm{l}}^{B}}{\mathrm{d}t} = -\frac{1}{\rho_{\mathrm{l}}}f_{\mathrm{l}}^{B-M} \tag{14.18}$$

Local source terms of heat and liquid flow rate per unit of volume and time are included in the balance equations to take into account the dissipation of heat and liquid pressure between matrix and embedded elements. Flow rates are defined in Eqs. 14.19 and 14.20 and are proportional to the difference between band and matrix temperature and liquid pressure.

$$f_{\mathrm{l}}^{B-M} = \psi_{\mathrm{l}}\left(p_{\mathrm{l}}^{B} - p_{\mathrm{l}}^{M}\right) \tag{14.19}$$

$$f_{\Theta}^{B-M} = \psi_{\Theta}\left(\Theta^{B} - \Theta^{M}\right) \tag{14.20}$$

where ψ_{l} and ψ_{Θ} are defined as liquid and energy transfer coefficients, respectively.

Eq. 14.19, which is for the local liquid source term f_{l}^{B-M} (mass per unit volume per unit time), is approximated in terms of a difference in pressures to facilitate the calculation at the local level. From a physical perspective, it should be related to the gradient of pressures through some permeability coefficient. This consideration suggests that the transfer coefficient ψ_l can be expressed as Eq. 14.21.

$$\psi_l = \frac{\rho_l k_{\text{int}}}{\mu_l A_{\text{ref}}} \tag{14.21}$$

Eq. 14.21 is expressed in terms of liquid density ρ_l, liquid viscosity μ_l, intrinsic permeability k_{int}, and a reference area A_{ref}. This area relates directly to the expected distance of fluid transfer from the shear band, included in a MP, and the surrounding clay matrix associated with the MP under consideration. Therefore, A_{ref} is made equal to the element area contributing to the MP (quarter of the element area if 4 MPs/element are adopted). A similar argument, in the case of local heat interchanged leads to Eq. 14.22.

$$\psi_\Theta = \frac{\Gamma_\Theta}{A_{\text{ref}}} \tag{14.22}$$

where Γ_Θ is the Fourier's coefficient for heat transfer. Computations run for the reference case described below to show the effect of varying A_{ref} from $A_{\text{ref}}/4$ to $4A_{\text{ref}}$ resulted in minimum changes in results.

The convective and advective terms in Eqs. 14.15 and 14.18 are defined exclusively at the matrix scale to impose the interaction between embedded elements and matrix, and are defined locally at the scale of the reference volume.

The heat source term $\dot{H}^{\text{B}}$, included into the heat mass balance equation of the band, is supplied by energy generated into the embedded shear band. It is made equal to the frictional work rate dissipated assuming that the plastic strains localise in the shear band (Eq. 14.23).

$$\dot{H}^{\text{B}} = \vec{\sigma}' \cdot \vec{\dot{\varepsilon}}^{\text{p}} \frac{L_{\text{ref}}}{L_{\text{B}}} \tag{14.23}$$

The effective stress controlling the constitutive behaviour is governed by the maximum value of liquid pressure prevailing in the embedded elements or in the matrix (Eq. 14.24) as they occupy the same position of the MP under consideration.

$$\vec{\sigma}' = \vec{\sigma} - \max\left(p_l^{\text{B}}, p_l^{\text{M}}\right) \vec{m} \tag{14.24}$$

Appendices A.1 and A.2 include the embedded shear bands in the formulation described.

14.2.3   Sensitivity analysis. Effect of mesh size, shear band thickness and permeability

The reference case evaluated previously using different mesh sizes is analysed again, including the embedded shear band, with the aim of evaluating the capabilities of the methodology. In all the cases analysed, embedded shear bands are generated at elements in which the computed plastic deviatoric strain is different from zero. The distribution of excess pore water pressure

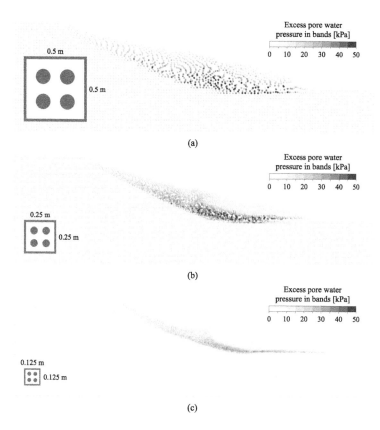

FIGURE 14.7: Excess pore water pressure at the end of the motion calculated for the reference case, including embedded shear bands, for a permeability value of $k = 10^{-11}$ m/s and for different mesh sizes: (a) 0.5×0.5 m, (b) 0.25×0.25 m, and (c) 0.125×0.125 m (after [226]).

and maximum displacement is plotted in Figs. 14.7 and 14.8. The effect of the mesh size is observed in minor details of the plots. However, the pathological dependence on the shear band thickness is removed. In fact, the calculated maximum pore pressure in the shearing band (Fig. 14.7) and the displacement (Fig. 14.8) are the same. In the three cases analysed, an embedded shear band thickness L_B of 1 cm was introduced as an input parameter.

The effect of the shear band thickness is shown in Fig. 14.9 in terms of maximum displacement. Band thickness varying between 10 cm and 0.5 cm are evaluated. Thinner shear bands lead to higher values of thermal induced pore water pressure and the associated reduction of the shear strength results in higher values of maximum velocity and displacement.

The effect of the permeability is evaluated with $L_B = 1$ cm, for all cases. Fig. 14.10 (b) shows the evolution of displacements of a point located initially

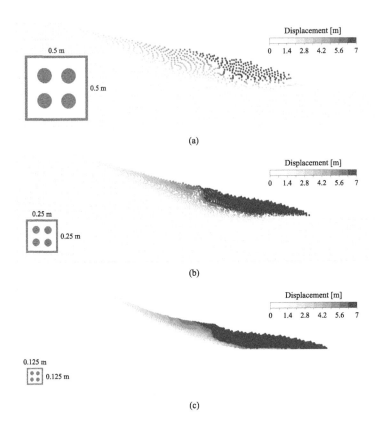

FIGURE 14.8: Maximum displacement calculated for the reference case, including embedded shear bands, for a permeability value of $k = 10^{-11}$ m/s and for different mesh sizes: (a) 0.5×0.5 m, (b) 0.25×0.25 m, and (c) 0.125×0.125 m, after [226].

at the surface and in the middle of the slope (Fig. 14.10a) for different values of permeability ranging from 10^{-3} m/s to 10^{-11} m/s. Fig. 14.10 (c) shows the calculated pore pressure records on a central point of the sliding surface. Zero values of pore water pressure are calculated in the most previous case ($k = 10^{-3}$ m/s). When permeability decreases to $k = 10^{-5}$ m/s, low values of excess pore pressure are accumulated. For $k = 10^{-7}$ m/s, pore pressure initially increases and reduces during the motion. As the values of permeability decrease further, the maximum velocity and the displacements increase due to additional excess pore water pressure accumulating in the shear band (Fig. 14.10 b).

The distributions of plastic strains, excess pore water pressures and temperature for the case with $k = 10^{-11}$ m/s are shown in Fig. 14.11. The shear strain increments localise in a band affecting one or two elements. The gen-

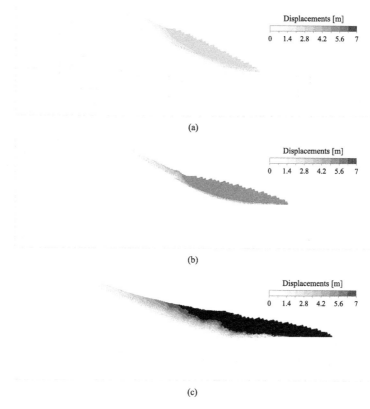

FIGURE 14.9: Maximum displacement calculated for the reference case for a permeability value of $k = 10^{-11}$ m/s and for different values of shear band thickness: (a) 10 cm, (b) 5 cm, and (c) 0.5 cm (after [226]).

eration of the excess pore water pressure concentrates at those elements and, due to their low permeability and the short time of the motion, pore pressures do not dissipate. Temperature increments are generated in all cases as a consequence of the frictional work dissipated during the motion (Fig. 14.11 c). Calculated temperature increments are small but their effect is significant. This is induced by changes in the thermal dilation coefficient of water and solids.

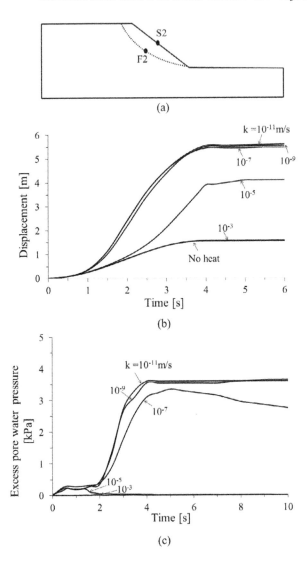

FIGURE 14.10: (a) Position of points analysed. (b) Displacement of point S2 and (c) excess pore water pressure evolution of point F2 for the reference case and different values of saturated permeability k (after [226]).

14.3   Vajont landslide

An ancient slide in the left bank of the Vajont river (Italy), under creeping motion, evolved into a rapid failure on 9th October 1963 when the dam

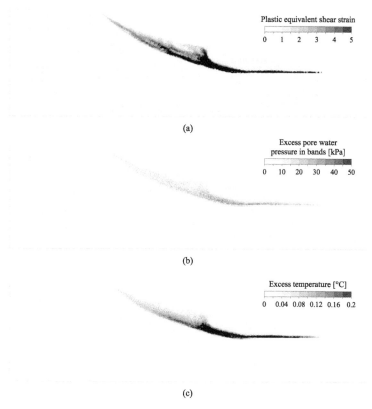

(a)

(b)

(c)

FIGURE 14.11: Distribution of (a) accumulated deviatoric plastic strains, (b) excess pore water pressures, and (c) temperature for the reference case at the end of the motion. Saturated permeability $k = 10^{-11}$ m/s and shear band thickness of 1 cm (after [226]).

reservoir reached the maximum level. According to geological observations, a long peripheral crack developed immediately after the first phase of the reservoir filling. A creeping motion towards the reservoir developed and continued until the rapid failure took place in October 1963. Creeping velocities of 20-30 cm/day were registered before the final failure.

Semenza [265] reports a tentative reconstruction of the past history of the slide. Two representative cross sections of the slide, located 400 m and 600 m upstream of the dam at distances, respectively, are reproduced in Fig. 14.12. Hendron and Patton [126] indicate that the surface of rupture is located in a thin continuous layer of high plasticity clay. Other authors suggest the existence of a more complex geometry defined by multiple sliding surfaces or a thick sliding area. This topic is discussed in the next section.

14.3.1 Model description

The landslide is triggered by the elevation of the water level in the reservoir. The impoundment of the dam and reservoir filling are modelled as a continuous increase in water level. Modelling the creeping stage prior to the failure would involve the consideration of additional features. In particular, Alonso et al. [8, 10] and Veveakis et al. [319] invoke rate effects on the strengthening of the behaviour of the material located in the basal surface to explain the slow motion observed in the Vajont slope. In addition, the internal shearing of the mobilised mass during the motion [7, 336] has probably an important role and contributed to maintain the slope at a slow motion. However, these aspects fall out of the scope of this chapter, which focuses on the thermal pressurisation phenomenon and its consequences on the landslide run-out.

The two-dimensional (2D) cross sections selected for the analysis follows the work of Hendron and Patton [126] as shown in Fig. 14.13. The mobilised mass is characterised by a unique homogeneous rock, which is modelled with a Mohr-Coulomb strain softening model with $c'_{peak} = 2,800$ kPa, $\varphi'_{peak} = 43°$, $c'_{res} = 200$ kPa, $\varphi'_{res} = 34°$ and $\psi = 0°$. The softening shape factor ($\zeta = 100$) leads to a drop of strength from peak to residual value in 30 mm of relative displacement. The selected values are accepted as a rough approximations of

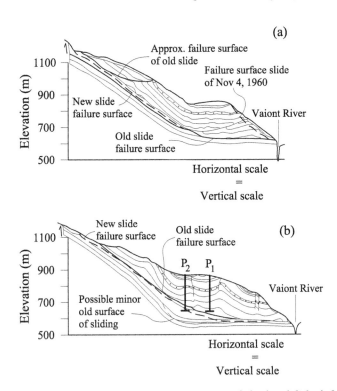

FIGURE 14.12: Two representative cross sections of the landslide (after [126]).

the complex stratification of limestone and marl layers [265] with different degrees of fracturing. They are also consistent with the range of average strength parameters suggested by Alonso and Pinyol [7] for the Vajont rock above the sliding surface. A porosity equal to $n = 0.2$, a Young modulus $E = 5,000$ MPa and a Poisson ratio $\nu = 0.33$ are estimated for the rock mass.

The model is able to simulate the progressive failure along shearing planes of the rock. This is expected because of the kinematics of the motion force internal shearing of the rock in the direction crossing the sedimentary surfaces at high angles. This is shown in the plots of shearing bands given below. These bands act as a rock degradation mechanism as the slide moves forward. A progressive failure develops along the shear bands during all stages of motion. It was also shown that the internal development of shearing bands in the rock mass is controlled by the geometry of the basal sliding surface. Progressive failure of the brittle rock mass alone was not capable of explaining the high sliding velocities of the slide. An apparent basal friction angle of $0°$ was required to match the high velocities of the slide.

An important aspect of the analysis is the nature of the basal sliding surface. In their comprehensive report [126], even if they accept the dominant role of the residual clay strength, they recognise the existence of areas in which shearing is across bedding planes, the absence of clay in some sections, areas where the clay is squeezed into rock voids and brecciated rock fragments within the clay beds. Parnouzzi and Bolla [222] characterised the basal "detachment" surface by a stepped pattern involving a variety of materials – limestone and marly limestone strata, clay interbeds, clay lenses and angular gravels. Wolter et al. [330] describe in detail the basal surface by means of terrestrial photogrammetry. The exposed surface is characterised by a complex geometry of undulations and ridges, which probably entered into rock-to-rock contact during sliding. They introduce four roughness classes to describe the sliding surface. They conclude that in some areas the asperities would imply the shearing of rock mass or a dilatant behaviour. Fig. 14.14 shows the aspect of the basal surface in the spring of 2007, immediately above the displaced

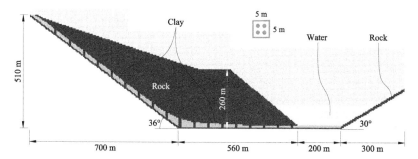

FIGURE 14.13: Computational model, mesh and initial distribution of MP (after [226]).

FIGURE 14.14: Exposed basal surface of Vajont landslide in 2007. Photographs taken in the lower part of the exposed sliding surface (after [226]).

rock mass. Some of the mentioned features could be observed in the photographs. The set of observations, summarised above, indicate that the basal sliding motion was far from being a smooth planar uniform shearing across a layer of clay. Rock shearing and rock-to-rock friction are present during the landslide.

The available information is insufficient for a precise model of the basal surface and it was decided to represent it by a number of brittle rock bridges separated by a clay areas (Fig. 14.13). The rock bridges are characterised by the rock properties adopted for the sliding rock mass.

Residual strength of the high plasticity clay at the basal surface, which is measured in shearing tests [302], suggest a value of $\varphi'_{\text{res}} = 11°$. A low permeability ($k = 10^{-11}$ m/s) is adopted for the clay in view of its high plasticity.

The underlying stable bedrock below the clayey layer is given the same properties as the overlying rock. This material is also present in the opposite valley slope. The river canyon on the bottom of the valley is not represented. The calculations indicate that the failure surfaces did not affect the lower boundary rock because of the significantly lower strength of the clayey layer with respect to the rest of the rock. Therefore, the thickness of the material below the clayey layer is small in order to reduce the dimension of the model and the computational cost.

The model includes also the reservoir water as an elastic material characterised by its real volumetric compressibility coefficient and imposing a shearing modulus close to zero. This procedure allows the application of the weight of the water and the corresponding water pressure on the slope surface at the initial time and also during the motion. The dynamic water forces induced during the motion are also included automatically in the calculation. This is a better option than imposing the effect of the reservoir water as a supporting force because these conditions should be applied on the slope surface, which is not fixed, and the position of the nodes or MP defining the slope surface are *a priori* unknown.

Fig. 14.13 also shows the computational mesh used, which defines the entire domain of the problem. Four MPs are initially distributed within the elements representing the materials. They are located at the corresponding integration points of a four-point Gaussian quadrature. The rectangular elements are 5×5 m. This size is limited by the computational cost of the calculation. Using such element size, it is not possible to define accurately the sedimentary layers of the Vajont slope. The lower heterogeneous clayey layer is defined by four elements across its thickness. The purpose of this discretisation is to avoid the inner elements of the clay band sharing nodes with the hard rock material.

Following Section 14.2, the mechanical work dissipated in heat is scaled by assuming a reference clay band thickness of 3 cm within the embedded localisation bands. No excess pore pressures are allowed in the rock material because of its high permeability. The initial stress state is the result of applying the gravity loading.

14.3.2   Numerical results

The application of weight before the rise of the reservoir level leads to the mobilisation of the strength of the clay material as shown in Fig. 14.15 (a) in which the equivalent plastic shear strains are plotted. The slope remains in static equilibrium due to the strength provided by the rock bridges and mass. The increase of the water reservoir level leads to the rise of pore water pressures on the sliding surface. The supporting action of the water is also accounted for. Failure initiates when the water level is 100 m above the toe. This corresponds to an elevation of 700 m according to the cross section 2 from [126]. At this time, equivalent plastic shear strains localise as shown in Fig. 14.15 (b). Shear surfaces grow from the basal sliding surface at the points where the geometry of the sliding surface, conditioned by the geometry of the clay layer, exhibit a change in curvature. In fact, the infinite curvature radius of the planar sliding surfaces reduces to a radius of approximately 200 m in the curved zone around the kink created at the junction between the planar surfaces. The localised shear bands are curved and tend to join inside the rock mass. The associated strain softening results in a weak fracture surface which is visible in the plots of Figs. 14.15 and 14.17.

The damage experienced by the rock mass at the end of the motion, once the landslide recovers a new static equilibrium, is shown in Fig. 14.15 (c). Fig. 14.16 shows the displacement and velocity records calculated as the average of several MPs located in the five elements of the computational mesh forming the landslide toe. The maximum velocity reached by the landslide is reported by several authors. Müller [209] mentions 25 m/s to 30 m/s. Ciabatta [68], cited by Nonveiller [215], performed a dynamic analysis to find 17 m/s. Nonveiller [215] calculates a maximum velocity of 15 m/s. Voight and Faust [320], in his pioneering contribution, found a maximum velocity of

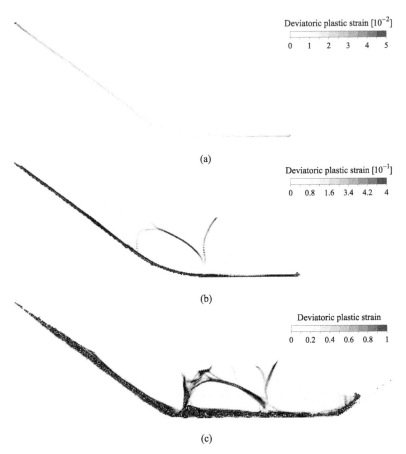

FIGURE 14.15: Equivalent plastic shear strain: (a) before reservoir impoundment upon application of gravity loading, (b) at the initiation of the motion when reservoir level is at 100 m, and (c) at the end of the motion (after [226]).

26 m/s (for a planar geometry). Hendron and Platton [126] estimate values in the range 20 m/s to 25 m/s. These are estimations based on a particular model for the dynamics of the motion and therefore they should be regarded as approximations of the actual field velocity. The maximum velocity of 20 m/s calculated here (Fig. 14.16 b) fits into the set of values mentioned. The calculated run-out distance of 320 m (Fig. 14.16 a) is close, but somewhat smaller, than other estimated values based on field observation [126].

Temperature, heat-induced pore water pressure increment and displacement distributions are given in Fig. 14.17. Increments of temperature around 50°C are computed in the basal shear band. Maximum temperature increments of around 40 °C are computed inside the rock mass. Excess pore water pressures concentrate in the impervious clayey soil, where the work has been

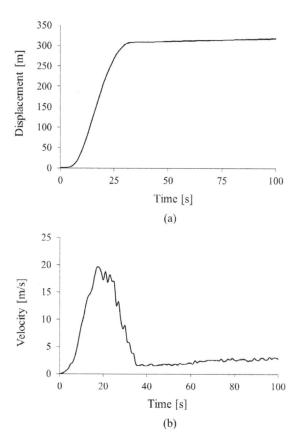

FIGURE 14.16: Calculated (a) displacement and (b) velocity of the Vajont landslide (after [226]).

scaled. Pore water pressure dissipates outside of the shear band affecting the rock overlying the clay layer. The presence of the previous rock bridges accelerates the pore water pressure dissipation. Fig. 14.18 shows the calculated final positions of the landslide materials. What is actually plotted in the figure is the position of all MPs. The basal clay layer has extended along most of the failure surface during the motion. In some places, the clay has opened its way through the rock. Some field observations mentioned before [126, 222] support this result. Note also that a horizontal layer of clay was located beyond the toe of the slide to avoid a direct rock to rock contact during the motion. Part of this clay was dragged by the landslide and is visible in Figs. 14.15 (c), 14.17 and 14.18.

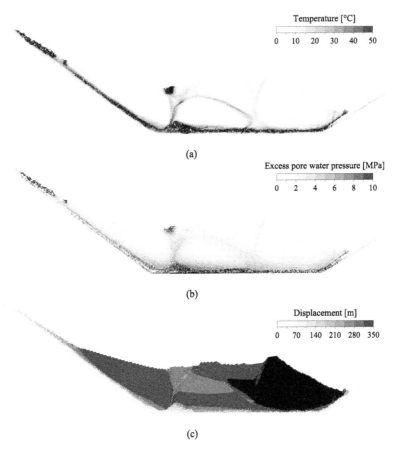

FIGURE 14.17: (a) Temperature, (b) excess pore water pressures, and (c) displacements at the end of the motion (after [226]).

FIGURE 14.18: Distribution of landslide materials at the end of the motion. The plot indicates the position of MP (after [226]).

14.4 Closure

The $\vec{u} - p_l - \Theta$ formulation for saturated porous media (excluding evaporation) developed in this chapter and implemented in the framework of MPM is applied to the analysis of landslides. The problem, strongly dependent on the shear band thickness, exhibits a pathological dependence on the computational mesh when the size of the elements is larger than the actual thickness of the shear bands (around few millimetres in fine soils). In order to be able to discretise the domain of the problem (usually several meters or kilometres long in case of landslides) with elements large enough to keep the calculation within tractable limits taking into account the computational cost, a numerical procedure is presented in the chapter. The proposed method is elegant for its simple implementation and requires introducing as an input parameter the thickness of the shear band, generally dependent on the grain size distribution. It does not require any additional parameters without physical meaning. A sensitivity analysis in a reference case shows the capabilities of the procedure and the numerical tool developed.

The well-known Vajont case is selected as a real case to check the model. Vajont, for the purpose of calibrating the model, offers essentially two key items of information – (1) the estimated sliding velocity and (2) the run-out distance. The model built is based on a representative 2D cross section. The calculation successfully reproduces the Vajont landslide and the calculated velocity and run-out match well the estimated values reported in previous publications.

The model developed is believed to offer an advanced tool to systematically incorporate thermal effects into landslides modelling. It contributes to an increased understanding of phenomena leading to landslide acceleration after failure. The model is capable of reproducing the initial pre-failure state, the onset of failure and the subsequent motion.

15

Excavation-Induced Instabilities

N.M. Pinyol and G. Di Carluccio

15.1   Introduction

A proper numerical analysis of geotechnical problems dealing with soil excavation is a useful tool for design and interpretation of civil engineering projects. It allows the prediction of the variation of stress and strain fields induced by excavation and its impact on the surrounding environment. The procedure adopted to simulate soil excavation with MPM consists in defining the range of calculation steps in which the material points (MPs) belonging to a certain volume have to be removed, once the geometry and the computational mesh are created. In order to contribute to the evaluation of this feature, numerical analyses of two excavation problems are carried out. In the first example, the bottom heave stability of a supported excavation in clay is analysed. This stability problem has been widely investigated theoretically and numerically by using different approaches. A series of undrained total stress analyses with decreasing value of soil strength are performed until failure conditions are reached. The aim is to discuss capabilities and limitations of MPM in simulating this kind of problem and to compare the results with the literature. A second simulation refers to a well-documented real case of a large landslide reactivated by quarry excavation [6]. The actual response, which induced a few centimetres of run-out, is well reproduced by the MPM model.

15.2   Strutted excavations in clay

There are typically three methods to evaluate the stability conditions for strutted excavations – (1) limit equilibrium methods (LEM), (2) limit analysis methods (LAM) and (3) finite element methods (FEM). Terzaghi [296] used the conventional LEM in order to analyse the stability of the bottom of an excavation on the basis of bearing capacity theory. The clayey soil is supposed to be homogeneous and isotropic under undrained conditions. The shearing resistance of soil is equal to its undrained strength c_u. Fig. 15.1 shows the geometry of the problem and the failure mechanism assumed. The weight of the clay located on both sides of the excavation produce a uniformly distributed load on the horizontal section passing through the bottom. During the excavation, this load may exceed the bearing capacity of the soil located below the bottom of the excavation. The surrounding soil moves downward and induces the heave of the base. According to Terzaghi [296], the critical depth D_{crit} for an excavation under plain strain conditions is given by Eq. 15.1.

$$D_{crit} = \frac{5.7c_u}{\gamma_{sat} - \dfrac{c_u\sqrt{2}}{B}} \tag{15.1}$$

where γ_{sat} is the saturated unit weight of soil, B is the excavation width and c_u is the undrained shear strength.

Bjerrum and Eide [37] showed based on field studies and for deep excavations or in presence of non-homogeneous clay that the bottom heave occurs at smaller depths than indicated by Eq. (15.1), and this without fully mobilising the shear strength of the upper clay layers. They suggested a more localised failure mechanism that is approximated by the bearing capacity factor N_c for deep foundations, as proposed by Skempton [270]. This stability factor depends on the shape and the depth of the foundation and is expressed as given in Eq. 15.2.

$$N_c = \frac{\gamma_{sat} D_{crit}}{c_u} \tag{15.2}$$

These solutions are widely used in the initial phases of the excavation design for its relative simplicity. However, they strongly depend on some assumptions

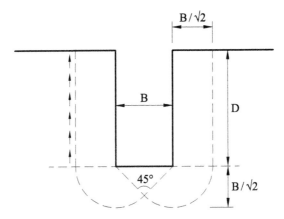

FIGURE 15.1: Failure mechanism of a 2D excavation in a clayey soil, according to Terzaghi [296].

regarding the shape and the location of the failure surface or the values of the bearing capacity factor. Limit analysis theory provides another method to evaluate this stability problem. It allows the calculation of an upper and lower bound for the bearing capacity factor in order to restrict the range of solutions for the basal stability calculations. FEM are usually used in advanced stages of the excavation design but the results depend on the constitutive model and the input parameters chosen. In this case, the influence of the support system on ground movements is taken into account. The stability of strutted excavations can be evaluated by using a strength-reduction procedure, which consists in reducing gradually the strength parameters of the soil until lack of convergence occurs in non-linear FE analysis [44,97]. Alternatively, FEM can be combined with the upper and lower bound limit analyses [305].

In order to investigate the stability of strutted excavations in clay with MPM, a numerical total stress analysis is conducted. In the following sections, the numerical model is described and the simulation results are discussed and compared with the literature. The support system of the excavation is not modelled and the main reference for the comparison of results will be the limit equilibrium approaches, which remain the fundamental methods for evaluating basal heave stability in braced excavations.

15.2.1 Model description

The geometry of the problem after the excavation process is given in Fig. 15.2. The dimensions of the model analysed are indicated in Table 15.1. The model is plane strain and is analysed by a 3D model having a 0.1 m thickness. Plane strain conditions are imposed by restricting horizontal displacements along the vertical contours. According to Terzaghi's assumptions, the thickness of the soil below the excavation base is greater than $B/\sqrt{2}$ in order to neglect the effect of an underlying stiff layer.

The mesh is composed of thin 3D mesh of tetrahedral elements. The thickness of the model is considered to be the same as the element size of the computational mesh for mesh quality reasons. Fig. 15.3 shows the discretised domain and the boundary conditions for the case of $D/B = 2$. The horizontal

TABLE 15.1
Dimensions of strutted excavation for different ratios of depth to width.

Parameter	Symbol	Unit	Case		
			$D/B = 1$	$D/B = 2$	$D/B = 4$
Box height	H	m	3.5	4.5	6.5
Box width	W	m	9	9	9
Soil height	S	m	3	4	6
Model thickness	T	m	0.1	0.1	0.1
Excavation width	B	m	1	1	1
Excavation depth	D	m	1	2	4

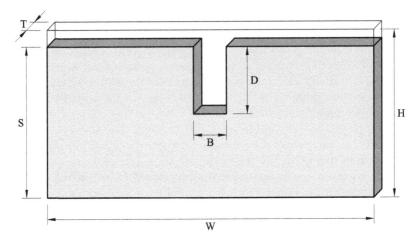

FIGURE 15.2: Geometry of strutted excavation.

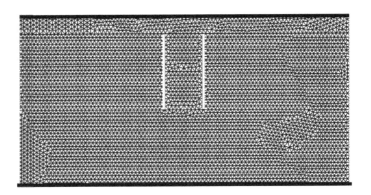

FIGURE 15.3: Computational mesh and and boundary fixities in x (grey) and y (black) direction.

displacements are prevented at the left and right boundaries. The displacements are constrained in the vertical direction at the top of the model, whilst no displacements are allowed at the bottom. The assumption of strutted excavation is ensured by constraining horizontal displacements at the vertical contours of the excavation.

The soil is assumed to be a purely cohesive material and modelled with a Tresca failure criterion; that is a Mohr-Coulomb failure criterion with $\varphi_u = 0°$. The stability against failure of the excavation base is evaluated by using a shear strength reduction procedure. Several simulations with decreasing value of the undrained shear strength c_u are performed until a sudden increase of upward vertical displacements at the excavation bottom is computed. The two-phase single-point MPM formulation (see Chapter 2) is used for this set

TABLE 15.2

Mohr-Coulomb material parameters.

Model parameter	Symbol	Unit	Value
Initial porosity	n	-	0.2
Density solid	ρ_s	kg/m³	2,650
Density liquid	ρ_l	kg/m³	1,000
Saturated unit weight	γ_{sat}	kPa	22.8
Undrained Young modulus	E_u	kPa	10,000
Undrained Poisson ratio	ν_u	-	0.49
Undrained friction angle	φ_u	°	0

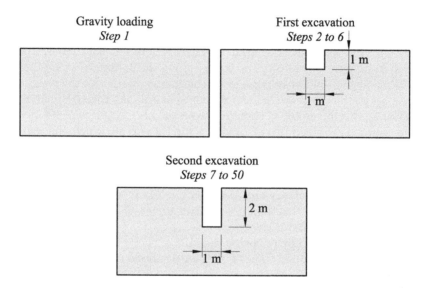

FIGURE 15.4: Calculation stages for $D/B = 2$.

of simulations. The input parameters for the material are summarised in Table 15.2.

The simulations are performed in two phases: (1) stress initialisation with quasi-static gravity loading and (2) the excavation process by consecutively removing soil volumes. Stresses are initialised by applying a gravity loading at the first calculation step. A local damping factor of 0.75 is applied to reach a quasi-static equilibrium state in a faster way allowing a considerable reduction in the computational time (see Chapter 3). In the second phase, the excavation process is simulated in several stages by removing the MPs belonging to volumes of 1 m height. The excavation of each volume is simulated instantaneously at the beginning of each excavation stage and then enough time is left in order to reach a new equilibrium. In this case, the full dynamic behaviour of the soil is analysed. Intermediate excavation stages are performed in five calculation steps of 0.2 seconds while, for the last one, additional calculation steps are simulated in order to evaluate the maximum upward vertical displacement at the excavation bottom. Fig. 15.4 shows the calculation stages for the case $D/B = 2$.

15.2.2 Numerical results

Fig. 15.5 shows the final maximum values of soil heaving movement for the case $D/B = 2$ and for each simulation in terms of the undrained shear strength c_u. They refer to the MP located at the middle of the excavation base. Two different ways of simulating the failure mechanism are considered. In the first case (Case 1), the soil is allowed to accumulate into the braced excavation during the failure mechanism development. The second approach (Case 2) respects an implicit hypothesis of classical solutions that consists in continuously removing soil as it rises inside the excavation. For both cases, upward vertical displacements are very small when using the undrained strength value of 8 kPa and the soil behaviour can be considered elastic. As the cohesion is reduced, a gradual increase in vertical displacements is expected at first.

Then a lower bound of c_u should be reached when displacements increase considerably. This trend is not observed for Case 1, in which the material accumulates into the excavation and the maximum vertical displacements are significantly lower in comparison with Case 2. This behaviour can be explained considering the accumulation of soil leading to a new stable configuration. Geometric conditions change and a sudden increase of soil heaving movement cannot be observed. In this case, it is difficult to detect an undrained strength at which failure occurs.

On the contrary and by removing the MPs as they rise into the excavation, a more evident lower bound value of undrained cohesion is reached. This result

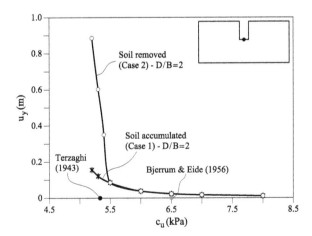

FIGURE 15.5: Maximum soil heaving movement for different values of undrained cohesion for $D/B = 2$.

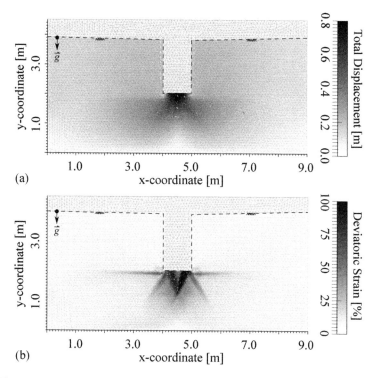

FIGURE 15.6: Results after about 10 seconds from last excavation for $c_u =5.2$ kPa and $D/B = 2$ (Case 2): (a) total displacements [m], (b) deviatoric strains [0% to 100%].

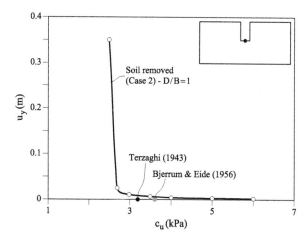

FIGURE 15.7: Maximum soil heaving movement for different values of undrained cohesion for $D/B = 1$.

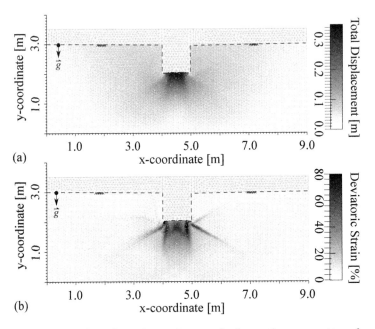

(a)

(b)

FIGURE 15.8: Results after about 2 seconds from the excavation for $c_u = 2.5$ kPa and $D/B = 1$ (Case 2): (a) total displacements [m], (b) deviatoric strains [0% to 80%].

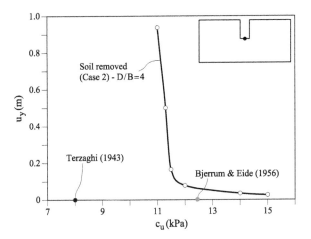

FIGURE 15.9: Maximum soil heaving movement for different values of undrained cohesion for $D/B = 4$.

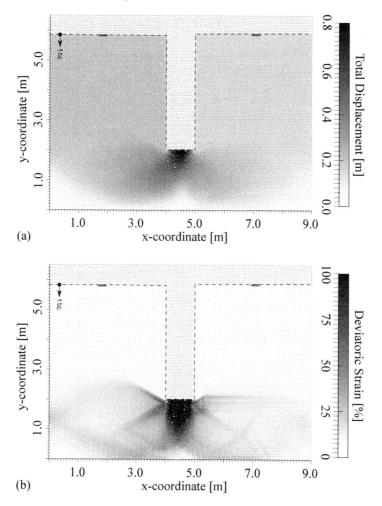

FIGURE 15.10: Results after about 15 seconds from last excavation for $c_u = 11$ kPa and $D/B = 4$ (Case 2): (a) total displacements [m], (b) deviatoric strains [0% to 100%].

is very close to Terzaghi's solution (Eq. 15.1) for a critical depth of $D_{crit} = 2$ m. As commented in Section 15.2, the reference solution is the one proposed by Bjerrum and Eide [37] when $D/B > 1$, based on the bearing capacity factors calculated by Skempton [270]. Accepting the value for strip foundations and $D/B = 2$ ($N_c = 7$), Eq. 15.2 gives a critical undrained strength of 6.5 kPa, significantly higher than the value obtained by the MPM simulation. However, the accumulated deviatoric strains are localised at the excavation base, as expected for deep foundations (Fig. 15.6).

The results for $D/B = 1$ are shown in Fig. 15.7. In this case, MPM modelling provides a critical value of c_u, when the computed vertical displacement rate increases suddenly, slightly lower than the theoretical solutions given by Terzaghi and Skempton. Fig. 15.8 shows the failure mechanism computed.

For a deep excavation ($D/B = 4$), Terzaghi's method provides unreliable results. The sudden increase of displacements is observed, when using MPM, for values of strength slightly lower than the critical undrained strength calculated with Eq. 15.2 (Fig. 15.9).

According to the observations based on Bjerrum and Eide [37], in the case of deep excavation the sliding surface beneath the base does not extend up to the ground surface. As shown in Fig. 15.10, the accumulated deviatoric strains are localised at the excavation base.

15.3 The Cortes de Pallas landslide

A large reactivated landslide was identified in 1985 in Valencia, Spain, as described in Alonso et al. [6]. The instability affected the left bank of a river upstream of the Cortes arch-gravity dam. The bank, constituted by limestone and marl strata, was part of a large isocline in which the river canyon carved its way in. The instability was observed in some silos and retaining walls founded on the upper part of the slope, which were used in the construction of the dam.

15.3.1 Site characterisation

The site characterisation included boreholes, inclinometers, and topographic measurements. The geological survey found the existence of an ancient landslide affecting the upper limestone layers, which are severely broken at the surface. The failure surface in located is a 2-m thick marl dominated by illite and dolomite crystals. No indication of water levels was found in any of the boreholes. The topographic marks were installed after the identification of the failure and indicate a large area moving downwards in a direction, approximately parallel to the average dip direction of the strata. The inclinometer readings (Fig. 15.11) show that the mobilised mass was sliding on the thin marl stratum. The total volume of the landslide is estimated to be around 5 million m^3.

Fig. 15.12 shows a representative cross section of the slide. A quarry provided the granular aggregate for the concrete used in the dam construction and was excavated in the lower part of the old landslide at the position indicated in Fig. 15.12. This excavation of the toe was identified due to the reactivation of the ancient landslide.

Laboratory shear tests were carried out on samples from the marl layer as well as from block samples taken in places where the marl layer outcrops.

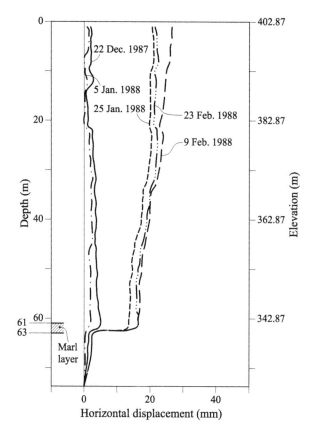

FIGURE 15.11: Inclinometer measurements in Cortes landslides (after [6]).

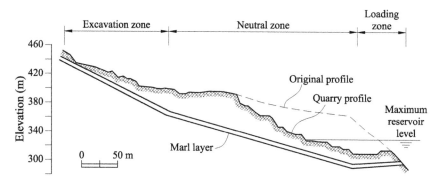

FIGURE 15.12: Representative cross-section of Cortes landslide (after [6]).

The marl material is characterised as a low plasticity clay ($w_L = 20\% -$ 28%, $w_P = 13\% - 14\%$) with high consistency ($w = 10\% - 13.6\%$) and low porosity ($n = 0.25$). The peak and residual frictional angles are determined from direct shear tests to be $\varphi_{max} = 23°$ and $\varphi_{res} = 22°$, respectively. Similar values ranging between between $\varphi_{max} = 20°$ and $\varphi_{res} = 21°$ are obtained in an undrained triaxial test with pore water pressure measurements under relatively large effective confining stress of 0.2, 0.45 and 0.7 MPa.

Due to the reactivation of the ancient landslide, the marl layer is assumed to have residual strength which was characterized by $\varphi_{res} = 21°$ according to shear tests carried out on reconstituted samples. This value is significantly higher than the strength obtained by the back-analysis ($\varphi = 17.7°$) [6]. This discrepancy was explained during the stabilisations work when undisturbed samples, taken from the sliding surface, were analysed.

To guarantee security conditions during the reservoir operation, which will partially submerge the unstable slope, it is decided to stabilise the hillside by reducing the weight of the upper part and increasing the weight of the lower part by filling in part the quarry excavation. This excavation uncovered the failure surface and the clayey soil located in the shear band of the landslide could be identified. Mineralogical differences are observed between closely located samples taken from the shear band and in the marl stratum. A block sample, including the sliding surface, was taken and tested in the laboratory in direct shear apparatus. The actual sliding surface was aligned carefully with the middle plan of the shear box. It is observed that the softening behaviour, from a peak value to a residual one observed in the previously tested sample, was absent. A moderately non-linear strength-strain curve is observed. The average frictional angle for the range of stresses considered is $\varphi' = 18°$. For higher values of normal stress, corresponding to the average normal stress on the sliding surface in the central part of the landslide (around 0.8 MPa), the representative frictional angle is lower ($\varphi' = 17°$). The obtained values in the shear box testing agree very well with the value derived from the back-analysis.

15.3.2 Model description and material parameters

A one-phase analysis MPM mixed calculation (see Chapter 3) is carried out to simulate the observed failure of the Cortes landslide due to the excavation of part of the bank toe. No groundwater is included in the modelling taking into account that no groundwater table above the sliding surface is observed [6].

The representative cross section presented by Alonso et al. [6] (Fig. 15.12) was simplified for the MPM calculations as indicated in Fig. 15.13 distinguishing two materials and different volumes to simulate the excavation process. The domain discretisation is shown in Fig. 15.14 and four particles per element are initially defined in the elements of the slope. The materials (marl layer and limestone below and above the marl layer) are modelled with a Mohr-Coloumb model with the parameters indicated in Table 15.3. The limestone is modelled as a brittle material with Mohr-Coulomb strain-softening model

TABLE 15.3

Material parameters.

Model parameter	Symbol	Unit	Limestone	Marl layer
Young modulus	E	MPa	1,000	200
Poisson ratio	ν	-	0.33	0.33
Peak effective cohesion	c'_{max}	kPa	100	-
Residual effective cohesion	c'_{res}	kPa	20	-
Peak friction angle	φ'_{max}	°	35	-
Residual friction angle	φ'_{res}	°	30	17.5
Porosity	n	-	0.3	0.3

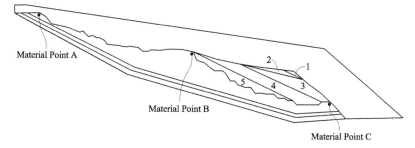

FIGURE 15.13: Cross section defined for MPM calculation.

(see Chapter 8). The marl layer is assumed to have residual strength at the surface of rupture with $\varphi'_{res} = 17.5°$ and in accordance with laboratory tests (Section 15.3.1). The other model parameters (i.e. stiffness, Poisson ratio) and initial state (i.e. porosity) are estimated.

Displacements are restricted in the three directions along the lower boundary surface, including in out-of-plane z-direction (Fig. 15.14) to simulate two-dimensional (2D) plane strain conditions.

The excavation volumes and the time interval for their excavation are indicated as input data. The MPs in the excavation volume are instantaneously removed at the beginning of the step. However, the excavation process of a given volume actually carried out in the field is significantly different from being an instantaneous unload. In order to simulate the excavation process in a more realistic way, the total volume excavated is divided into five sub-volumes that are excavated in a sequenced way. Since no pore water pressure is included and the constitutive model is not time dependent, the duration of the steps is not relevant. However, it is ensured that the slope is at rest prior to any excavation.

The calculation is divided into 7 stages. The first stage is composed of five loading steps of 0.5 seconds each. The initial stresses are calculated as quasi-static in which the gravity loading is applied gradually and a local damping of 0.75 is applied (see Chapter 6). The initial stress distribution of the slope is not easy to estimate as no field measurements are available. It is assumed that

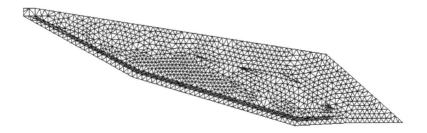

FIGURE 15.14: Computational mesh of 8,264 tetrahedral elements.

FIGURE 15.15: Accumulated deviatoric strain distribution after gravity loading (initial stress state before excavation process).

the stress distribution resulting by imposing the weight of the materials is a good approximation. Then, five stages loading steps of 0.5 seconds are defined to simulate the excavation of the volumes (1 to 5) as indicated in Fig. 15.13. For these stages, the local damping is reduced to a low value of 0.05. Finally, the last stage of 75 load steps of 0.5 seconds is defined to evaluate the response of the slope without applying any change in the model.

15.3.3   Numerical results

The initial state, caused by the gravity loading prior to any excavation, induces some deviatoric strain as shown in Fig. 15.15. Deviatoric strains localise along the soft layer of marl. Fig. 15.16 shows the accumulated deviatoric strain and accumulated displacement at the end of calculation. The excavation process induces displacements of around 20 cm at the slope surface and the slope reaches a new equilibrium.

Fig. 15.17 shows the displacement curve versus the time step calculation for three MPs of the slope, as indicated in the Fig. 15.13. Note that for each stage of excavation, the accumulated displacement increases abruptly and the motion is interrupted quickly after a few centimetres when a new equilibrium

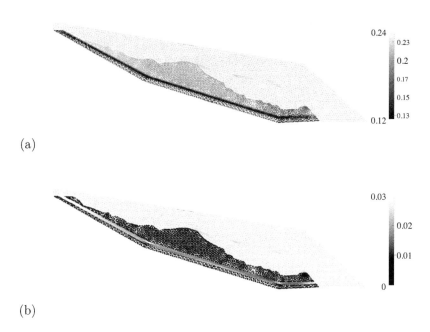

(a)

(b)

FIGURE 15.16: (a) Total displacements [m] and (b) accumulated deviatoric strains at the end of calculation, after excavation.

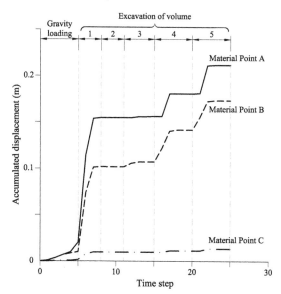

FIGURE 15.17: Displacement of individual MPs.

is reached. The displacement observed is not homogeneous along the slope. Maximum accumulated displacement of 24 cm is computed in the upper part of the landslides, probably favoured by a local failure of the upper part due to the geometry of the slide cross section. The minimum accumulated displacement calculated is located in the more stable part of the toe of the slope due to the shape of the marl layer.

15.4   Closure

The capabilities of MPM to model large displacement provide a realistic interpretation of the excavation problems. The analysis of deep excavations protected by lateral walls is a good example. Failure is understood as an increasing soils displacement rate. However, the resulting accumulation of soil mass on the excavation bottom has the positive effect of retarding the motion. The realistic interpretation of failure offered by MPM is compared with classical solutions based on simplified assumptions.

The second example solved refers to a well-documented case history: the Cortes Landslide. Again, the concept of failure closely related with the accumulation of landslide displacements provides some insight, which cannot be obtained with limit analyses. In the case of the Cortes Landslide, the excavation resulted in significant displacements but it did not result in a catastrophic accelerated motion. This information is outside the formulation of Limit Analysis. The MPM analysis was consistent with field observation: the landslide displaced downward response to the excavation but it reached a new stable configuration at the expense of severe straining of the sliding surface and the rock cover.

16

Slope Reliability and Failure Analysis using Random Fields

G. Remmerswaal, M.A. Hicks and P.J. Vardon

16.1 Introduction

The need for reliability based design (RBD) has increased over the years, due to the necessity for more complex and efficient designs [138]. The uncertainty, which must be accounted for during RBD, can be divided into two categories; namely, aleatoric uncertainty and epistemic uncertainty [144]. Aleatoric uncertainty or natural variability, such as variability of soil properties, cannot be reduced with increasing investigation; i.e. the property has a natural randomness. On the other hand, epistemic uncertainty, consisting of model and measurement uncertainty, can be reduced by acquiring more knowledge. Due to soil variability, removing measurement uncertainty is impossible without an infinite amount of site investigation.

Therefore, a stochastic analysis is required to take account of both the natural variability and epistemic uncertainty. The random finite element method (RFEM) uses random fields, i.e. property fields generated from the measured statistics, for modelling the natural (spatial) variability of soil properties, within a finite element framework [119, 130]. A random field is one (of an infinite number) of possible property fields based on the material statistics. Measurement uncertainty is included in RFEM with the generation of multiple random fields based on the available knowledge. A Monte Carlo analysis is performed with these random fields in RFEM to compute the reliability of a structure. Note that recent studies have confirmed that conditioned random fields can be used to reduce the uncertainty when additional information is acquired [169] and small failure probability levels can be accessed with reduced computational effort using subset simulation [166, 307].

Analysing the impact of soil variability with RFEM has been proven useful in geotechnics, for example in: deep and shallow foundation settlements [99, 211]; slope stability [119, 129, 130, 168, 347]; retaining walls [85] and groundwater flow [279].

Even though RFEM has been applied in many parts of geotechnical design, it is limited by the applicability of FEM. Specifically for this chapter, RFEM, just as FEM, is incapable of modelling large deformations without the occurrence of excessive mesh distortions. One solution to this problem is to replace FEM with the material point method (MPM) for modelling the structural response in each realisation of the Monte Carlo simulation, so giving the random MPM (RMPM) first outlined by Wang et al. [324]. This chapter presents an overview of RMPM, including two simple examples of its use.

16.2   Theoretical background

RMPM is comparable with RFEM, in that it combines random field theory with a numerical simulation method. This section explains the four steps involved in an RMPM analysis:

1. Quantifying variability in materials.

2. Numerical representation of variability using an ensemble of random fields.

3. Simulation using MPM.

4. Quantifying the outcome.

16.2.1   Quantifying variability in geomaterials

In practice only, a limited amount of site investigation is performed to quantify soil parameters and their spatial variability. Cone penetration tests (CPTs) are typically a good candidate for the quantification of spatial variability, due to their continuous profile in the vertical direction, as shown in Fig. 16.1, where a 'real' soil is shown with CPT measurements superimposed. For simplicity a single soil layer is considered here; discrete layers, lenses or inclusions are also possible in reality.

Two types of statistical measures must be obtained from CPT measurements, namely the point and spatial statistics. As indicated by Fig. 16.2, the point statistics can be represented by a probability density function, which is often chosen as either a normal or lognormal distribution defined by the mean μ and standard deviation σ. For soils, the standard deviation is generally dependent on the mean property value. Therefore, the coefficient of variation $V = \sigma/\mu$ of a soil layer is often referred to instead of (or in addition to) the standard deviation.

Spatial statistics, represented by a vertical θ_v and horizontal θ_h scale of fluctuation, can also be obtained from CPT data. The scale of fluctuation θ

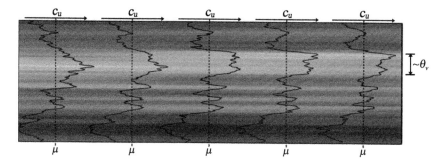

FIGURE 16.1: Spatial variability of undrained shear strength c_u, with light and dark areas indicating high and low c_u respectively, and CPT measurements to quantify the variability.

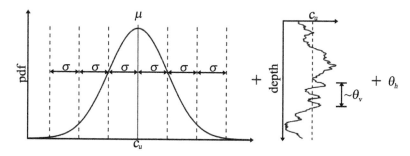

FIGURE 16.2: Statistical measures of soil properties given by a probability density function, and vertical θ_v and horizontal θ_h scales of fluctuation.

defines the degree of spatial correlation and is a function of the distance between adjacent weak or strong zones [130]. A larger θ is the result of a material being more strongly correlated over greater distances. θ_v can be accurately obtained from CPT data, due to the continuous profile in the vertical direction as shown in Fig. 16.2. However, it is more difficult to obtain θ_h from CPTs, as multiple CPT profiles must be combined from which the horizontal correlation can be approximated. More detailed descriptions for obtaining both scales of fluctuation can be found in [78].

Due to the deposition process of soils, θ_h is generally several times larger (at least) than θ_v. The amount of layering of the soil is often defined by the degree anisotropy of the heterogeneity $\xi = \theta_h/\theta_v$. In 3D, three scales of fluctuation are present; two horizontal scales and one vertical scale. In many soil deposits the horizontal scales may be taken as equal, i.e. $\theta_x = \theta_y = \theta_h = \theta_v\xi$, thereby allowing the quantification of the statistics with one horizontal scale of fluctuation.

16.2.2   Numerical representation of variability using random fields

After the quantification of the soil variability, the point and spatial statistics are used to a generate random fields (see Fig. 16.3). For example, local average subdivision (LAS) [100] can be used to generate the random fields in both 2D and 3D RMPM. A random field is one of many (infinite) possible outcomes of the statistics and is not an exact representation of the actual soil properties at a site. Some differences can be observed between the soil properties (Fig. 16.1) and the generated random fields (Fig. 16.3):

- Resolution is limited to the random field discretisation, thereby averaging the finer details.

- Strong and weak zones in the random fields can be at different locations than in reality.

A random field is therefore a representation of the characteristic of the soil variability, as it is generated using only μ, σ, θ_v and θ_h. Furthermore, conditioned random fields can be used to reduce the possible outcomes of the random fields [169]. For example, the CPT data in Fig. 16.1 can be used to condition the random fields and reduce the differences between them and the real soil properties; i.e. the soil property values would be correct at the CPT locations.

As a logical extension of RFEM, in which the random field is mapped to the Gauss points, the random field is mapped to the material points (MPs) in RMPM, see Fig. 16.3. Random fields can be used for each material property and correlations between properties can be applied. However, in this chapter, only the undrained shear strength of an MP is generated using a random field, while other properties have been assigned their mean value.

16.2.3   MPM solution adopted

Once the material properties of each MP have been assigned from the random field, a simulation using MPM is performed. As indicated in Chapters 2 and 3 , several MPM formulations exist, for example, implicit, explicit, single-phase, multi-phase, which all have different applications. RMPM can be used to include variability in any MPM formulation. At the time of writing only purely mechanical, single-phase, MPM formulations have been used in RMPM [241, 324]. For the 2D analysis in Section 16.3.1 an implicit MPM version is used, whereas an explicit MPM version is used for the 3D analysis in Section 16.3.2. The reader is referred to Chapter 3 for further details on integration schemes.

Besides different formulations, many improvement techniques have been developed for MPM to tackle well known problems such as stress oscillation. Heterogeneity studies often involve heterogeneity in strength parameters and, due to this heterogeneity, weaker zones are present. Numerical stress oscillations are more likely to cause failure in weaker zones compared to a ho-

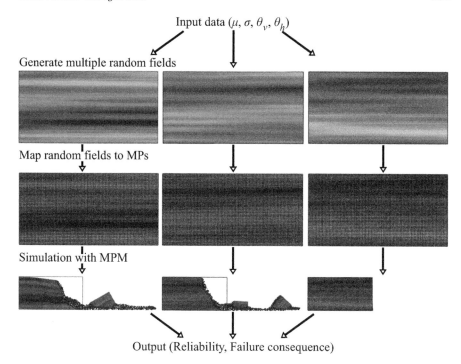

FIGURE 16.3: Flow chart of RMPM.

mogeneous material. Therefore, oscillation reduction techniques might even be more valuable in RMPM compared to standard MPM. GIMP [24] and CMPM [115] have been used in the 2D RMPM applications described in Section 16.3.1. GIMP reduces the grid crossing error, whereas CMPM reduces the error in the stress recovery, due to MPs generally not being at the optimal locations, by using a higher order stress approximation.

16.2.4 Quantifying the outcome

Two types of analyses can be performed with RMPM. In a single realisation analysis, one random field is generated and simulated using MPM. A single realisation can be used to observe the processes and general failure mechanism which may occur. Due to the computational cost of 3D MPM, this single realisation approach is used in Section 16.3.2. However, a single realisation cannot be used to assess the reliability of the slope, because this is only one of many possible solutions. As indicated by Fig. 16.3, multiple realisations are used in a Monte Carlo analysis to compute the probability of different types of failure mechanism. These probabilities can be used to compute the reliability of a structure as well as the likelihood of a specific consequence. Therefore,

RMPM is capable of computing both the likelihood of failure as well as its consequence, and is thereby well-suited for risk assessments.

Different applications can require the quantification of different consequences. For example, in slope stability modelling, run-out distance and failure volume can be used to quantify the required maintenance to repair a failed slope. Height reduction can be measured in, for example, dyke stability modelling, which can give a better prediction of the probability of flooding. For landslides or avalanches, the impact force on downhill structures can be predicted, thereby better quantifying the risks of a slide.

16.3 Applications of RMPM

RMPM can be used to model any problem involving large deformation under the influence of spatial variability. Due to the large effect of spatial variability in soils, because failures are attracted to weaker zones, the application of RMPM in geotechnical designs is vast. Both slope and liquefaction failure are influenced greatly by spatial variability and studying the effects of these failures requires the use of large deformation modelling. Besides failure modelling, pile installation and cone penetration testing involve large deformations during their intended use. Modelling the influence of spatial variability on these penetration problems is another application of RMPM. In this section the focus is on retrogressive slope failure, which is the initial step on dyke reliability modelling with RMPM.

16.3.1 2D embankment reliability

An idealised 2D boundary value problem has been analysed to study the effect of spatial variability within an embankment. The 5 m high embankment is composed of a strain-softening clay with a peak to residual shear strength ratio of 4.5, and has been modelled with an implicit version of RMPM. In this simple example, only the peak undrained shear strength is considered to be random. Due to the fixed ratio between the peak and residual shear strengths, the residual undrained shear strength is also spatially variable. A total of 8,850 MPs have been used, onto which a random field is mapped for the peak undrained shear strength based on a mean of 18 kPa, $V = 0.2$ and $\theta_v = 1$ m, as shown in Figs. 16.4 (a), 16.5 (a), 16.6 (a). In subsequent sub-figures, the current undrained shear strength is shown, i.e. with a reduced value in the shear bands, due to the strain-softening material model. Six Monte Carlo simulations with $\theta_h = 1$, 2, 3, 6, 12 and 48 m have been performed to study the effect of θ_h on a retrogressive failure mechanism, because an accurate determination of θ_h is difficult, as mentioned in Section 16.2.1.

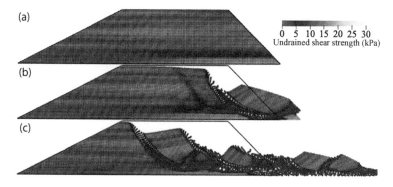

FIGURE 16.4: Retrogressive slope failure of the steep slope under influence of soil heterogeneity ($\theta_h = 48$ m).

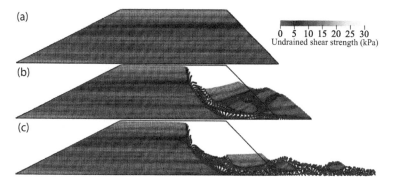

FIGURE 16.5: Stable slope after initial failure mechanism of the steep slope under influence of soil heterogeneity ($\theta_h = 48$ m).

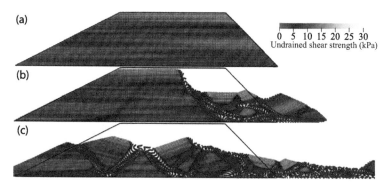

FIGURE 16.6: Failure of steep and gentle slopes under influence of soil heterogeneity ($\theta_h = 48$ m).

The embankment is located on top of a frictional bottom boundary condition, i.e. friction in the horizontal direction and fixed in the vertical direction, and the vertical boundary conditions, which are fixed in the horizontal direction and free in the vertical direction, have only a small influence on the run-out of the embankment. As the consequence of initial failure was the objective of the study, most slopes in this study were unstable under their own weight after the in-situ stresses were generated by gravity loading. The factor of safety of the slope with the mean material properties is $FOS \approx 0.9$.

To indicate the effect of spatial variability on slope failure, the embankments shown in Figs. 16.4 to 16.6 are for the case of $\theta_h = 48$ m. These three realisations have been chosen from the thousand realisations in total, because they are representative of the different failure mechanisms occurring. As expected, and similar to a homogeneous slope, the steeper sloping side tends to fail first. Fig. 16.4 (b) indicates that slope failure can initiate in soft layers above the toe of the slope. However, due to the fact that the mean undrained shear strength is constant with depth, deep failures are still far more likely, as shown in Figs. 16.5 (b) and 16.6 (b).

After the initial failure, the sensitivity of the clay causes a retrogressive failure mechanism within the weak layer of the slope shown in Fig. 16.4 (c). Even though the steepness of the slope after the initial failure mechanism in Fig. 16.5 is higher than in Fig. 16.4, retrogressive failure does not occur within the former cross-section due to the absence of a weak layer at the bottom of the embankment scarp. A small amount of softening is still visible within the embankment, especially in the weak layer close to the crest of the embankment, seen via the darker (weaker) colour in the weak band approximately one third of the slope height from the top. Enough residual strength is present within the embankment to prevent a further collapse. Fig. 16.6 shows that failure of the complete slope is another possible outcome. Due to the presence of a relatively weak layer along the entire bottom of the embankment, a complete collapse eventually occurs on both sides of the embankment. As expected the steeper slope experiences a quicker failure mechanism compared to the gentler slope.

Combining the results of the six Monte Carlo simulations gives rise to the reliability curves shown in Fig. 16.7. Fig. 16.7 (a) presents the reliability of the slope against an initial failure. At most, 20% of the embankments remain completely stable over the 12 seconds of the simulation. As the amount of layering increases, i.e. as θ_h increases, more embankments remain stable, due to the fact that a strong layer close to the bottom of the slope prevents an initial failure from occurring (i.e. due to the larger θ_h making it harder for potential failure surfaces to avoid the stronger layer).

After the initial failure, secondary failures can occur which can cause failure through the entire crest, i.e. complete failure of the embankment. From the three slope failures shown earlier, only the slope in Fig. 16.6 has a complete slope failure. Fig. 16.4 is close to a complete failure, whereas Fig. 16.5 has to overcome a significant amount of residual strength before a complete failure would be reached. Fig. 16.7 (b) shows the reliability of the various embank-

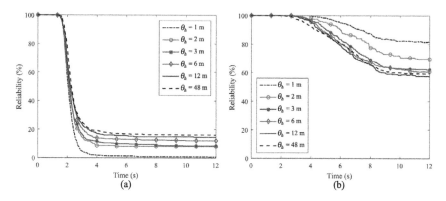

FIGURE 16.7: Reliability of 2D embankments with different horizontal scales of fluctuation θ_h. (a) Reliability against initial failure. (b) Reliability against complete failure.

ments against a complete failure. Over 50% of the slopes prevent a complete collapse, even though only 20% prevent an initial failure. This indicates that the residual strength of these embankments is generally high enough. Somewhat counter intuitively, the reliability against complete failure decreases with a larger θ_h even though the reliability for initial failure increases. This is likely to be due to the occurrence of an initial failure mechanism in a weak layer. Retrogressive failure in this weak layer is more likely for a larger θ_h, leading to an increase in the number of complete collapses.

16.3.2   3D embankment

The work on 3D RMPM has only recently started, but already appears to be an important tool in the modelling of heterogeneous slopes. A 3D homogeneous slope stability analysis produces similar results compared to a 2D analysis, with the exception of edge effects. However, for a heterogeneous slope a 2D analysis inherently assumes an infinite horizontal scale of fluctuation in the out-of-plane direction; i.e. it represents a failure mechanism along the entire length of the embankment. In practice, initial slope failures are often small compared to the slope dimensions and retrogressive failure mechanisms can cause further collapse. As shown in Fig. 16.8, 3D RMPM is capable of producing an initial localised failure mechanism, which then leads to collapse in other parts of the slope. However, only a single realisation has been analysed due to the computational cost of the 3D analysis. In total, 248,000 MPs have been used to model a 20 m wide, 5 m tall and 50 m long slope, composed of a strain-softening soil with a peak to residual cohesive strength ratio of 4. The material properties are: mean peak undrained shear strength of 20 kPa, $V = 0.3$, $\theta_v = 2.5$ m and $\theta_h = 10$ m. The embankment has been placed on a fixed horizontal boundary. Moreover, explicit MPM has been used without

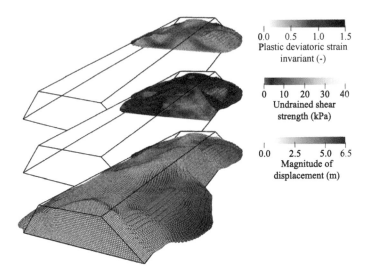

FIGURE 16.8: 3D slope failure mechanism of a heterogeneous slope with $\theta_h = 4\theta_v = L/5$. (Top) Cut through the slope with plastic deviatoric strain invariant represented by red contour scale. (Middle) Cut through the slope with mobilised undrained shear strength represented by blue contour scale. (Bottom) Full domain with displacement magnitude represented by green contour scale.

any improvement techniques. Due to the relatively crude discretisation, as well as the absence of stress oscillation and regularisation techniques, the failure bands are rather thick. The model, therefore, most likely over-predicts the strength of the slope, as is common in all mesh-based strain-softening models. The shear band, which may in practice be thin, is limited to being at least one element thick, providing additional strength to the model.

16.4   Closure

An overview of the random material point method (RMPM) is presented together with 2D and 3D examples of slope failure and reliability. The random finite element method (RFEM) proves to be a useful reliability analysis tool, with the examples shown here indicating that RMPM is a good large-deformation extension of RFEM. RMPM allows for a risk assessment analysis as both failure probability and the consequence of failure can be modelled.

17

Jacked Pile Installation in Sand

A. Rohe and P. Nguyen

17.1 Introduction

During installation, jacked displacement piles are pushed into the ground causing severe distortion of the surrounding soil. As a result, large shear strains and significant increase of stresses surrounding the pile tip and shaft can be observed (e.g. [255]). This complex installation process influences the bearing capacity of the pile foundation substantially.

In engineering practice, the bearing capacity of simple foundation structures is often estimated using empirical methods without taking installation effects into account (e.g. [238]). If a more reliable prediction of the load-settlement behaviour is required, empirical methods are limited in use (see Chapter 1). In these cases, the finite element method (FEM) can be an alternative approach. It has been used by many authors (e.g. [117, 274]) to assess the ultimate bearing capacity of foundations assuming small strain analysis. Although those models provide good basic techniques to determine the pile capacity, the influence of the installation process and occurring large deformations is not taken into account. Installation effects have been included in the analyses of Baligh [18] and Broere and Tol [46] by applying prescribed stresses or strains to the soil along the pile shaft. However, the installation process itself cannot be modelled as classical FEM suffers from severe mesh distortion when dealing with large deformations. Engin et al. [94] presented a simplified FEM technique to model a jacked pile, i.e. the so-called press-replace technique, which does not suffer from mesh distortion. However, the continuous flow mechanism of soil being pushed from below the pile tip towards the shaft is not properly captured, leading to an underestimation of predicted pile capacity. Current FEM approaches for determining pile capacity do not take into account the change of soil state and properties due to installation, or do so in a simplified manner only. Consequently, the accuracy in predicting the pile capacity by FEM is rather limited.

Choosing a suitable soil constitutive model plays a crucial role in numerical modelling of pile installation. The hypoplastic (HP) constitutive model in the formulation of Von Wolffersdorff [322] is chosen as it is incorporating dilation, contraction and the dependence of stiffness on stress and density. Centrifuge tests (e.g. [84, 156]) showed that very high stresses can occur at the pile tip during the installation process. The stress range may increase by up to 100 times its initial value. Such a large stress increase in the soil leads to dilation, grain crushing and a significant reduction of soil strength (e.g. [182]). These effects have to be addressed by adapting the HP model parameters accordingly in order to simulate the pile installation process accurately.

Numerical simulations are presented for modelling the installation process of jacked displacement piles in sand using *Anura3D* [14] and the HP model of Von Wolffersdorff [322]. Two different initial densities of sand are investigated, namely loose sand and medium-dense sand. Simulations of a static load test following the pile installation process are carried out. The numerical simulation results are compared with experimental results of centrifuge tests of jacked pile installation and static load tests.

17.2   Numerical model

Centrifuge tests [137] are used for the validation of the numerical simulations. The tests were carried out in a steel container of 0.6 m diameter and 0.79 m height, which was filled with saturated sand up to a height of 0.46 m. Due to a very low loading rate, excess pore water pressures can be assumed negligible during pile installation and static load tests (SLT). The model pile was a steel pile with a length of 0.3 m and a diameter of 0.0113 m. During preparation at normal gravity (1 g), the pile was initially embedded at ten times its diameter (10D). The installation of the model pile started after the centrifuge had been accelerated to a level of 40 g. During the installation phase, the pile was installed in-flight to a final depth of 20D and with a velocity of $1.67 \cdot 10^{-4}$ m/s. After the installation, a series of SLTs with a velocity of $1.67 \cdot 10^{-6}$ m/s and an additional displacement of 0.1D were performed to determine the pile capacity.

The numerical simulations are performed using a geometry at prototype scale. The geometry of the numerical model is shown in Fig. 17.1. The right side boundary is at a distance of 26D from the pile centre, which is identical to the size of the sample container in the centrifuge experiment. All boundaries are fixed in normal direction and free in lateral directions. The pile is modelled as a rigid body penetrating the soil. The shape of the pile tip is flat, which is identical to the centrifuge test. However, the edge of the pile tip in the simulation is slightly curved to avoid numerical difficulties due to locking (Fig. 17.1 a). The simulations are carried out with one-phase drained material behaviour (see Chapter 2) since no changes in pore pressure were recorded in the centrifuge tests due to the low applied loading rates.

A contact algorithm in combination with the moving mesh algorithm (see Chapter 6) is used to model the frictional contact between pile and soil. The friction angle of sand against polished steel of the model pile is determined to be 11° as suggested by Murray and Geddes [210]. The mobilised wall friction angle may be different along the mantle due to varying stress level. However, for simplification reasons it is reasonable to assume a constant wall friction coefficient along the pile (e.g. [116]). The chosen wall friction coefficient for all simulations in this chapter is assumed based on the characteristics of a polished steel pile surface, which is $\mu = \tan \varphi = 0.194$. The influence of μ on the load-displacement curve is investigated and shown in Fig. 17.11.

The mesh (Fig. 17.1 a) is formed of 4-noded tetrahedral elements with linear interpolation of displacements. It has a total of 26,826 elements, including the initially inactive elements, for a total 152,020 material points (MPs). The mesh is refined near the pile. The inactive elements above the soil surface can be activated during the calculation process as MPs are entering. Although the considered problem is axisymmetric, the simulations are 3D due to the nature of the MPM implementation. A 20° section of the axisymmetric problem is considered for discretisation. For smaller angles elements near the vertical axis can generate numerical inaccuracies due to the extreme aspect ratios.

The K_0-procedure (see Chapter 6) is used to initialise the stresses in the soil. The pile is initially embedded at 10D below the soil surface and is then

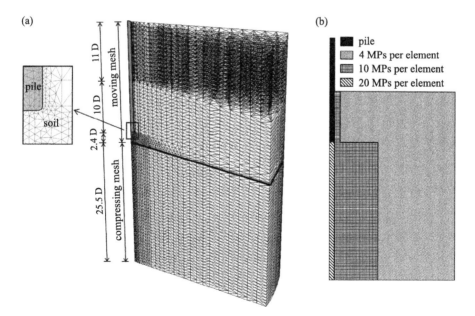

FIGURE 17.1: (a) Mesh discretisation with moving and compressing mesh (20° wedge) and detail of rounded pile tip. (b) Distribution of MPs per element (front view).

pushed an additional 10D into the soil with a velocity of 0.02 m/s. The penetration of the pile into the soil is modelled by applying a prescribed velocity on the top of the pile. A relaxation phase follows the installation phase. During the relaxation phase, the pile is slowly unloaded until the pile head force becomes zero. Finally, a SLT is performed with a velocity of 0.002 m/s. The velocities used in the numerical analyses are higher than the values used in the centrifuge test to optimise the calculation time. It has been confirmed in a parametric study that the increased loading rate does not significantly influence the results for the drained type of material behaviour.

The evolution of shear bands in granular material is strongly related to its micro-properties. Shear band thickness is influenced by the soil grain size, which is difficult to model with a continuum model. In order to properly describe the behaviour of granular materials with shear localisation, MPM may need regularisation by using micro-polar or non-local terms, for example, but this topic falls out the scope of this chapter.

17.3   Constitutive model

During the installation process the soil around the pile tip is exposed to (very) high stress levels. The centrifuge tests show a maximum stress at the pile tip towards the end of the installation process of about 8.5 MPa for medium-dense sand (Fig. 17.3 a) and 5.5 MPa for loose sand (Fig. 17.3 b). Leung and Touati [182] performed several triaxial tests on very dense Fontainebleau sand and indicated that with increasing cell pressure the peak friction angle and dilatancy decrease significantly. No dilation at all is observed at very high cell pressures. This stress dependency of the friction and dilation angle for sands is also confirmed by Bolton [40]. Therefore, the HP model used for the numerical simulations must be able to account for an appropriate soil response over a wide range of stress levels to avoid an overestimation of the pile force in the numerical simulation of the centrifuge tests.

In the centrifuge tests, Baskarp sand with a mean grain size diameter of $d_{50} = 130\mu$m was used. The tests were conducted using medium-dense sand with an initial void ratio of $e_0 = 0.68$ ($I_D = 54\%$) and loose sand with an initial void ratio of $e_0 = 0.75$ ($I_D = 36\%$). Both tests were simulated using a HP constitutive model in the formulation of Von Wolffersdorff [322] to model the soil behaviour at different initial densities. Several simulations of triaxial element tests at confining stresses below 1 MPa were done for both initially loose and dense sands to calibrate the HP parameters as given in Table 17.1 and referred to as *HP-original*.

For element test simulations with increasing cell pressure and using the original HP parameter set, the observed friction and dilation angles are only slightly reducing. [128] indicated that it is possible to control the peak state in a triaxial compression test by considering the exponent α, which reduces significantly with decreasing friction angle. Combination with the stress-dependency of the friction allows determining a corresponding value of α for each stress level. Another set of triaxial element tests using the newly proposed HP parameter set in which a value of $\alpha = 0.02$ is used show a reduced contractive behaviour. This set of parameters is referred to as *HP-modified* and the parameters are given in Table 17.1. The comparison for both parameter sets for a cell pressure of 10 MPa is shown in Fig. 17.2.

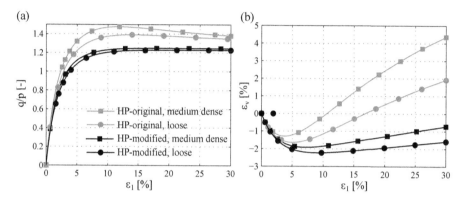

FIGURE 17.2: Comparison of triaxial element test simulations at 10 MPa cell pressure on Baskarp sand using the original and modified hypoplastic (HP) parameter set. (a) Stress ratio q/p' vs. axial strain, and (b) volumetric strain vs. axial strain.

TABLE 17.1

Hypoplastic parameters for Baskarp sand.

Parameter	Symbol	Unit	HP-original	HP-modified
critical state friction angle	φ_{cs}	°	31	31
granular stiffness	h_s	MPa	4,000	4,000
compression exponent	n	-	0.42	0.42
minimum void ratio	e_{d0}	-	0.548	0.548
critical void ratio	e_{c0}	-	0.929	0.929
maximum void ratio	e_{i0}	-	1.08	1.08
densification exponent	α	-	0.12	0.02
densification exponent	β	-	0.96	0.96

17.4   Load-displacement curve during installation phase

The numerical simulation results of the installation phase are compared with the experimental results of the centrifuge test. Fig. 17.3 shows the evolution of the axial force on the pile head during the installation phase for initially medium-dense (left) and loose sand (right). The simulation using the original HP parameter set significantly overestimates the pile head force after 10D penetration compared to the experimental results. Simulations using the modified HP parameter set show a much better agreement with the experiments. Note, however, that using the modified HP parameter set the response at the very beginning of the installation process (up to 0.5D penetration) is less stiff than observed in the centrifuge tests. Here, the original parameter set is performing much better, which makes sense, as they were calibrated for low stress levels.

Some oscillations in the numerical results can be observed in Fig. 17.3. They can occur when MPs cross element boundaries. These could be reduced by applying one of the recent extensions to MPM (see Chapter 3) but this falls out of the scope of this chapter.

17.5   Load-displacement curve during static load test

After the pile has been installed to a depth of 20D, it is unloaded and a pile load test is carried out. During the unloading phase, the pile head is slowly pulled out at a prescribed velocity of 0.0001 m/s until the pile head force

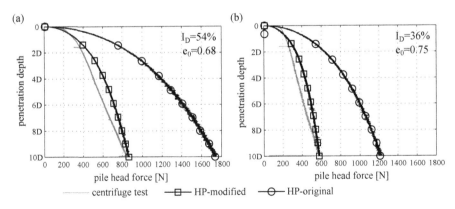

FIGURE 17.3: Load-displacement curve of the pile head during installation phase for (a) medium-dense and (b) loose sand. Comparison of experimental results and MPM simulations with different hypoplastic (HP) parameter sets.

becomes nearly zero. The displacement of the pile head during this phase is less than 0.02D, which corresponds to the centrifuge test (Fig. 17.4).

The load-displacement curves of the calculated SLT compared with the centrifuge tests are plotted in Fig. 17.5. The base bearing capacity in the simulation is in good agreement with the centrifuge test results. Although the stiffness of the load-displacement curves during SLT between the simulation and centrifuge test is slightly different, the ultimate bearing capacity at 0.1D penetration is corresponding well, for both loose and medium-dense sand.

Simulations, in which the SLT is performed after pre-embedding the pile at 20D, a so-called wished-in-place pile, are also shown in Fig. 17.5. Clearly, without considering installation effects the calculated bearing capacity of the pile is significantly lower than the test results. This emphasises the importance of accounting for installation effects when simulating the SLT. The ratio of base capacities of non-displacement and displacement piles ranges from 0.18 for loose sand to 0.33 for dense sand according to a data base in [110]. The ratios in Fig. 17.5 are between 0.31 and 0.34 for the base capacity after penetration of 10D.

The SLT results are compared with the prescribed load-displacement curves in NEN 9997-2011 [1], the Dutch application document of Eurocode 7 which is, to our knowledge, the only Eurocode adaptation that provides (dimensionless) numerical values for load-displacement behaviour. The normalised plots given in Fig. 17.6 show the load-displacement curves from the numerical simulations of a medium-dense sand in comparison to the design curves suggested in NEN 9997-2011. For a reliable design using this code, the ultimate base capacity is determined at 0.1D displacement for a driven pile and at 0.2D displacement for a bored pile. The resulting normalised base resistance curve for the simulation of the jacked pile SLT is in good agreement with curve 1 for driven piles from the NEN 9997-2011 code. This demonstrates

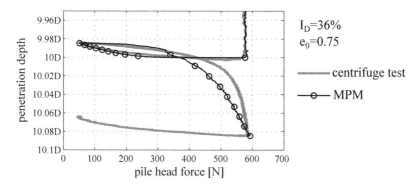

FIGURE 17.4: Load-displacement curve of the pile head during unloading and reloading phase for loose sand. Comparison of experimental results and MPM simulations.

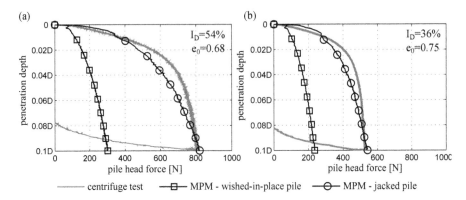

FIGURE 17.5: Load-displacement curve of the pile base during SLT for (a) medium-dense sand, and (b) loose sand. Comparison of experimental results and MPM simulations for wished-in-place pile and jacked pile.

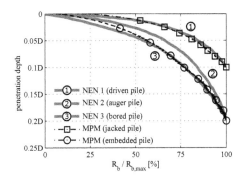

FIGURE 17.6: Comparison of the design load test suggested by Dutch code (NEN 9997-2011) with MPM simulations for jacked and embedded pile SLT. R_b is effective stress of pile tip during SLT and $R_{b,max}$ is effective stress of pile tip after SLT.

the importance of including the pile installation in the simulation and using an advanced constitutive model, e.g. HP model, for modelling pile load tests. The curve that simulates the pre-embedded pile shows a good correspondence with the curve suggested by curve 3 for a bored pile in NEN 9997-2011 code.

17.6   Stress state after pile installation

For both loose and medium-dense sand, a significant increase of radial stresses due to pile installation is observed after 10D penetration (Fig. 17.7). The

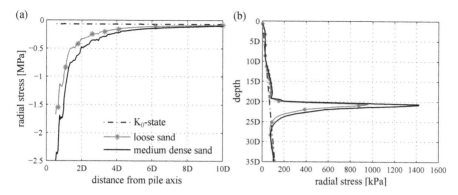

FIGURE 17.7: Radial stresses after 10D penetration for medium-dense and loose sand at (a) horizontal cross section AA', and (b) vertical cross section BB' close to the pile shaft.

radial stresses are increased along the pile shaft and peak near the pile tip. Two cross sections, AA' (close to the pile tip) and BB' (along the pile shaft), are investigated in detail for the change in radial stresses (see Fig. 17.8 for the location of cross sections). The radial stresses after 10D penetration at a horizontal cross section AA' are plotted in Fig. 17.7 (a). Close to the pile, the initial radial stresses are increased by a factor of almost 75 for medium-dense sand and a factor of 50 for loose sand. At a distance from the pile centre larger than 10D, no stress changes can be observed for either case. That indicates that a container diameter of 26D for the centrifuge tests and the numerical simulations can be considered as sufficiently large to exclude boundary effects.

Fig. 17.7 (b) shows the radial stress distribution along the vertical cross section BB'. It should be noted that the position of the pile in Fig. 17.7 (b) is initially at $z/D = -10$ and -20 at the end of the penetration process. The radial stresses along the pile shaft after 10D installation are only 1.5 to 2 times the initial K_0 value. This may be caused by a loosening zone observed along the pile shaft in the HP model simulations (as can be seen in the next section), and a relaxation of the radial stress at the shaft can be expected. The radial stress reaches a peak value at pile tip level and drops down to a value closer to the K_0-state below the pile tip. The observed change of radial stresses along a vertical cross section near the pile shaft is in good agreement with [187].

17.7 Density change after installation

The density distribution after 10D penetration is shown in Fig. 17.8. Significant densification of the soil around the pile after installation is observed for both medium-dense and loose sand, except in a small dilative zone close to the

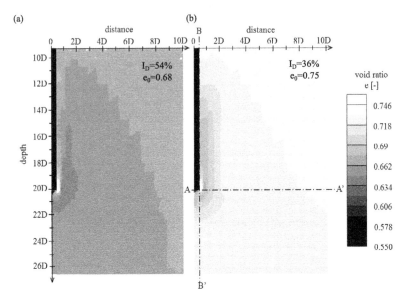

FIGURE 17.8: Void ratio change after 10D penetration. MPM simulation results for (a) medium-dense sand, and (b) loose sand.

end of the pile shaft and around the corner of the pile tip. The dilative zone may be caused by the high shear strains in the soil surrounding the pile corner. Hence, the compaction of the soil close to the pile corner is superimposed by the shearing process, while at a larger distance from the pile corner the compaction is dominant, which is in line with the findings of other researchers (e.g. [187]).

In Fig. 17.9, the distribution of void ratio in vertical cross sections at different radial distances from the centre of the pile after 10D penetration is shown for both medium-dense sand (Fig. 17.9 a) and loose sand (Fig. 17.9 b). For medium-dense sand at a distance of 0.5D, the increase in void ratio is observed very close to the pile shaft, whereas for loose sand it is only seen at the edge of the pile tip. For both medium-dense and loose sands, the lowest void ratio value is found just under the pile tip, which is considered as the zone with highest densification. The densification reduces gradually with increasing distance from the pile. At a distance of 10D from the pile centre and 5D to 7D below the pile tip, no change in void ratio is observed.

To better understand the changing void ratio during pile installation, the void ratio at a vertical cross section $r/D = 0.5$ is plotted at different stages of the installation process as shown in Fig. 17.10. At the beginning of the installation, the soil under the pile tip is pushed downwards and compacted, and a loosening zone forms around the corner of the pile which causes a strong increase of the void ratio. The peak void ratio (in the dilation zone) increases and moves onwards together with the pile tip as it penetrates. This

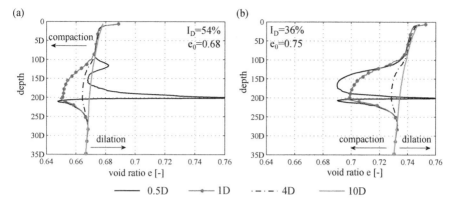

FIGURE 17.9: Void ratio distribution along the pile shaft at different distances after 10D penetration for (a) medium-dense sand, and (b) loose sand.

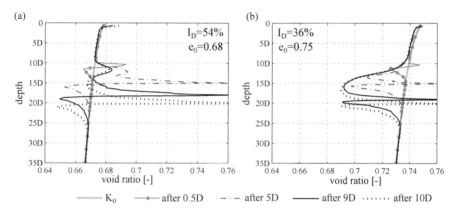

FIGURE 17.10: Void ratio distribution along the pile shaft at distances of 0.5D by time after 0.5D, 5D, 9D and 10D penetration for (a) medium-dense sand, and (b) loose sand.

explains the sudden increase of void ratio at $z/D = 20$. The oscillations are the remnants of the high peak void ratios from on-going penetration, which decrease by relaxation due to compaction in the far field.

17.8 Parametric study

Several simulations are performed to study the influence of contact properties, domain size and number of MPs on the evolution of the load-displacement curve during penetration.

Fig. 17.11 shows the influence of the coefficient of friction $\mu = \tan\varphi$ between pile and soil on the load-displacement curve during installation. The value of μ has a larger influence on the evolution of total pile shaft force (Fig. 17.11 a), rather than the evolution of total pile tip force F_t (Fig. 17.11 b).

The influence of the radial boundary distance r/D is shown in Fig. 17.12 (a). Due to the boundary effect, the simulation 10D results in a larger pile load force compared to 35D. However, this difference is quite small, about 8% of the total load value. The simulation for 25D, which is the domain applied for all simulations, gives quite similar result to the one for 35D. Thus, there is no or only negligible influence of domain size on the calculation results.

In order to investigate the influence of number of MPs, the initial number of MPs per element is varied between 4, 8 and 10 for all elements in the mesh. Results of the load-displacement curve are shown in Fig. 17.12 (b) which only slightly change for a different number of MPs. Hence, the amount of MPs per element in this simulation has no or only negligible influence on the load-displacement curve.

17.9   Closure

Several numerical analyses were performed using MPM to model the installation process and static load capacity of a jacked displacement pile in sand. The geometry and soil parameters of the simulations were determined based on a centrifuge test in which the pile was installed in-flight at 40 g. The results of the simulations were compared with the centrifuge experiments for validation.

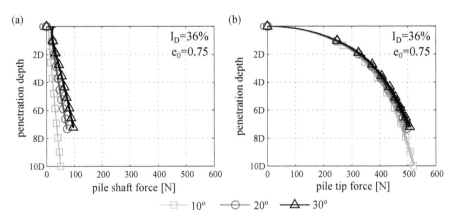

FIGURE 17.11: Influence of wall friction on the load-displacement curve for loose sand. (a) Pile shaft force, and (b) pile head force.

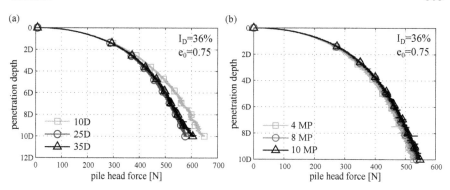

FIGURE 17.12: Load-displacement for loose sand. (a) Influence of boundary distance, and (b) influence of number of MPs per element.

The results of this chapter denote the importance of considering installation effects in numerical simulations to determine the pile capacity. Moreover, it shows the capability of the used numerical scheme to simulate the installation process of jacked piles in sand. MPM, instead of classical FEM, is well suited to model the large deformations and high stresses encountered during the pile installation process. In order to successfully model the centrifuge test, it is necessary to account for a reduction of the friction and dilation angle of the soil at very high stress levels. With this reduction, the MPM simulations show good agreement with the centrifuge test results for both the installation process, as well as the SLT performed after the installation. The numerical analyses of the pile installation show significant differences in the soil stresses and strains around the pile after installation compared with the initial K_0-state. During installation, the soil is pushed downwards and alongside the pile, which leads to significant densification around the pile and very high lateral stress at the pile tip. Such influences of the installation extend to about eight times pile diameter in the lateral direction from the centre of the pile and seven times pile diameter below the pile tip. As a consequence of the change in soil state after installation, a significantly higher pile bearing capacity is observed during SLT, as compared to simulations without installation effects. Using modified HP model parameters for the determination of the pile capacity results in a good agreement with the load-displacement curve suggested in NEN 9997-2011 code.

18

Cone Penetration Tests

F. Ceccato and P. Simonini

18.1 Introduction

The numerical simulation of penetration processes is challenging because it involves large deformations, soil-structure interaction, non-linear constitutive behaviour of soil, and the interaction between the solid skeleton and the pore fluid. Standard Lagrangian FEM is not suitable for these problems because of element distortions, but Arbitrary Lagrangian-Eulerian methods (ALE) [142, 186, 236] and meshless methods such as MPM [58, 224] have been successfully applied.

This chapter shows the applicability of the two-phase single-point MPM (see Chapter 2) to the simulation of soil penetration problems such as the cone penetration test (CPT).

18.2 Cone penetration test

CPT is a widely used in situ soil testing technique, which consists in pushing a conical tip into the ground at a constant rate while measuring the resistance offered by the soil to the advancement of the instrument. According to the ISSMFE IRTP standard, the rate of penetration should be 20 ± 5 mm/s. The reference test equipment consists of a $60°$ cone, with 10 cm^2 base area and 150 cm^2 friction sleeve located above the cone. During penetration, the resistance at the cone tip q_c and the resistance of the friction sleeve f_s are recorded with specific sensors. The device can be equipped with a pore pressure transducer (piezocone or CPTu) that is usually placed at the cone shoulder (u_2 position). However, it can also be placed in the cone face (u_1 position) or above the friction sleeve (u_3 position). Extra sensors are available in the market to measure other soil properties such as the shear wave velocity (sCPT) and the dielectrical conductivity. Fig. 18.1 shows a schematic representation of the device.

CPT is mainly used to determine the sub-surface stratigraphy and to estimate the geotechnical parameters of the soil for geotechnical site characterisation. The results are often directly applied for geotechnical design such as calculation of pile bearing capacity or settlement estimation. CPT measurements depend on the in situ mechanical response of the soil. Robertson [244] suggests the following soil behaviour type classification.

- Clay-like: typical of fine-grained plastic soils such as clay, in which the penetration occurs in undrained conditions and excess pore pressures are generated and do not have time to dissipate.

- Sand-like: typical of coarse grained soil such as sand, in which the penetration occurs in drained conditions and excess pore pressures may be generated but quickly dissipate.

- Transitional: typical of intermediate materials such as silt, sand-clay mixtures, tailings and are characterised by a complex penetration process and often the excess pore pressure partially dissipates during penetration (partially drained conditions).

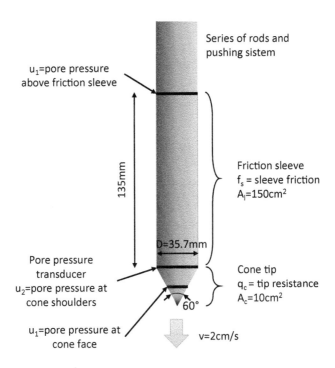

FIGURE 18.1: Schematic representation of the cone penetration test.

Most of the existing empirical or theoretical correlations between CPT or CPTu measurements and soil properties have been developed for undrained or drained conditions, and are thus inappropriate in intermediate soils. For this reason, it is important to gain a better understanding of the penetration process in these conditions. The problem has been mainly studied experimentally with centrifuge tests, calibration chamber tests, and in situ tests [154, 186, 218, 240, 259]. Only a few numerical studies investigated partial drainage in installation problems [123, 269, 338].

The CPTu device is used to estimate the soil consolidation coefficient by performing a dissipation test. It consists in stopping the penetration and monitoring pore pressure evolution along time. The results are interpreted with the aid of specific charts that correlate the time at which 50% of the initial pore pressure have been dissipated t_{50} to the consolidation coefficient [245]. The initial pore pressure distribution around the cone is a consequence of the installation process and it influences the following dissipation process. However, this is neglected in the simplified theories applied to interpret the results.

Although the entire range of drainage conditions can be simulated with the MPM two-phase formulation in drained and undrained conditions (see Chapter 2), the presence of the water can be taken into account in a simplified way applying the one-phase MPM formulation (see Chapter 2) to reduce the computational cost. In the first case, the pressure dissipates nearly instantaneously. Therefore, the presence of water can be neglected. In the latter, the soil-water mixture behaves as a single-phase material because the solid and the liquid phase move with the same velocity.

18.3   Numerical model

Cone penetration is an axisymmetric problem. However, only three-dimensional (3D) meshes could be used with the applied MPM code when this study was carried out. Therefore, the presented results are obtained with a full 3D implementation. In order to reduce the computational cost and taking advantage of the rotational symmetry of the problem, only a 20° slice is considered. The geometry and discretisation of the numerical model are illustrated in Fig. 18.2.

The dimensions of the cone correspond to the specification of ASTM D-5778 with apex angle $\alpha = 60°$ and cone diameter $D = 35.7$ mm. The standard device has a discontinuous edge at the base of the cone. At this location, boundary conditions are not uniquely defined, thus causing numerical problems which are circumvented by using a slightly rounded shape of the cone shoulders.

The dimensions of the model and the mesh size are determined through preliminary calculations as a compromise between computational cost and

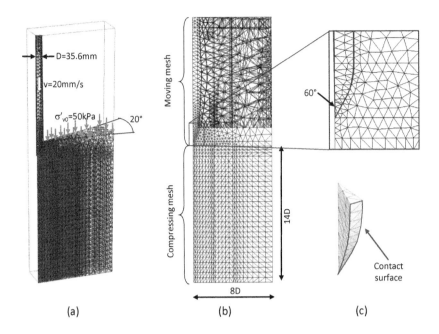

FIGURE 18.2: Geometry and discretisation of the CPT problem (a) MP discretisation, (b) connectivity plot, (c) detail of cone discretisation.

accuracy [52,60]. It extends 14D below the tip at the beginning of the computation, and 8D in the radial direction. These dimensions prove to be sufficient to prevent boundary effects. Like in FEM, a finer grid is necessary where high gradients of stress and strain are expected. That is near the tip, for instance. The mesh has 13,221 tetrahedral elements and 105,634 material points (MPs) are located in the initially active elements. Each element below and near the cone contains 20 MPs. The high density of MPs prevents small elements in the vicinity of the cone tip from becoming empty. Elements further away from the cone have 4 or 10 MPs.

In order to reduce the stress oscillations due to grid-crossing errors, Gauss integration is applied in fully filled elements (see Chapter 3). Those are elements in which the sum of the volumes of the contained MPs is greater than 90% of the element volume. Issues with volumetric locking, typical of low-order finite elements, are prevented with the use of a nodal mixed discretisation technique (see Chapter 6).

Displacements are constrained in the normal direction at the lateral surfaces and the bottom of the mesh is fully fixed. The side boundaries of the 20° slice are impermeable because they correspond to symmetry axes of the problem. The bottom and the lateral boundaries are permeable. A prescribed velocity of 2 cm/s is applied to the MPs and to the nodes of the structure, which thus move downward as a rigid body.

The contact between the soil and the piezocone is modelled with a contact algorithm presented in Chapter 6. The soil-cone interface is characterised by a friction coefficient $\mu = 0.3$, which is a typical value for clay in contact with steel [234].

The moving mesh concept illustrated in Chapter 6 is used to avoid difficulties with the application of the prescribed cone velocity. Because of this procedure, the fine mesh is always kept around the cone. The moving mesh zone is attached to the penetrometer and moves with the same displacement as the cone. The elements of this zone keep the same shape throughout the computation whereas the elements in the compressed zone reduce their vertical length as shown in Fig. 18.3. The elements of the compressed zone must keep a reasonable aspect ratio until the end of the simulation. Thus, the initial discretisation should be cautiously defined.

The use of this strategy is also beneficial in conjunction with the contact algorithm because the need for identifying the new soil-structure interface and the unit normal vectors during the computation is eliminated. Hence, the inaccuracy related to recomputing them is prevented.

In CPT, the penetration rate is very slow. Thus, the problem is almost quasi-static, which might make the simulation with an explicit dynamic code inefficient. In order to increase the computational efficiency, the mass scaling approach illustrated in Chapter 6 is used. It consists in multiplying the mass

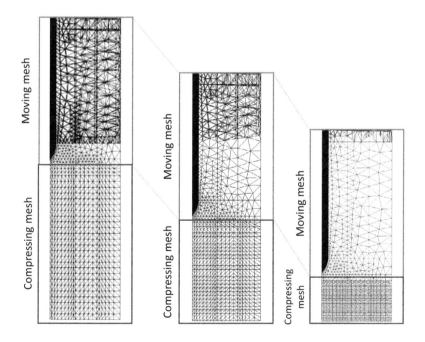

FIGURE 18.3: Moving mesh applied to CPT simulation.

matrices by a factor β and, thus, increasing the critical time step by a factor $\sqrt{\beta}$ (see Chapter 3).

An additional issue to be considered in dynamic analyses is the presence of oscillations due to wave reflections at mesh boundaries. A possible solution to this problem is the introduction of a local damping (see Chapter 6). The optimal mass scaling factor $\beta = 400$ and local damping factor $\alpha = 0.05$ are determined with preliminary analyses. Further details can be found in [52, 58].

For simplicity reasons, gravity is neglected. Therefore, the initial stresses are constant with depth and the pore pressure is zero. The initial vertical and horizontal effective stress are $\sigma'_{v0} = 50$ kPa and $\sigma'_{h0} = 34$ kPa, respectively.

The soil response is modelled with the Modified Cam-Clay model [248] (see Chapter 8). The material parameters assumed for the numerical analyses are summarised in Table 18.1. The soil is assumed normally consolidated and is typical of kaolin.

18.3.1   Simulation of cone penetration

If the pressure dissipation rate is relatively high compared to the penetration rate, the soil in the vicinity of the advancing cone consolidates during penetration thereby developing larger shear strength and stiffness compared to undrained conditions. This results in higher tip resistances. Finnie and Randolph [105] show that the effect of soil drainage can be taken into account by introducing a normalised penetration rate, also called normalised velocity (Eq. 18.1).

$$V = \frac{vD}{c_v} \tag{18.1}$$

where V is the normalised velocity, v the penetration rate, D the cone diameter and c_v the soil vertical consolidation coefficient.

The consolidation process near the advancing cone is not only affected by the permeability, but also the compressibility of the soil, the probe diameter and penetration rate play a significant role.

It is assumed that the soil compressibility during penetration is well described by the virgin compression index λ, and the consolidation coefficient

TABLE 18.1
Material properties used in CPT analyses

Parameter	Symbol	Unit	Value
Virgin compression index	λ	-	0.205
Swelling modulus	κ	-	0.044
Poisson ratio	ν	-	0.25
Critical state stress ratio	M	-	0.92
Initial void ratio	e_0	-	1.41
Bulk modulus of water	K_L	kPa	36,600
Overconsolidation ratio	OCR	-	1

can be estimated with Eq. 18.2. In the presented numerical simulations, the variation of V is obtained by changing the soil permeability k while keeping a penetration rate $v = 20$ mm/s.

$$c_v = \frac{k(1 + e_0)\sigma'_{v0}}{\lambda \gamma_w} \qquad (18.2)$$

Fig. 18.4 shows how the tip stress increases with the vertical displacement of the cone for different drainage conditions. As expected, the tip resistance increases by decreasing V that is moving from undrained to drained conditions. In case of $V = 1.2$, the tip resistance is 7% lower than for drained conditions, and in case of $V = 12$ the tip resistance is 5% higher than for undrained conditions. The steady state tip stress, which corresponds to the tip resistance q_c, is reached after a penetration depth which ranges from 5D in drained conditions to 7D in undrained conditions.

Figs. 18.5 and 18.6 show the distribution of deviatoric stress and excess pore pressure around the cone after a penetration of 10D, respectively. Reducing the normalised penetration velocity, the excess pore pressures also reduce while the effective stresses increase. The lowest deviatoric stress around the cone is observed in undrained conditions (Fig. 18.5 a), which corresponds to the undrained shear strength of the soil. In partially drained conditions, it is about 40 kPa in case of $V = 12$ (Fig. 18.5 b) and 90 kPa in case of $V = 1.2$ (Fig. 18.5 c). The highest deviatoric stress is about 110 kPa and is encountered in drained conditions (Fig. 18.5 d).

Approximately undrained behaviour is observed for $V = 12$ at which the

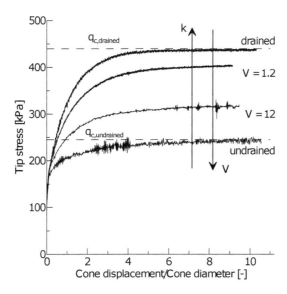

FIGURE 18.4: Evolution of tip stress with normalised cone displacement (after [54]).

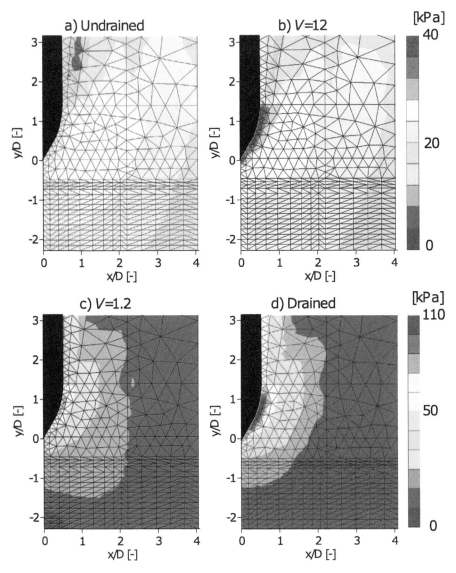

FIGURE 18.5: Deviatoric stress contour in (a) undrained conditions, partially drained conditions with (b) $V = 12$ and (c) $V = 1.2$, (d) drained conditions.

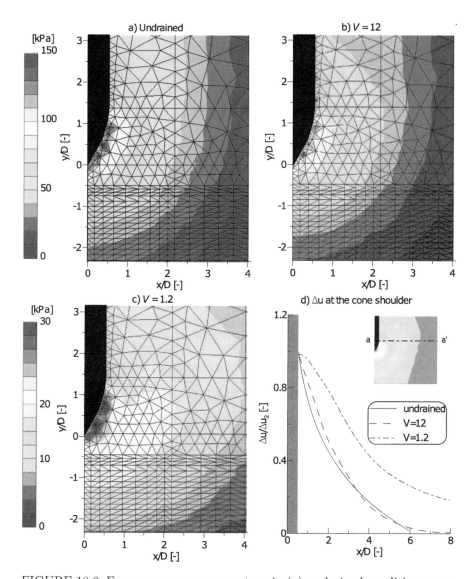

FIGURE 18.6: Excess pore pressure contour in (a) undrained conditions, partially drained conditions with (b) $V = 12$ and (c) $V = 1.2$. (d) Normalised distribution of excess pore pressure in a horizontal section at the shoulder of the cone.

excess pore pressure next to the tip is about 110 kPa (Fig. 18.6 b). For $V = 1.2$ the behaviour is nearly drained and the excess pore pressure is about 30 kPa (Fig. 18.6 c). The distribution of excess pore pressure along the radial co-ordinate x at the level of the cone shoulder is plotted in Fig. 18.6 d. Note that the pressure is normalised by Δu_2 (excess pore pressure at the cone shoulder) to emphasize that the gradient of excess pore pressure is lower in partially drained conditions than in undrained condition. The gradients of the excess pore pressure distribution after CPTu penetration govern the dissipation response when performing dissipation tests to estimate the consolidation coefficient as explained in the next section.

Assuming the undrained penetration as a reference condition, the normalised resistance and the normalised pore pressure are defined by Eqs. 18.3 and 18.4, respectively.

$$\frac{q_{net}}{q_{net,\text{ref}}} = \frac{q_{\text{c}} - \sigma'_{\text{v0}}}{q_{\text{c,undrained}} - \sigma'_{\text{v0}}} \tag{18.3}$$

$$\frac{\Delta u}{\Delta u_{\text{ref}}} = \frac{\Delta u}{\Delta u_{\text{undrained}}} \tag{18.4}$$

where $q_{\text{c,undrained}}$ and $\Delta u_{\text{undrained}}$ are the tip resistance and the excess pore pressure in undrained conditions, respectively.

Figs. 18.7 and 18.8 plot the normalised resistance and the normalised pressure as function of normalised velocity, and compare the results obtained between the MPM simulations and the experimental results for normally consolidated kaolin [186, 240, 259]. Constant values of these normalised parameters are obtained for $V \leq 0.2$ in drained conditions and $V \geq 60$ in undrained conditions.

Silty deposits are characterised by a consolidation coefficient that can vary from 10^{-3} m^2/s to 10^{-4} m^2/s for silty sand to 10^{-6} m^2/s to 10^{-7} m^2/s for silty clay. For the standard penetration rate, the normalised velocity varies approximately between 1 and more than 500, which corresponds to the range of partially drained conditions. For this reason, it is important to control the penetration velocity in these types of soil because cone measurements vary significantly with V and this can lead to misinterpretation of the results.

Correlations between cone measurements and shear strength parameters originally proposed for undrained or drained conditions are inappropriate for intermediate soils. Thus, numerical modelling offers a strategy to determine the material characteristics and study its response.

18.4   Simulation of pore pressure dissipation

In fine-grained soils, dissipation tests are often carried out with the CPTu device to estimate the consolidation coefficient. It consists in stopping the

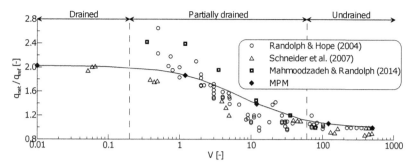

FIGURE 18.7: Normalised resistance as function of normalised velocity (after [60]).

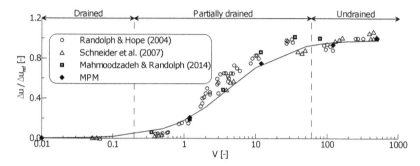

FIGURE 18.8: Normalised pressure as function of normalised velocity (after [60]).

penetration and monitoring pore pressure evolution along time. The results are interpreted with the aid of specific charts that correlate the time at which 50% of the initial pore pressure have been dissipated t_{50} to the consolidation coefficient [245]. These charts are based on analytical solutions of the consolidation problem assuming that the cone penetration occurs in undrained conditions [165, 293].

If the penetration occurs in partially drained conditions, the distribution of excess pore pressures around the cone differs form the undrained case and the shape of the pressure-time curve deviates from the theoretical solution thus leading to an incorrect estimate of the consolidation coefficient in intermediate soils [259, 269].

The problem is investigated with the two-phase single-point MPM (see Chapter 2). After a cone penetration of 10D, the penetrometer is stopped and the evolution of the consolidation process is monitored.

The time required for excess pore pressure dissipation around the piezocone is a function of the spatial extent of the excess pore pressure field which, in turn, depends on the rigidity index $I_r = G/c_u$ of the soil. Assuming an undrained cone penetration, Teh and Houlsby [293] suggest using a modified time factor T^* that is inversely proportional to the square root of I_r and validated in the range of $25 < I_r < 500$.

$$T^* = \frac{c_h t}{r^2 \sqrt{I_r}} \tag{18.5}$$

where $r = D/2$ is the radius of the penetrometer and c_h is the horizontal consolidation coefficient.

Fig. 18.9 shows the dissipation of excess pore pressure with normalised time after a 10D penetration at the normalised rate $V = 1.2$. The shape of the dissipation curve is significantly influenced by the position of the filter for the pore pressure measurements.

Fig. 18.10 shows $\Delta u / \Delta u_{\text{init}}$ (Δu_{init} = pressure before the start of dissipation test) over T^* measured at u_2 and u_1 positions (cone shoulder and middle of cone, respectively) for the normalised velocities of $V = 1.2$ (nearly drained penetration), 12 and 121 (nearly undrained penetration). Note that the curves shift to the right by increasing V. This means that the normalised pressure dissipates at a slower rate when the cone penetrates in partially drained conditions.

For the pore pressure filter at u_2 position, the time required to dissipate

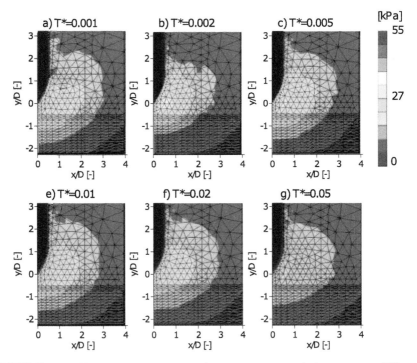

FIGURE 18.9: Excess pore pressure dissipation around the cone at different normalised time (V=1.2).

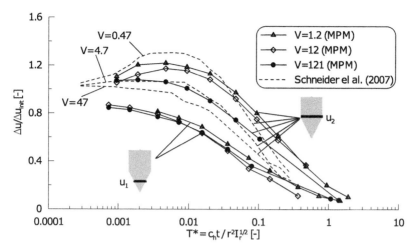

FIGURE 18.10: Excess pore pressure dissipation at different positions of the cone surface.

50% of the initial excess pore pressure for $V = 1.2$ is approximately 2.5 times that required in the case of $V = 121$. This difference is smaller, about 1.4, for the u_1 position. This leads to an underestimation of the consolidation coefficient when interpreting the dissipation test in soils where the cone penetrated in partially drained conditions.

At the u_1 position the pore pressure decreases monotonically. On the contrary, at the cone shoulder the pore pressure slightly increases at the beginning of the process. This is due to a pore pressure redistribution around the cone, as a consequence of the three-dimensional water flow. This effect is predicted analytically by Levadoux and Baligh [165] using the uncoupled consolidation theory.

The value of the pore pressure at the peak is a function of the normalised penetration velocity. In particular, it increases with the decrease of V. This behaviour is also observed during centrifuge tests on kaolin. There is a reasonable agreement between the numerical results obtained here and the experimental data by Schneider et al. [259].

18.5   Closure

In this chapter, the two-phase single-point MPM is applied to the study of cone penetration considering the effect of the drainage conditions. Moreover, dissipation tests are simulated. Large deformations, soil-cone contact, soil-water interaction and non-linear soil constitutive behaviour are taken into account.

Since the cone penetrates the soil at a slow rate, the mass scaling procedure ($\beta = 400$) is used to improve computational efficiency, and a small amount of local damping ($\alpha = 0.05$) is able to reduce oscillations of the tip stress due to dynamic waves without affecting the results.

The tip resistance increases moving from undrained to drained conditions because decreasing the penetration rate V the pressure dissipates and the soil consolidates, therefore, developing higher shear strength and stiffness.

Lower pore pressures and lower gradients of pressure are observed when the penetration occurs in partially drained conditions. This produces a shift toward the right of the normalised dissipation curves when increasing V and, thus, in an underestimation of the consolidation coefficient.

The shape of the dissipation curve depends on the location of the pore pressure transducer. The MPM simulations capture very well the experimental results on kaolin published by Randolph and Hope [240] and Schneider et al. [259] confirming the validity of the method.

19

Dynamic Compaction

A. Chmelnizkij and J. Grabe

19.1 Introduction

Dynamic compaction is a method used to increase the density of soil making the soil more stiff and increasing the peak strength (see Chapter 7). The compaction can be of several metres in depth and is carried out by inferring large amounts of kinetic energy to the ground. Two different techniques exist and are depicted in Fig. 19.1. The first technique is the free fall system, which consists in lifting a pounder with a crane and letting it free fall on the ground. The amount of energy transferred to the ground corresponds to the elevation height of the crane. The second technique follows the same principles as the first but the pounder is connected to the crane by a cable. Friction and braking forces affect the trajectory of the pounder and less kinetic energy is induced into the ground. However, the second technique is more practical as the pounder can be lifted and dropped again directly after impact without having to be reconnected.

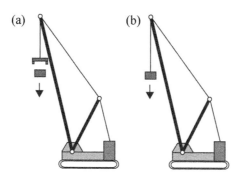

FIGURE 19.1: Compaction techniques: (a) free fall system and (b) rope-guided system.

The impact of the pounder induces plastic deformations, displacements and waves, which propagate in the ground and on the surface as shown in Fig. 19.2 (a). Plastic deformations and displacements cause a crater during the procedure. As the soil reaches a maximum density, the crater continues growing only due to displacement which can be observed in the elevation of the surface around it.

The compaction process is carried out multiple times until an entire region reaches the desired density. However, this is difficult to achieve. Once a point, or location, is sufficiently compacted, the procedure is repeated at another point. This sequential process spans a grid of compaction points over the whole site. It is common to use several grids, referred to as phases, as they are completed one after each other starting with a coarse grid with bigger spacing between the compaction points and closing the gaps with finer grids. Compacting neighbouring points leads to interactions between them and this can cause loosening at already-compacted points. This can be prevented by filling the craters with additional material, which is again compacted until the crater is closed. The compaction of water-saturated soils is especially challenging as the pore-water pressure might prevent the soil from compacting. Drains can be applied in such cases to accelerate the consolidation process but, due to the impulse load induced by the pounder, a big amount of the kinetic energy dissipates in displacement and not compaction. This chapter presents a case study of dynamic compaction and compares the numerical predictions with experimental data.

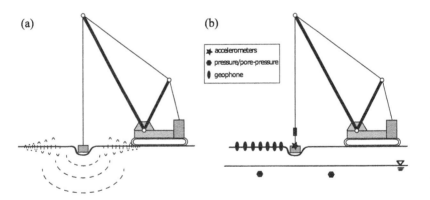

FIGURE 19.2: Schematic representation of (a) wave propagation during the impact and (b) measurements of dynamic compaction.

19.2 Experimental data

A field monitoring programme was set up in order to measure the process of dynamic compaction and is used for the validation of the numerical prediction presented in the next section. The results of the field measurements at the site are described in detail in [64] and [155]. The measurements were performed on a system with a permanent connection between pounder and crane. The movement of the pounder is captured by triaxial accelerometers attached to the pounder. The signal was recorded at a sampling rate of 20 kHz to cover the influence of higher frequencies during the impact. The range of the accelerometers was 600 g whereas the maximum measured vertical acceleration was around 110 g. The two measured locations were chosen to be as far as possible from other already-compacted points in order to mitigate the influence of the distorted soil. At both locations, the soil conditions were approximately homogeneous consisting of fine up to medium-coarse grains, and loose up to medium-dense sands. At each location, the pounder was dropped seven times. Both locations were part of the first compaction phase performed on a coarse compaction grid and, therefore, on approximately undisturbed soil.

As the movement of the pounder was almost only in the vertical direction during the fall and the impact, the vertical acceleration is a characteristic value describing the dynamics of the system. It includes the free fall behaviour as well as the interaction of the pounder and soil during the impact. Therefore, it is desirable to deduce information about the current soil state from those values. The slope up to the maximum vertical acceleration indicates the soil stiffness. As for softer soil, it is expected to be lower than for stiffer ones. The oscillation after the impact indicate the elastic and attenuated behaviour of the soil. Fig. 19.4 shows the measured vertical acceleration for two neighbouring locations, which were compacted by seven blows each. The distance between both locations was approximately 6 m.

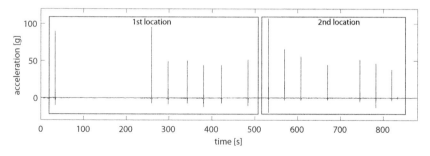

FIGURE 19.3: Measured vertical acceleration of the pounder at the first and second location.

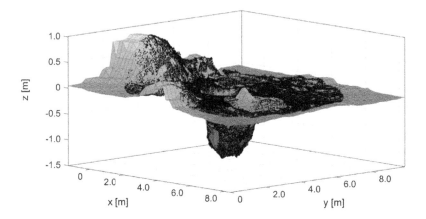

FIGURE 19.4: 3D topographic scan of the ground surface during dynamic compaction.

For reasons aforementioned, this data is mainly used for the validation of the numerical model presented in Section 19.3. Apart from the accelerometers, other measurements are performed to better estimate the amount of energy in the system such as the amount of energy dissipated by frictional contacts and brake forces. Strain gauges are attached between the pounder and the crane in order to measure the forces. The wave propagation at the surface, which is induced by the impact of the pounder on the ground, can be monitored with a series of geophones placed at equidistance on the ground surface. This allows calculating the propagation speed as well as the damping of the oscillation amplitude with distance. Pressure transducers can be used to monitor the compression wave within the soil. Topographic scans are also carried out in order to quantify the size of the crater and the surface heave. Fig. 19.4 shows an example of a 3D scan in which the crater can be observed. Fig. 19.2 (b) shows a schematic representation of the experiment layout performed as in [64] and [155]. In the following sections the focus will be on the accelerometer and three-dimensional (3D) scans only. Nevertheless, the additional data is also used for further validation of numerical models.

19.3   Numerical model

The motivation for simulation dynamic compaction is the necessity of optimising and improving the efficiency of the construction method. During the progressive impacts of the pounder, the completeness of the compaction is unclear and only estimated by the depth of the crater. This crater depth is estimated by measuring the length of the steel rope at the winch after each drop, which is not accurate and cannot be done with a free-fall system (Fig. 19.1 a). The measured acceleration of the pounder gives the opportunity to deduce more information and create a simple model [198, 264] giving an instant response on the compaction process. Simple replacement models consisting of spring-dashpot-elements were introduced to predict and estimate the compaction success [297]. The question arises whether focussing on only one dimension is sufficient to give satisfactory results. The investigation of this question can be carried out in many field tests by varying soil conditions in validated numerical models [206, 220], which allow changing all conditions without big expenses of time consumption.

In this section, a numerical model is introduced and is capable of approximately reproducing the measured vertical accelerations presented in Section 19.2. The advantage of such a numerical model is that, once it is validated, parametric studies can be carried out to investigate, for example, the influence of different soil and pounder parameters on the behaviour of the vertical acceleration of the pounder and conclude improvements for the method and its success. In addition the influence of one or more neighbouring compacted points is of interest during the compaction process. Therefore, a 3D model is necessary to take into account the influence of other compaction points as no symmetries can be applied. The presented model is considering two locations as in the field measurements in Section 19.1 compacted by a cubic shaped pounder.

As described in Section 19.1, different phenomena take place during the impact of the pounder and wave propagate beneath the soil surface. The study of propagating stress waves requires techniques which mitigate the reflections at the boundaries. Alternatively, a sufficiently big computational domain is required in which the waves are almost completely attenuated when reaching the boundary. Mitigation techniques are for example absorbing boundaries of infinite elements. However, those techniques are not straightforward for 3D problems in which the directions of the stress waves are changing during the simulation and unstructured meshes are used. The propagating waves can still be measured at a big distance from the compaction location and, therefore, can be seen as a far field phenomena. The displacement and compaction of soil are taking place at a distance of several metres from the compacted location and, therefore, can be regarded as a near-field phenomena. As previously mentioned, the objective of this study lies in the vertical acceleration of the pounder. Considering that the mass of the pounder is between 15-35 t, the influence of reflected stress waves on the vertical acceleration of the pounder is rather small. To further mitigate this influence, a sufficiently big computational domain was chosen.

The geometry of the numerical model is shown in Fig. 19.5. To avoid boundary effects and still perform an efficient computation the height of the

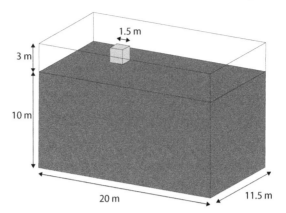

FIGURE 19.5: Geometry and dimension of the numerical model.

soil was chosen to be 10 m, the width 20 m and the depth 11.5 m. The boundary of the computational domain is fixed everywhere in the normal direction to prevent material points (MPs) from leaving the mesh. The model is discretised by 42,448 tetrahedral elements connecting 60,759 nodes containing 1 MPs/cell. The element size of the partially structured mesh ranged from 0.2 to 1.4 m. The pounder is cubic shaped with a side length if 1.5 m. It is initially placed 0.5 m above the soil surface. A material free void domain of 3-m height is defined above the soil surface. This domain represents the space where material points can move during the simulation. Once a material point reaches the boundary of this domain the boundary conditions prevent it from leaving the computational domain.

The mass assigned to a MP in the initialisation phase depends on the volume of the element where this MP is located. In the presented situation, 1 MP/cell was used for the calculation. Therefore, the weight of each MP is equal to the weight of the initial element volume. If the element size increases with depth, the surface elements are much lighter than other MPs located deeper in the ground. These MPs might be subjected to move further than the ones laying in bigger elements. During the simulation of dynamic compaction, a big crater is created by the impacts of the pounder and, therefore, MPs initially located few metres under the surface at a certain point will form the new surface of the crater together with lighter MPs, which were representing the surface before the impacts. In such a case, it is important to distinguish between the lighter and heavier MPs as they are now located in the same region but have a different inertia and, therefore, different movement.

The results of a calculation for two compaction locations are shown in Fig. 19.6. The left point was compacted first and the right point afterwards. It can be seen that some MPs are floating over the surface. The radius of those MPs indicates that they possess a small mass. The floating can be caused by a small upwards resistance force preventing the MPs from settling. This kind

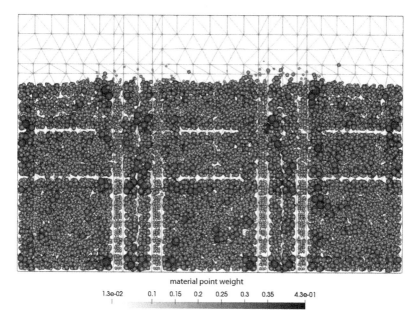

FIGURE 19.6: Representation of material point weight using weighted radius for the spheres.

of resistance force can be produced by MPs at the surface as they immediately interact with a floating MP as soon as it enters their element. It can also be caused by numerical errors leading to a small artificial resistance force strong enough to keep the light material points floating. As long as only a few small-mass MPs float around the crater without causing numerical issues or influencing the overall solution, the results are acceptable for further investigation. Nevertheless, MPs with too small masses can cause a non-physical behaviour and lead to an unstable calculation.

If the soil is represented by MPs with different weight, it is cumbersome to interpret the movement of the soil. The heave and settlement of the soil surface due to the impacts observed in Fig. 19.6 depend on the weight of those MPs laying at the surface. Lighter MPs will show a bigger heave in the result than the heavier ones. It is not a numerical issue but the nature of the discrete representation of MPM results. The point here is that the same continuum solution of an MPM simulation can be represented in many ways by MPs depending on their size.

A moving mesh (see Chapter 6) is applied to the model. The moving part moves together with the pounder. The initial velocity of the pounder is prescribed according to the accelerometer measurements and is set to 18 m/s. After each impact, the pounder is lifted up together with the moving mesh. This procedure is repeated 7 times before the second compaction point is

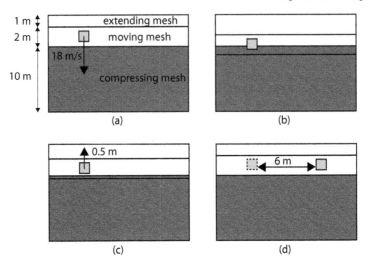

FIGURE 19.7: Calculation steps: (a) the pounder is released 0.5 m above the surface with a vertical velocity of 18 m/s; (b) the "moving mesh" follows the movement of the pounder without mesh-deformations, the "extending mesh" extends and the "compressing mesh" compresses in vertical direction respectively; (c) after the impact the pounder is lifted 0.5 m above the surface, the "extending mesh" now compresses and the "compressing mesh" extends. Steps (a)-(c) are repeated seven times; (d) after seven impacts, the pounder is shifted horizontally to the next location and steps (a)-(c) are repeated respectively.

compacted in the same way. Fig. 19.7 shows the computational steps and the moving mesh is shown in more detail.

19.3.1 Constitutive model

The behaviour of the soil was simulated by using the hypoplastic constitutive model [214]. The constitutive equation can be written as function H depending on the current stress state, void ratio and strain rate and expressed with the Jaumann stress rate $\dot{\sigma}_{ij}$ (see Chapter 2).

The function H can be decomposed in a linear part L and a non-linear part N (Eq. 19.1).

$$\dot{\sigma}_{ij} = L_{ijkl}\dot{\epsilon}_{kl} + N_{ij}\sqrt{\dot{\epsilon}_{ij}\dot{\epsilon}_{ij}} \tag{19.1}$$

L and N are in turn functions of stress and void ratio (Eq. 19.2).

$$L_{ijkl} = f_b f_e (f^2 \delta_{ik}\delta_{il} + a^2 \hat{\sigma}_{ik}\hat{\sigma}_{ji})$$

$$N_{ij} = f_d f_b f_e \frac{af}{\hat{\sigma}_{ij}\hat{\sigma}_{ij}}(\hat{\sigma}_{ij} + \hat{\sigma}_{ij} - \frac{1}{3}\delta_{ij}) \tag{19.2}$$

where $\hat{\sigma}_{ij}$ is a stress tensor defined as $\hat{\sigma}_{ij} = \sigma_{ij}/(\sigma_{11} + \sigma_{22} + \sigma_{33})$. The parameters a and f depend on the critical state friction angle φ_{cs} the stress state (Eq. 19.3).

$$a = \frac{\sqrt{3}(3 - \sin \varphi_{cs})}{2\sqrt{2} \sin \varphi_{cs}} \tag{19.3a}$$

$$s = \sqrt{\frac{1}{8} \tan^2 \psi + \frac{2 - \tan^2 \psi}{2 + 2\sqrt{2}} \tan \psi \cos 3\upsilon - \frac{1}{2\sqrt{2}} \tan \psi} \tag{19.3b}$$

with

$$\tan \psi = \sqrt{3(\hat{\sigma} - \frac{1}{3}\delta_{ij})(\hat{\sigma} - \frac{1}{3}\delta_{ij})} \tag{19.4a}$$

$$\cos 3\upsilon = -\sqrt{6}\frac{\delta_{ji}\left[(\hat{\sigma} - \frac{1}{3}\delta_{ik})(\hat{\sigma} - \frac{1}{3}\delta_{kl})(\hat{\sigma} - \frac{1}{3}\delta_{lj})\right]}{\left[(\hat{\sigma} - \frac{1}{3}\delta_{ij})(\hat{\sigma} - \frac{1}{3}\delta_{ij})\right]^{\frac{3}{2}}} \tag{19.4b}$$

Three characteristic values for the void ratio are defined, the maximum void ratio e_i defining an upper limit, the minimum void ratio e_d defining the lower limit and the critical void ratio e_{cs}. These characteristic values evolve over time depending on the trace of the stress tensor (Eq. 19.5).

$$\frac{e_i}{e_{i0}} = \frac{e_{cs}}{e_{c0}} = \frac{e_d}{e_{d0}} = \exp\left[-\left(\frac{-(\sigma_{11} + \sigma_{22} + \sigma_{33})}{h_s}\right)^n\right] \tag{19.5}$$

The void ratio e is derived from the differential equation Eq. 19.6.

$$\dot{e} = (1 + e)\dot{\epsilon}_{kk} \tag{19.6}$$

where ϵ_{kk} represents the change of the volumetric strain, e.g. $\epsilon_{kk} = \sum_i \epsilon_{ii}$. The subscript 0 indicates initial values and h_s is a constant representing the granular stiffness. The exponent n is the non-physical input parameter.

The calculated void ratio (Eq. 19.6) is checked to be within the region defined by Eq. 19.5. If the void ratio is bigger than e_i, it is set equal to e_i. If it is smaller than e_d, it is set equal to e_d. Note that e_i and e_d are not constant (Eq. 19.5).

The values of the factors f_e, f_d and f_b can now be defined in terms of the characteristic void ratios (Eq. 19.7).

$$f_e = \left(\frac{e_c}{e}\right)^\beta \qquad f_d = \left(\frac{e - e_d}{e_c - e_d}\right)^\alpha \tag{19.7a}$$

$$f_b = \frac{h_s}{n}\left(\frac{1 + e_i}{e_i}\right)\left(\frac{e_{i0}}{e_{c0}}\right)^\beta \left(\frac{-\sum_i \sigma_{ii}}{h_s}\right)^{1-n}\left[3 + a^2 - \sqrt{a}\left(\frac{e_{i0} - e_{d0}}{e_{c0} - e_{d0}}\right)^\alpha\right]^{-1} \tag{19.7b}$$

The hypoplastic model was extended by Niemunis and Herle [214] for small strain stiffness and a new tensor S, called the intergranular strain tensor, was introduced. The rate of intergular strain depends on the direction $\hat{S}_{ij} = S_{ij}/\sqrt{S_{ij}S_{ij}}$ of S and the strain rate $\dot{\epsilon}$ (Eq. 19.8).

$$\dot{S}_{ij} = \begin{cases} (\delta_{ik}\delta_{il} - \hat{S}_{ik}\hat{S}_{il}r^{\beta}) & \text{for } (\hat{S}_{ij}\dot{\epsilon}_{ij} > 0) \\ \dot{\epsilon}_{ij} & \text{for } (\hat{S}_{ij}\dot{\epsilon}_{ij} \leq 0) \end{cases} \tag{19.8}$$

where β is a material parameter describing the degradation of the stiffness and r is the normalised magnitude, which is calculated as $r = \sqrt{S_{ij}S_{ij}}/R$ with R the size of the elastic domain in the strain space.

Two additional scalar m_T and m_R are introduced to define the extended hypoplastic function and depend on the normalised magnitude r and $\hat{S}_{ij}\dot{\epsilon}_{ij}$ (Eq. 19.9).

$$D_{ijkl} = [r^{\chi}m_T + (1 - r^{\chi})m_R]\,L_{ijkl} +$$
$$\begin{cases} r^{\chi}(1 - m_T)L_{ijmn}(\hat{S}_{mk}\hat{S}_{nl}) + r^{\chi}N_{ik}\hat{S}_{jl} & \text{for } (\hat{S}_{ij}\dot{\epsilon}_{ij} > 0) \\ r^{\chi}(m_R - m_T)L_{ijmn}(\hat{S}_{mk}\hat{S}_{nl}) & \text{for } (\hat{S}_{ij}\dot{\epsilon}_{ij} \leq 0) \end{cases} \tag{19.9}$$

where χ affects the stiffness degradation and has no physical meaning. Finally, the stress rate can be calculated as given in Eq. 19.10.

$$\dot{\sigma}_{ij} = D_{ijkl}\dot{\epsilon}_{ij} \tag{19.10}$$

The soil parameters for the model were determined in the laboratory by oedometer and triaxial tests and are given in Table 19.1.

TABLE 19.1
Hypoplastic parameters for soil

Description	Symbol	Unit	Value
Critical friction	φ_{cs}	$^\circ$	33.5
Granular stiffness	h_s	MPa	93
Porosity	n	–	0.39
Minimum void ratio	e_{d0}	–	0.341
Maximum void ratio	e_{c0}	–	0.761
Critical void ratio	e_c	–	0.662
Model parameter	α	–	0.013
Model parameter	β	–	3.59
Model parameter	m_T	–	2.0
Model parameter	m_R	–	5.0
Model parameter	R	–	0.0001
Model parameter	β_R	–	0.95
Model parameter	χ	–	1.0

19.4   Results and comparison

Reference points were chosen in the numerical model to compare the measurements with the simulation. A MP located at the top of the pounder was chosen to record all accelerations during the impacts. Fig. 19.8 shows the simulated acceleration of the pounder for all 14 impacts. It can be observed that the peak acceleration at the second location, i.e. impact 8-14, is slightly higher than at the first location, i.e. impact 1-7. This indicated a higher stiffness at the second location due to the compaction of the first one.

Fig. 19.9 shows both the measured and calculated result for the 5th impact and the data shows good agreement in the slope and maximum value of the vertical acceleration. After the impact, the measured values show a strongly damped behaviour while the simulated values tend to oscillate around zero. The averaged value of these oscillations matches approximately the measured acceleration. This kind of behaviour was observed for every impact and might be reduced if boundary effects of the model would be removed by applying absorbing boundaries. In both cases, a second peak after the decrease behind the maximum acceleration can be observed. This behaviour describes that after a maximum resistance and, therefore, deceleration of the pounder at a certain instant the pounder experiences less resistance before the second peak indicates a new growing resistance. The moment directly after the maximum acceleration can be identified with the displacement of the soil after the first contact. The second peak indicates the end of the soil displacement process.

The comparison of the maximum vertical acceleration of all impacts for both locations is shown in Fig. 19.10 and gives an idea of the quality of the numerical estimation. The maximum vertical acceleration is an important value as it indicates the stiff response of the soil and the amount of energy induced to propagating waves. It was observed that in general higher values of vertical acceleration at the pounder were leading to bigger surface waves measured by the geophones. The results presented in Fig. 19.10 show a different quality of agreement depending on considered impact. The averaged absolute deviation of the results is around 13 g, which is acceptable regarding the sensor accuracy of $\pm$ 6 g. In both cases, the first impact of the second location showed a significant higher acceleration compared to the first impact at the first location indicating a pre-compacted soil at the second location due to the first seven impacts. Some points are in a very good agreement with the measurements as for example the impacts 4, 5, 7, 9 and 12. Also the shape of the curves is similar as was shown for the 5th impact in Fig. 19.9. It is difficult to find a particular reason for this deviation in such a complex simulation but again the boundary effects could cause this kind of behaviour as the movement of the soil in unnaturally restricted. The advantage of the numerical model is the possibility to visualise different variables at any location, which is almost impossible to do with measurements. Fig. 19.11 shows the field of vertical

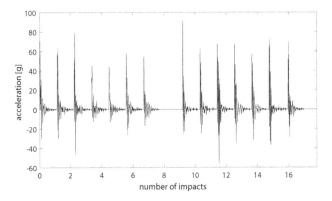

FIGURE 19.8: Vertical acceleration in [g] at a material point on top of the pounder for all 14 impacts.

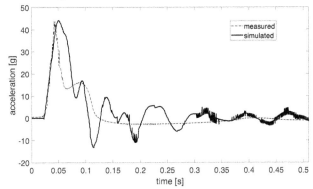

FIGURE 19.9: Comparison of the measured and simulated values for the vertical acceleration of the pounder.

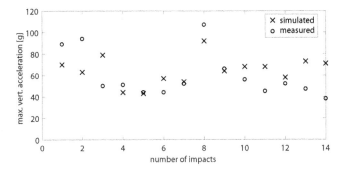

FIGURE 19.10: Maximum vertical acceleration for 14 impacts at both locations.

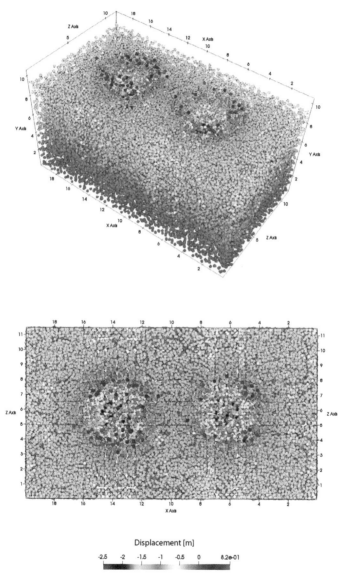

FIGURE 19.11: Vertical displacement of material points for the whole computational domain.

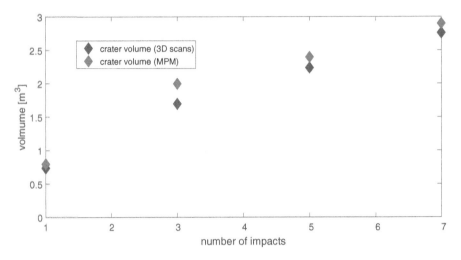

FIGURE 19.12: Estimated crater volume for the first seven impacts.

displacements from different perspectives. It is easy to estimate interesting information such as the depth or the width of the craters and calculate their volumes, which is shown later.

The red circles around the crater in Fig. 19.11 indicate an elevation of the soil surface. Such kind of elevations are created by the displacement of soil. Once the soil is maximally compressed only displacement leads to the growth of the crater. Therefore it is inefficient to continue dynamic compaction after this moment. In practise the size of the crater will continue increasing even when the compaction has been finished long before. As already discussed above the representation of simulated field results must be handled with care as each material point might represent a different amount of soil. The dark red material points in Fig. 19.11 are exactly those floating ones shown in Fig. 19.6. Even though the comparison of the simulated heave of the surface with measured ones is difficult the estimation of the crater size can be done. The determination of the crater volume is performed approximately by fitting cylinders with different radii at different depths inside the crater and calculating the volumes of those resulting cylinders. Fig. 19.12 (a) compares the calculated volumes with and the measured ones. As the 3D scans were only performed for the first location, only one crater is considered.

The results presented in Fig. 19.12 are in good agreement with the experimental data. Both the predicted and measured deformations show a big settlement during the first three impacts and smaller ones afterwards.

19.5 Closure

In general, it is challenging to realistically reproduce complex three-dimensional problems involving several bodies with complex material behaviours. Therefore, the interpretation of calculated results must be done with caution. As explained in Section 19.4, some values are in good agreement such as the maximum vertical acceleration (Fig. 19.9) and are helpful for a prediction or deeper study. However, other parameters deviate from the experimental data such as vertical acceleration after the impact (Fig. 19.9). It is obvious that both systems, the numerical model and the real world, are for certainly behaving differently and the task to identify which similarities between those two systems can be useful for the user is always present. The choice of the MPM as a numerical method of dynamic compaction seems reasonable as large deformations occur during the formation of the crater. The MPM can deal with such deformation without numerical issues such as mesh distortion. The presented numerical simulation still shows shortcomings such as the treatment of boundary effects, which can be improved to properly capture the wave propagation induced by the impacts; the issue of wave propagation is discussed in Chapter 20. The application of absorbing boundaries can be considered but also introduces new difficulties and complexities to the model. A bigger computational domain can be defined but it increases the computational cost. New techniques in MPM now permit simulating the filling of the crater after compaction. Also, a case with more than two compaction locations should be considered. This is especially important when the compaction is performed in many stages. In the presented simulation, this would lead to a third compaction location between the first and the second, which would be compacted after the first two are done.

20

Installation of Geocontainers

B. Zuada Coelho

20.1 Introduction

The modelling of large deformations is of great importance for engineering problems. One example that illustrates the need for taking large deformations into account is the installation of geocontainers. Geocontainers are sand-filled bags encapsulated by a geotextile membrane. Due to their resistance to erosion, they are often used for coastal applications such as revetments and breakwaters. However, the modelling of geocontainer installation is a challenge as it combines large deformations with complex soil and water interaction behaviour.

The recent development of MPM combined with the development of a fully coupled formulation that is able to capture the state transition for both liquid and solid phases, i.e. two-phase double-point formulation (see Chapter 2), has established the basic framework to model the installation of geocontainers.

This chapter presents some first steps towards the modelling of geocontainers with MPM. Despite the practical interest on this topic, very little experimental research has been conducted. Therefore, a series of elementary problems, for which an analytical solution exists, is presented in order to validate MPM before discussing the modelling of geocontainers. The elementary problems consist of one-dimensional (1D) wave propagation in dry soil (Section 20.2), 1D wave propagation in saturated soil (Section 20.3) and 1D poroelastic flow (Section 20.4). Section 20.5 presents the modelling of a simplified geocontainer.

20.2   One-dimensional wave propagation in dry soil

The first problem is a 1D wave propagation analysis of a wave travelling through an elastic and dry solid. Fig. 20.1 presents the geometry of the problem. The solid column has a finite length L, and it is subjected to a compressive load p_0, which is suddenly applied at the top $(x = L)$. The column is assumed to be fixed at the bottom $(x = 0)$. The column has a constant cross section A, Young modulus E, Poisson ratio ν and density ρ. The governing equation of motion is given by Eq. 20.1.

$$\frac{\partial p}{\partial x} A = \rho A \frac{\partial^2 \vec{u}}{\partial t^2} \tag{20.1}$$

where $\vec{u}$ is the displacement, x the position, and t the time. After some elaborations, the solution of Eq. 20.1 gives Eq. 20.2 [67].

$$\vec{u}(x,t) = \frac{p_0}{K} \left[x + \frac{8L}{\pi^2} \sum_{n=1}^{\infty} \frac{(-1)^n}{(2n-1)^2} \sin(\lambda x) \cos(\lambda ct) \right] \tag{20.2a}$$

$$\vec{v}(x,t) = \frac{p_0}{K} \frac{8L}{\pi^2} \sum_{n=1}^{\infty} -\frac{(-1)^n}{(2n-1)^2} \lambda c \sin(\lambda x) \sin(\lambda ct) \tag{20.2b}$$

$$p(x,t) = p_0 \left[1 + \frac{8L}{\pi^2} \sum_{n=1}^{\infty} \frac{(-1)^n}{(2n-1)^2} \cos(\lambda x) \cos(\lambda ct) \right] \tag{20.2c}$$

where $c = \sqrt{K/\rho}$ is the wave velocity of the solid. The wave velocity is a function of the solid density ρ and solid bulk modulus K (Eq. 20.3).

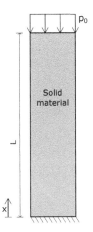

FIGURE 20.1: Geometry of the solid column and loading conditions.

$$K = \frac{E(1-\nu)}{(1+\nu)(1-2\nu)} \tag{20.3}$$

The parameter λ in Eq. (20.2) is given by Eq. 20.4.

$$\lambda = \frac{(2n-1)\pi}{2L} \tag{20.4}$$

The solutions of Eq. 20.2 are written as a Fourier series of n terms. The more terms considered in the Fourier expansion, the more accurate the solution is.

Fig. 20.2 shows the velocity and stress at three different points along the solid column and for a solid with the properties given in Table 20.1, and a load $p_0 = 10$ kPa. The analytical solution is computed with $n = 100$ terms for the Fourier series. A clear propagation of the compression wave is identified. The wave starts travelling at $t = 0$ until it reaches the bottom boundary where it is reflected, creating a tension wave travelling in the opposite direction. As there is no damping in the system, the wave keeps on travelling through the solid column indefinitely.

The MPM calculation is performed with *Anura3D* [14] and it is conducted for a 1-m column and an element size of 0.01 m. Low-order tetrahedral elements with 4 MPs/cell are used. A structured mesh is used in the analyses.

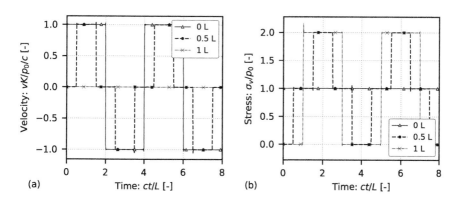

FIGURE 20.2: Analytical solution of the wave propagation in an elastic solid: (a) velocity and (b) stress.

TABLE 20.1
Material properties for the dry solid column.

Parameter	Symbol	Unit	Value
Density	ρ	kg/m^3	2,700
Young modulus	E	kPa	5,000
Poisson ratio	ν	–	0.30

A load of 10 kPa is applied at the top of the solid column. The nodes at the bottom of the column are fixed in the vertical direction. The solid is modelled as a linear elastic material with the parameters given in Table 20.1. The numerical analysis is performed using the one-phase single-point formulation (see Chapter 2).

Fig. 20.3 shows the comparison between the numerical and analytical results. The time required for the wave to travel along the solid column is correctly captured as well as the average amplitude of the stress and velocity. The maximum value of the stress and velocity is not constant, as in the analytical solution, but oscillates around the exact value. This oscillation is related to the numerical discretisation and the integration of the equations of motion. As discussed in Chapter 6, these oscillations can be reduced by using bulk viscosity damping.

20.3   One-dimensional wave propagation in saturated soil

The second problem is a 1D wave propagation analysis through a fully saturated poroelastic solid material. Fig. 20.4 presents the geometry of the problem. The saturated solid column has an infinite length, and it is submitted to a compressive load p_0 at the top. Consider a column of poroelastic material with a porosity n, a tortuosity τ and a Biot coefficient α. The solid material has a density ρ_S and compressibility m_v, while the incompressible liquid has a density ρ_L, a storage coefficient S_p, an intrinsic permeability κ and a viscosity μ. The equations for the propagation of plane waves in a porous medium can

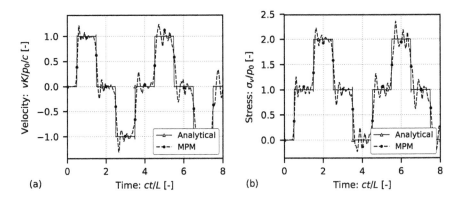

FIGURE 20.3: MPM solution of wave propagation in an elastic solid: (a) velocity and (b) stress.

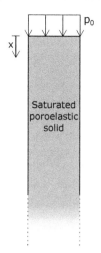

FIGURE 20.4: Geometry of the infinite solid column and loading conditions.

be formulated as Eq. 20.5.

$$\alpha \frac{\partial \vec{w}}{\partial x} - S_p \frac{\partial p}{\partial t} = -\frac{\partial n \left(\vec{u} - \vec{w} \right)}{\partial x} \tag{20.5a}$$

$$m_v \frac{\partial \boldsymbol{\sigma}'}{\partial t} = -\frac{\partial \vec{w}}{\partial x} \tag{20.5b}$$

$$n \rho_L \frac{\partial \vec{v}}{\partial t} + \left(1 - n \right) \rho_S \frac{\partial \vec{w}}{\partial t} = -\frac{\partial \boldsymbol{\sigma}'}{\partial x} - \alpha \frac{\partial p}{\partial x} \tag{20.5c}$$

$$n \rho_L \frac{\partial \vec{v}}{\partial t} + \tau n \rho_L \frac{\partial \left(\vec{u} - \vec{w} \right)}{\partial t} = -n \frac{\partial p}{\partial x} - \frac{n^2 \mu}{\kappa} \left(\vec{u} - \vec{w} \right) \tag{20.5d}$$

where $\vec{w}$ and $\vec{v}$ are the solid and liquid velocity, respectively, $\boldsymbol{\sigma}'$ the effective stress and p the pore pressure.

Eq. 20.5a expresses the mass conservation of the liquid and the solid, Eq. 20.5b the linear elastic stress-strain relation of the solid, Eq. 20.5c the momentum conservation of the mixture, and Eq. 20.5d the momentum conservation of the pore liquid, i.e. the generalisation of Darcy's law for the dynamic case. For the case of a periodic load p_0 applied at the free surface, a general solution for Eq. 20.5 exists; the detailed derivation of the solution is given in [318].

The MPM calculation is performed with *Anura3D* [14] and it is conducted for a 10 m column and an element size of 0.01 m. A structured mesh of low-order tetrahedral elements with 4 MPs/cell is used in the analyses. A load of 10 kPa is applied at the top of the solid column. The nodes at the bottom of the column are fixed in the vertical direction. The solid is modelled as a linear elastic material with parameters given in Table 20.2. The choice of the column length depends on the fact that the analytical solution only exists for an infinite column. A length of 10 m proved sufficient to prevent the waves from reflecting and returning to the points of interest during the selected time of analysis. The numerical analysis is performed using a fully coupled two-phase single-point formulation (see Chapter 2).

Fig. 20.5 shows the comparison between the analytical and numerical results at three different locations. The time is normalised with respect to the bulk wave velocity c of the saturated materials (Eq. 6.19) and the point coordinate x. The analytical and numerical solutions are in good agreement. No effects of wave reflection are observed which shows the appropriateness of the modelling strategy. Some oscillation of the numerical solution exists, which is less significant than for the dry solid. This is related to the damping caused by the drag force.

Fig. 20.5 shows two compression waves – one corresponding to the velocity of the bulk wave (undrained wave), where the velocities of solid and liquid are the same ($t = 1$ s), and one with a lower velocity, which is approximately half of the velocity of the first wave. In the second wave, the solid and liquid move in different directions. It is, therefore, strongly attenuated in time and space due to the frictional forces that are generated between the liquid and solid. This means that the second wave is only visible within the vicinity of the stress boundary condition, immediately after the application of the stress. This is further illustrated in Fig. 20.6, which shows the development of the stresses along the column for different time instants. Immediately after application of the stress boundary condition, the two waves are clearly visible (Fig. 20.6 a). However, the second wave starts vanishing (Fig. 20.6 b) as time progresses. As soon as the stress pressure reaches a distance larger than 1 m the second wave is completely damped (Figs. 20.6 c and d).

The existence of the second compression wave was theoretically predicted by Biot [33], but it was only measured for the first time in 1980 [227] in a water saturated porous material composed of sintered glass spheres, and later for granular soil [212,308]. The difficulties related to the measurement of the second wave are related to its high damping, which makes it very hard to capture.

TABLE 20.2

Material properties for the solid column.

Parameter	Symbol	Unit	Value
Density solid	ρ_S	kg/m^3	2,700
Solid Young modulus	E_S	kPa	5,000
Solid Poisson ratio	ν	-	0.3
Solid initial porosity	n_0	-	0.4
Permeability	κ	m·s	0.001
Liquid density	ρ_L	kg/m^3	1,000
Liquid bulk modulus	K_L	kPa	$2 \cdot 10^6$
Liquid viscosity	μ_L	kPa·s	$1 \cdot 10^{-6}$

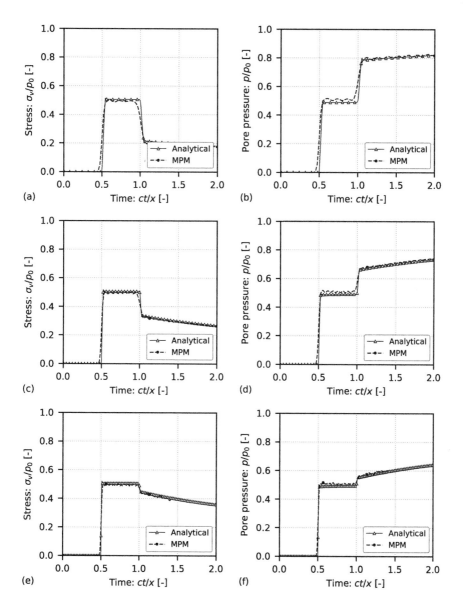

FIGURE 20.5: Comparison of the analytical and numerical solutions for the effective stress and pore pressure at different positions: (a–b) $x = 0.25$ m, (c–d) $x = 0.50$ m and (e–f) $x = 1.0$ m.

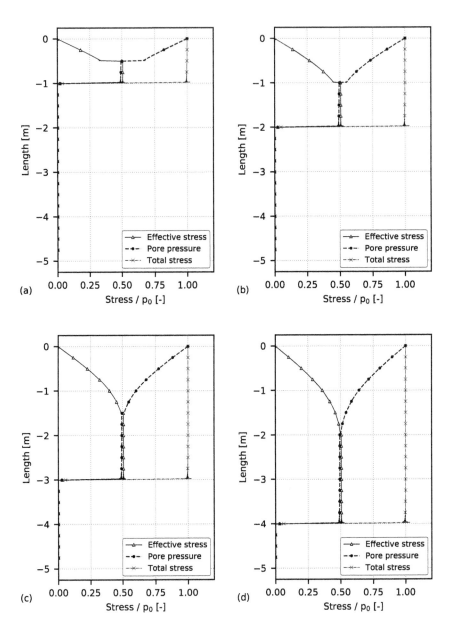

FIGURE 20.6: Stresses for the saturated soil column at different times: (a) $t =$ 0.44 ms, (b) $t = 0.88$ ms, (c) $t = 1.33$ ms and (d) $t = 1.77$ ms.

20.4 One-dimensional submerged poroelastic solid flow

In this section, the analysis of a 1D poroelastic solid falling by gravity through a Newtonian liquid is performed. This 1D problem is chosen as it offers the convenience of having a closed-form analytical solution for the steady-state velocity of the poroelastic solid flow. One of the challenges of modelling such a problem is related to the interaction between solid and liquid materials, and the change in state of the liquid as it transitions from a free surface flow to groundwater flow. In order to tackle these challenges, the two-phase double-point formulation has been developed as presented in Chapter 2.

The geometry of the problem is illustrated in Fig. 20.7. The poroelastic solid material is modelled as linear elastic, while the liquid material is

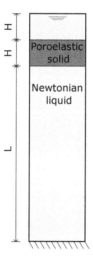

FIGURE 20.7: Geometry of the poroelastic solid and water column (after [70]).

TABLE 20.3
Material properties for the poroelastic solid flow.

Parameter	Symbol	Unit	Value
Solid density	ρ_S	kg/m^3	2,700
Solid Young modulus	E_S	kPa	$10 \cdot 10^3$
Solid Poisson ratio	ν	-	0.3
Solid initial porosity	n_S	-	0.4
Solid grain diameter	D_S	m	$2 \cdot 10^{-3}$
Liquid density	ρ_L	kg/m^3	1,000
Liquid bulk modulus	K_L	kPa	$20 \cdot 10^3$
Liquid viscosity	μ_L	kPa·s	$1 \cdot 10^{-6}$

assumed to be a compressible Newtonian liquid. Table 20.3 summarises the material properties. A poroelastic solid is considered with a submerged volumetric weight γ' and embedded in a Newtonian liquid. The steady-state velocity $\vec{v}_S$ of a rigid poroelastic solid falling by gravity through a liquid follows Eq. 20.6 [192, 352].

$$\vec{v}_S = -\frac{\gamma'}{\dfrac{\mu}{\kappa} + \rho_L \beta |\vec{v}_S|} \tag{20.6}$$

where κ is the intrinsic permeability and it is a function of the solid grain diameter (Eq. 2.116). β is the non-Darcy flow coefficient [95] (Eq. 2.115). It follows from Eq. 20.6 that the velocity calculation is implicit; i.e. it has the form $\vec{v}_S = f(\vec{v}_S)$, which requires an iterative method for its computation.

20.4.1   MPM solution

All numerical analyses are performed with *Anura3D* [14] using the two-phase double-point formulation (see Chapter 2). The domain is discretised with 780 low-order tetrahedral elements with 10 solid MPs and 10 liquid MPs for the saturated poroelastic solid elements, and 10 liquid MPs for the liquid elements. The nodes at the bottom of the column are fixed in the vertical direction. The dimensions H and L are assumed as 0.1 m and 1 m, respectively.

Fig. 20.8 shows the location of the poroelastic solid at different times. The poroelastic solid flows through the liquid due to the gravity force, exhibiting large deformations. Fig. 20.9 (a) presents the solid displacement and its comparison with the analytical solution. It follows that the analytical and numerical solution are initially in close agreement up to a normalised time of approximately 0.15 s. Then, the numerical solution starts deviating from the analytical. Due to the high permeability of the poroelastic solid (solid grain diameter D_S), the transient behaviour of the flow is not noticeable in Fig. 20.9. Fig. 20.9 (b) shows that the solid velocity oscillates periodically around the analytical solution, up to a time of 0.1 s, followed by a non-periodic oscillation.

The periodic oscillation is related to the liquid compressibility, as discussed in Zuada Coelho et al. [352] and can be mitigated by implementing a incompressible formulation of the liquid material. The non-periodic oscillation is caused by element crossing, due to the discontinuity in the liquid concentration ratio, at the boundary between the liquid and the solid. The fact that the velocity oscillates around the analytical solution is likely to be responsible for the deviation of the numerical displacement (Fig. 20.9 a) from the analytical solution.

20.4.2   Effect of solid Young modulus

The effect of the solid Young modulus on the poroelastic flow is investigated for a range of values between 50 kPa and 50 MPa. The solid grain diameter is assumed to be $5 \cdot 10^{-4}$ m, while the remaining material parameters are

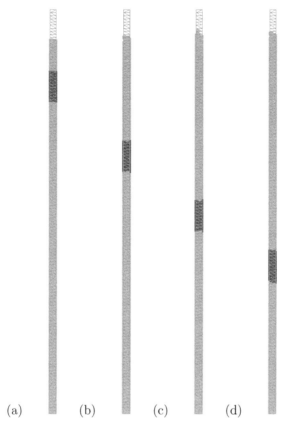

FIGURE 20.8: Poroelastic solid flow at different times: (a) $t = 0$ s, (b) $t = 10$ s, (c) $t = 20$ s, and (d) $t = 30$ s.

those given in Table 20.3. The grain size diameter was reduced in relation to the previous analyses in order to capture the transient behaviour of the poroelastic solid.

Fig. 20.10 (a) presents the displacement of the poroelastic solid for several Young moduli. A transient behaviour of the poroelastic solid is clearly identified for the analysis with a Young modulus of 50 kPa. Up to a normalised time of 0.004, the solid accelerates, after which it reaches the steady-state velocity. This transient behaviour is related to the consolidation that develops during the flow. The flow occurs as there is a pressure gradient between the top and bottom of the solid, which causes an excess pore water pressure at the bottom. This results in the occurrence of consolidation (i.e. dissipation of excess pore water pressure) and swelling of the poroelastic solid (extension). This is further illustrated in Fig. 20.10 (b) which shows the poroelastic solid extension during the flow. For the remaining Young moduli, the transient re-

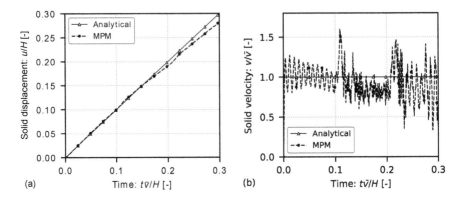

FIGURE 20.9: Poroelastic solid flow: (a) displacement and (b) velocity of all solid material points (after [70]).

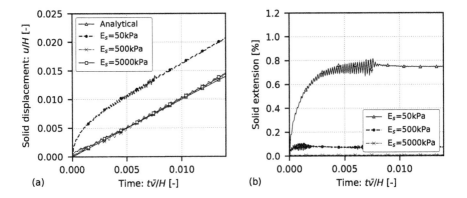

FIGURE 20.10: Effect of the solid Young modulus on (a) the poroelastic solid displacement and (b) the swelling of the poroelastic solid due to consolidation (after [70]).

sponse is less noticeable. This is caused by a faster consolidation and a smaller extension of the solid due to the larger Young moduli.

20.5   Two-dimensional submerged poroelastic solid flow

This section presents the modelling of a 2D poroelastic solid flow, which resembles a simplified geocontainer installation process. Previous work on the modelling of geocontainers with MPM has been conducted using the single-point formulation [121,122,328]. The analyses in this section use the two-phase double-point formulation (see Chapter 2). In order to model the installation of geocontainers, it is required to model large deformations, the interaction between the solid and liquid materials, and the change in flow conditions of the liquid between free surface and groundwater flow, and the liquid flows around and through the solid.

Fig. 20.11 shows the geometry of the problem with $H = 1$ m, $L = 2$ m and $B = 4$ m. The problem is assumed to be symmetric with the nodes at the bottom of the model fixed in the vertical direction and at the side in the horizontal direction. The model has a total of 6,300 elements with 10 MPs per element for the solid and liquid phase in the poroelastic material, respectively, and 20 MPs per element for the liquid phase in the liquid material. The solid material is assumed to be linear elastic and the liquid material is modelled as compressible Newtonian liquid. Table 20.4 presents the properties for the solid and liquid materials.

20.5.1 MPM solution

Fig. 20.12 shows the movement of the poroelastic solid through the liquid for different time instants. For the chosen parameter set (i.e. a stiff geocontainer with low permeability), the poroelastic solid moves through the liquid as a rigid body, i.e. without significant bending. It is visible in the figure that the liquid flows into the bottom of the solid. The grey MPs (initially liquid MPs) enter the domain of the solid MPs (dark MPs). The light grey MPs (initially groundwater liquid MPs) flow out of the top of the solid and leave the domain

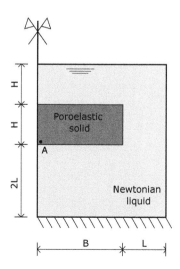

FIGURE 20.11: Geometry of the 2D poroelastic block (simplified geocontainer) (after [70]).

of the solid MPs. However, the movement of the solid in this case is mainly due to the liquid that flows around the solid rather than through it.

20.5.2 Effect of solid Young modulus

The effect of the solid Young modulus on the 2D poroelastic flow problem is investigated for three different values – 500 kPa, 5,000 kPa and 50,000 kPa. All remaining parameters are given in Table 20.4.

TABLE 20.4
Material properties for the 2D poroelastic solid flow.

Parameter	Symbol	Unit	Value
Solid density	ρ_S	kg/m^3	2,700
Solid Young modulus	E_S	kPa	$50 \cdot 10^3$
Solid Poisson ratio	ν	-	0.3
Solid initial porosity	n_0	-	0.5
Solid grain diameter	D_S	m	$2 \cdot 10^{-3}$
Liquid density	ρ_L	kg/m^3	1,000
Liquid bulk modulus	K_L	kPa	$20 \cdot 10^3$
Liquid viscosity	μ_L	kPa·s	$1 \cdot 10^{-6}$

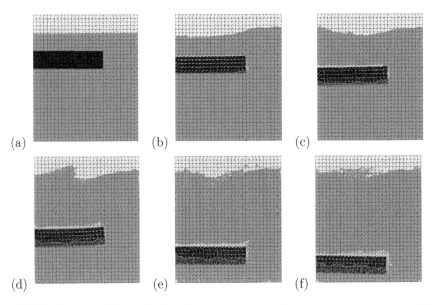

FIGURE 20.12: Flow of the 2D poroelastic solid at times: (a) $t = 1.25$ s, (b) $t = 2.5$ s, (c) $t = 3.75$ s, (d) $t = 5$ s, (e) $t = 6.25$ s, and (f) $t = 7.5$ s (after [70]).

Fig. 20.13 shows the effect of the Young modulus on the solid vertical displacement and the velocity for Point A (Fig. 20.11). The velocity increases with the reduction of the solid Young modulus. This increase in solid velocity is related to the bending deformation of the poroelastic solid. As the stiffness decreases, the solid exhibits higher deformation due to bending as the water flows around the solid. As the solid bends, the water flow around the block is facilitated and the solid velocity increases. This is illustrated in Fig. 20.14, which shows the poroelastic solid displacement for a Young modulus of 500 kPa at $t = 5$ s. Fig. 20.12 (d) shows the poroelastic solid displacement at the same time for the default case.

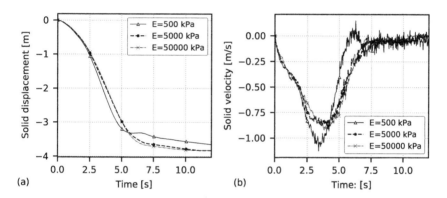

FIGURE 20.13: Effect of the solid Young modulus on the: (a) vertical displacement and (b) vertical velocity of point A of the poroelastic solid (after [70]).

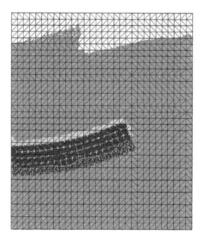

FIGURE 20.14: Deformation of the poroelastic solid for a Young modulus of 500 kPa at $t = 5$ s (after [70]).

20.6   Closure

The numerical modelling of geocontainer installation is challenging, as it involves large deformation analysis in combination with complex solid-liquid interaction. In this chapter, a first step is made by modelling the installation process in a simplified way. A series of 1D elementary problems for which analytical solutions are available are modelled with MPM: (1) wave propagation in a dry soil column, (2) wave propagation in an infinite saturated soil column, and (3) submerged poroelastic solid flow. These three problems required the use of distinct formulations of MPM: one-phase single-point, two-phase single-point and two-phase double-point formulation. The results show the appropriateness of the MPM solution in comparison to the analytical one.

Although the first two examples do not involve large deformations, they illustrate that the MPM formulation provides an accurate solution at low strain levels. Furthermore, they show that the fully-coupled two-phase MPM formulation accurately predicts the second compression wave in saturated materials. The submerged poroelastic solid flow problem, including its analytical solution for the steady state regime, provides the validation of the two-phase two-point formulation. Additionally, it is shown that there is a transient response of the poroelastic flow velocity, where the solid swells due to consolidation. The poroelastic solid velocity oscillates sinusoidally around the analytical flow velocity due to the liquid compressibility and element crossing. However, the average value of the velocity is in good agreement with the analytical solution.

The analysis of the 2D submerged poroelastic flow resembles the installation of a geocontainer. The analysis shows that the flow velocity of the falling poroelastic solid depends on the solid Young modulus. As the solid Young modulus reduces, the solid bending is larger. This accelerates the flow around the solid block and therefore the solid velocity is higher. In next steps for the modelling of geocontainer installation, the use of more advanced constitutive models for the solid material and the use of a membrane around the solid have to be incorporated.

21

Applications in Hydraulic Engineering

X. Zhao and D. Liang

21.1 Introduction

Free surface hydrodynamics is of significant industrial and environmental importance in hydraulic engineering. However, challenges arise in the implementation of surface boundary conditions on an arbitrarily moving water surface [268]. Due to the fact that the MPM combines the advantages of the mesh-based and mesh-free methods, it is an attractive method for computing free surface flows. MPM has many attractive advantages over other numerical methods [3]. It is convenient to incorporate time-dependent constitutive models because variables are carried by the moving MPs. This allows the spatial and temporal tracking of the history of the material motion. The use of a background mesh facilitates the definition of the boundary conditions offering a strong advantage over other methods. In addition, the MPM is free of the tensile instability that is evident in Smoothed Particle Hydrodynamics (SPH) [170, 185]. Despite these advantages, there has been little application of MPM in the field of hydraulic engineering.

This chapter presents some applications of MPM in the field of hydraulic engineering such as dam break flows, Scott Russell's wave generator, the free overfall flow and the scouring of underwater pipelines. The simulation results are compared with experimental data, other numerical methods as well as theoretical solutions. It is found that the MPM calculations show good agreement with the MPM simulation results. Through these studies, the MPM proves to be a useful tool for investigating hydrodynamic problems.

21.2   Dam break flow

The dam break flow is a well-established experiment in the field of hydraulic engineering for which extensive experimental data is available in the literature (e.g. [191]). The experiment consists in releasing a column of water on a flat surface, which is similar to the granular column collapse presented in Chapter 9. As water does not have any shear strength, the experiment consists in analysis of the travel speed of the water front, expressed in terms of position and time. The length and time are normalised according to the Froude scaling law (Eq. 21.1) in order to facilitate the comparison between different experiments and numerical predictions.

$$t^* = \frac{t}{\sqrt{h_0/g}} \quad \text{and} \quad x^* = \frac{x}{h_0} \tag{21.1}$$

where t^* is the normalised time, x^* the normalised position of the flow front, t time, x the position flow front, and g gravity. The initial aspect ratio a of the water column is defined in the same way as for the granular column collapse (Eq. 9.1) as the ratio between the initial height h_0 and the width r_0.

Shallow water equations (SWEs) are favoured for rapid flood simulations due to the ease in obtaining solutions in comparison to solving the Navier-Stokes equations. The SWEs description is accurate and convenient for nearly horizontal flows. Detailed discussions on the model are available in the literature (e.g. [106, 170, 171]). For cases where the flow changes abruptly, the underlying assumptions behind the SWEs (i.e. hydrostatic pressure distribution, smooth variation of the water surface, and negligible vertical acceleration) may not hold true so the accuracy of the results of the SWEs solver may be questionable [170]. The applicability may need to be evaluated before solving the SWEs for flood routing.

In this section, the MPM simulations are compared with experimental data and SPH simulations. Then, a parametric study of a two-dimensional (2D) dam break flow is conducted with MPM and the results are compared with the theoretical solution given by the SWEs. Viscous and turbulent effects are neglected as the behaviour of dam break flow are dominated by convection processes.

21.2.1   Model description and material parameters

The MPM model resembles the one of the granular column collapse (Fig. 9.4) with three differences. (1) The MPs represent a piece of water continuum. (2) The boundary conditions are applied to water MPs, and (3) all the contact surfaces are assumed to be frictionless and, hence, a base layer is no longer required. Fig. 21.1 shows a schematic description of a typical MPM model used for dam breaking flood simulations, which has a structured mesh with 3,880 elements with 11,640 MPs (20 MPs/cell).

The water is modelled as a compressible Newtonian fluid with mechanical parameters as given in Table 21.1. The bulk modulus of the water is reduced by a factor of 100 in order to increase the critical time step in order to gain computational efficiency with the explicit integration scheme. Liang [170] showed this practice has marginal effects on the results as the sound speed is over 10 times bigger than the maximum flow velocity. The water pressure is initiated from the water column surface by K_0-procedure. The calculations are run with Gauss integrations and a Courant number of 0.8 (see Chapter 3), together with strain smoothing and liquid pressure smoothing techniques (see Chapter 6).

21.2.2 Mesh size and material point discretisation

As demonstrated in Chapter 9 for the granular column collapse, MPM predictions are mesh sensitive and it is necessary to check if the mesh is sufficiently small. A series of simulations of dam break flow are carried out for a column with an initial aspect ratio of $a = 1$, for two different mesh sizes (coarse: 0.1

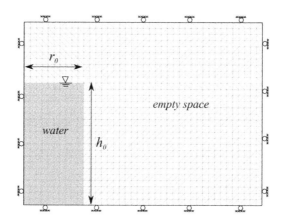

FIGURE 21.1: Schematic description of the MPM model for the dam break simulations.

TABLE 21.1

Material parameters for MPM simulations of dam break flows.

Parameter	Symbol	Unit	Value
Density	ρ_w	kg/m³	1,000
Bulk modulus	K_w	kPa	20,000
Dynamic viscosity	μ_w	kPa · s	10^{-6}

m, fine: 0.05 m) and 3 different MPs densities (4, 8 and 10 MPs/cell). The results of the simulations are given in Fig. 21.2 (a), and the results show no apparent mesh or MPs density dependency. Therefore, the fine mesh with 4 MPs/cell is used for small columns ($a = 0.2, 0.25$, and 0.5), and the coarse mesh with 4 MPs/cell for all other columns.

21.2.3   Comparing numerical predictions with experimental data

Fig. 21.2 compares the velocity of the flow front of the numerical predictions using MPM and the experimental data [191]. The experimental data shows that the water front accelerates up to a normalised time of $t^* = 0.5$, and then reaches a constant velocity. The results of the MPM simulation are in good agreement with the experimental data. Fig. 21.2 also compares the MPM predictions with other numerical methods. The MPM, SPH and volume of fluid (VOF) give results very similar to one another and in-line with the experimental data. However, the SWEs solver does not simulate the acceleration phase induced by the vertical momentum, but immediately predicts the maximum velocity of the water front. Therefore, the slope of the SWEs curve corresponds to those of the experimental data and the other numerical predictions but it is off-set due to the absence of the acceleration phase.

A series of MPM simulations is carried out with different initial aspect ratios ($a = 0.2, 0.25, 0.5, 1, 2, 3$ and 4). The models are generated by maintaining a width of $r_0 = 1$ m and increasing the initial height h_0. Two representative simulations with aspect ratios $a = 0.2$ (small column) and 4.0 (high column) are shown in Fig. 21.3 for an illustration. The contours represent the MPM simulations while the solid lines represent the SWEs predictions. The results

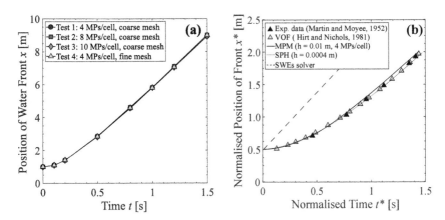

FIGURE 21.2: (a) Mesh and MPs-density sensitivities of dam break MPM simulations and (b) movement of the flood front position.

show that the MPM and SWEs simulations of the small column ($a = 0.2$) are in good agreement, but the results for the large column ($a = 4.0$) show large differences. However, this difference could be due to the absence of the acceleration phase in the SWEs. The results of the MPM simulations are then compared with the solution provided by the SWEs solver, and the difference between both solutions is computed with Eq. 21.2.

$$\Delta x = x_1 - x_2 \tag{21.2}$$

where Δx is the difference in position, x_1 is the solution provided by the SWEs, and x_2 the one by MPM.

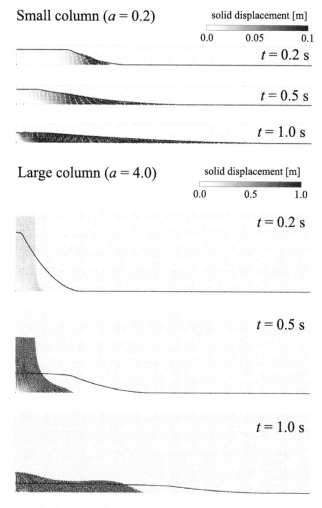

FIGURE 21.3: Evolution of water column collapse from MPM simulations (contour) and SWEs (solid line): (a) small column and (b) large column.

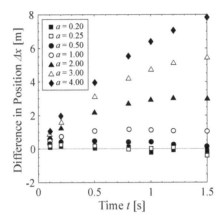

FIGURE 21.4: Difference between flood fronts predicted by the MPM and SWEs models.

Fig. 21.4 shows the results in terms of Δx for the different aspect ratios. The results show that the difference between the MPM and SWEs predictions is bigger for large columns ($a > 1$) than for small columns ($a \leq 1$). In the case of $a > 1$, the flow front difference gets larger and larger with time, while there is a limit for the difference of about ± 1.0 m when $a \leq 1$. This means, when the aspect ratio $a \leq 1$, the simulation results of MPM and SWEs agree well with each other, which is no longer true when the aspect ratio exceeds 1. For this reason, $a = 1$ can be regarded as the critical value for the applicability of SWEs.

21.3   Scott Russell's wave generator

The generation of solitary waves in flume tanks is an important problem in hydraulic engineering. These can be generated by pushing downwards a solid block into a tank full of water. Monaghan and Kos [203] investigated the generation of solitary waves by means of experimental investigations and numerical analysis using SPH. The experiment consists of a weighted box (length: $l = 0.3$ m, height: $h = 0.4$ m) dropped vertically into a wave tank. The case with a water depth of $d = 0.21$ m is simulated with MPM in order to demonstrate the ability of MPM to simulate wave generation and propagation.

21.3.1 Model description and material parameters

Fig. 21.5 shows the 2D model, which is composed of a solid box, a reservoir of water and an empty space in which the wave can propagate. The box is modelled as a porous solid continuum. The material parameters of the simulations are listed in Table 21.2.

A schematic description of the MPM model used for Scott Russell's wave generator is given in Fig. 21.5. It has a structured mesh with 13,656 elements with 7,200 solid MPs and 66,000 liquid MPs (solid: 4 MPs/cell, liquid: 10 MPs/cell). The initial condition and the boundary conditions of the problem are shown in Fig. 21.5. The length of the computational domain is reduced to 2 m from 9 m in the experiment for saving computational cost. All the contact surfaces are assumed to be frictionless. The water pressure is initiated by K_0-procedure from the initial water surface elevation (0.21 m) while the solid box stresses are initiated from the box top surface elevation (0.51 m). The calculations are run with a Courant number of 0.95 (see Chapter 3) without any strain or pressure smoothing techniques (see Chapter 6).

TABLE 21.2

Material parameters for MPM simulations for wave generator.

Parameter	Symbol	Unit	Value
Water density	ρ_w	kg/m^3	1000
Water bulk modulus	K_w	kPa	20000
Water dynamic viscosity	μ_w	kPa · s	$1 \cdot 10^{-6}$
Box specific density	ρ_s	kg/m^3	1,592
Box porosity	n	-	0.5
Box Young modulus	E	kPa	10,000
Box Poisson ratio	ν	-	0.3
Box intrinsic permeability	k	m^2	$1 \cdot 10^{-9}$

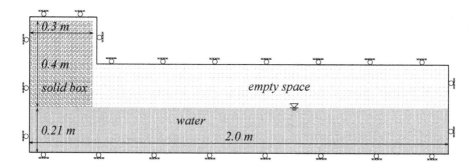

FIGURE 21.5: MPM model for Russell's wave generator test.

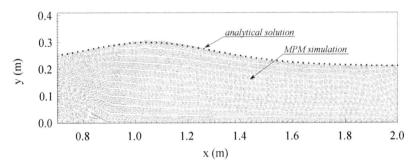

FIGURE 21.6: Wave profile at 0.7 s (MPM in grey and analytical in black dots).

21.3.2   Comparing numerical predictions with experimental data

By assuming that a fixed proportion of gravitational energy of the falling block turns into the energy of the solitary wave, Monaghan and Kos [203] established Eq. 21.3 for calculating the wave amplitude A_w.

$$A_\text{w} = 3d \left(\frac{m_b}{40\rho_w lwd} \right)^{2/3} \left(\frac{l}{d} \right)^{2/3} \tag{21.3}$$

where m_b and w are the block mass and width, respectively.

The profile of the solitary wave generated by the falling box into the water is given by Eq. 21.4.

$$\eta_\text{w}(x, t) = d + a\text{sech}^2 \left[\sqrt{\frac{3a}{4d^3}} (x - ct) \right] \tag{21.4}$$

where η_w is the water surface elevation, ρ_w is water density, x is the axis along which the wave propagates, and c is the solitary wave celerity given by $c = \sqrt{g(d + a)}$.

The falling velocity of the solid box is computed as given in Eq. 21.5.

$$\frac{v_\text{b}}{\sqrt{gd}} = \sqrt{\left(\frac{M}{\rho_W L^2 W} \right) \frac{Z}{d}} \sqrt{\left(1 - \frac{Z}{d} \right)} \tag{21.5}$$

where Z is the height of the bottom of the box above the bottom of the tank at time t and v_b is the falling vertical velocity of the box at time t.

Fig. 21.6 shows the solitary wave generated by the falling box and the profile of the wave is compared with Eq. 21.4. The agreement between the analytical profile of the wave and the simulated free surface is good. Fig. 21.7 compares the falling velocity of the solid box with the analytical solution (Eq. 21.5). The results show that the MPM predictions are in-line with the theoretical solution, albeit some oscillations may be attributed to the grid crossing errors (see Chapter 3).

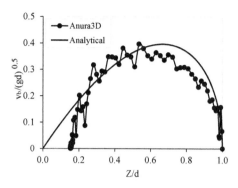

FIGURE 21.7: Solid box falling velocity.

21.4 Free overfall

The free overfall occurs where the bottom of a flat channel is discontinued, which is a particular case of hydraulic drop and is described by Chow [65]. According to the conservation law, the water surface will seek its lowest possible energy configuration providing that no external energy is added to the system. This means that the brink depth will not be less than the critical depth y_c. The theoretical water surface curve is as shown in Fig. 21.8. The critical depth y_c is expressed as given in Eq. 21.6 for flows with a rectangular cross-section.

$$y_c = \sqrt[3]{\frac{\alpha(vh_i)^2}{g}} \qquad (21.6)$$

where v is the mean velocity of the cross section, h_i the initial flow depth and α the velocity coefficient, which can be approximated as $\alpha \approx 1.0$.

Due to the fact that the flow at brink is a rapidly varied flow and Eq. 21.6 is only applicable to gradually varied flow, Rouse [252] found that the critical depth is approximately $y_c \approx 1.4y_0$. A comprehensive review of the free overfall flow in open channels can be found in [82].

21.4.1 Model description

A typical computational mesh of free overfall simulations is shown in Fig. 21.9. A quasi plane-strain condition is assumed for the simulations, considering only a strip of elements. The so-called inlet elements are used to generate water flows with various initial velocities and depths and the outlet elements are used to remove MPs from calculations. Gauss integration is adopted in the calculations where the Courant number is set as 0.98 (see Chapter 3) and strain smoothing and liquid pressure smoothing are employed. 8 MPs/cell is used for the calculations. As in other studies of this chapter, frictional effects are not considered. The material parameters are the same as listed in Table 21.1.

21.4.2   Brink depth and end pressure distribution

The theoretical brink depth y_c and the actual exit water depths y_0 for an open channel flow with an initial flow depth of 15 cm and for different mean flow velocities v are plotted in Fig. 21.10. The water depth at the brink h_e of numerical simulations with six different mean flow velocities ($v = 0.4, 0.6, 0.8, 1.0, 1.1$ and 1.2 m/s) are also plotted in Fig. 21.10. The critical flow depth y_c and the actual brink depth y_0 are obtained as in Sec. 21.4. It can be seen that the MPM predictions of the brink depth agree well with empirical solutions.

The end-pressure distribution is an extensively studied topic of free overfall such as notable studies detailed in [205, 237, 242, 252, 317]. An analytic solution is given in Montes [205] for the analysis of open channel flows near a discontinuity of the bed geometry (i.e. free outlets). The predictions of end-pressure distribution provided by the MPM are shown. From a qualitative point of view, they show good agreement with above mentioned laboratory measurements and analytical solutions. However, it is worth mentioning that the MPM simulations show slight over-prediction of the pressure near the bed

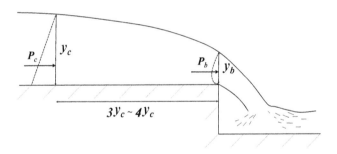

FIGURE 21.8: Sketch of free overfall.

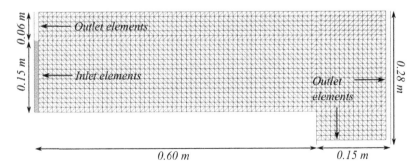

FIGURE 21.9: Computational mesh of an open channel flow.

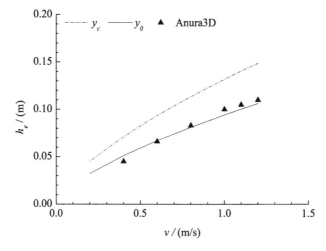

FIGURE 21.10: The comparison of simulation results of the brink depth with analytical solution.

region at the brink. This may be attributed to the fact that the elements around the brink are restricted by the vertical fixities applied on the meeting nodes at the brink causing the pressure increase. This assumption can be proved. If we select MPs an element away from the brink, the pressure distribution can be significantly improved near the bed.

21.5 Scour initiation beneath an embedded pipeline

Submarine pipelines are considered as the most efficient and safest way to transport gas and oil and are indispensable for the rapid development of the global economy. They are considered as lifelines between storage units and offshore fields. It is crucial to ensure the safety of the pipeline as their failures can lead to tremendous losses [288].

Scouring has been recognised as one of the leading causes of submarine pipeline failures, and numerous studies have been conducted in this area over the past decades [331]. Interactions between an exposed pipeline and the approach flow will lead to changes in the local flow field, often changing the shear stress. Consequently, the modified flow could entrain and remove sediment particles beneath the pipeline, causing pipeline scour. Scour is a threat to the stability of pipelines, which may lead to the suspension of the pipeline and is closely related to structural failure.

There has been considerable research on the onset of scours and the mechanics are well understood. According to Sumer and Fredsøe [288], scours are

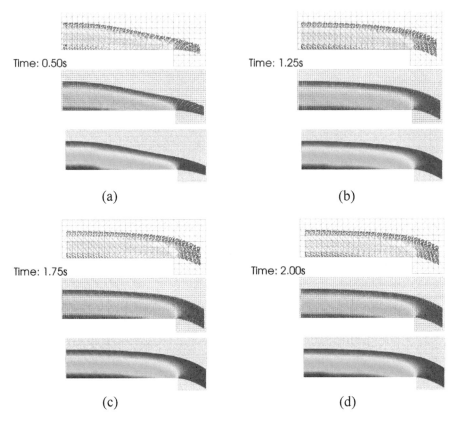

FIGURE 21.11: End pressure distribution of simulations with various mesh sizes (for each sub-figure the characteristic element length is: top 0.03 m, middle 0.01 m, and bottom 0.005 m).

initiated by piping caused by a pressure gradient between the upstream and downstream sides of the pipe. Chiew [63] investigated the initiation of scours artificially by creating a static pressure gradient across a pipe. It is observed that the values of the critical pressure gradient for soil failure are very close to the soil flotation gradient.

MPM is gaining popularity in recent years with double-point formulation (see Chapter 2). The soil-water interactions can be modelled through a drag force, which enables the MPM to be applied to various scenarios such as seepage flows [3, 4, 20, 21, 194, 344], submarine landslides [192], soil fluidisation [39, 194], water jetting on a soil bed [172]. It shows great potential for modelling scours around submarine pipelines.

Pipelines are circular objects, which require special boundary conditions in MPM in order to consider the rounded surface. However, very few MPM papers have discussed simulations with rounded shapes, and entirely fixed boundary conditions are usually applied on the round surface nodes. However, as the restrictions caused by no-slip boundary conditions are mapped to the MPs inside the elements containing the surface nodes during MPM calculations, unrealistic restrictions are therefore put on these MPs. This section focuses on developing the free-slip boundary conditions for modelling scour around pipelines.

21.5.1 Double-layer formulation for scour initiation

The description of the two-phase double-point MPM formulation is given in Chapter 2. The two sets of MPs represent the solid skeleton of the soil and water, respectively, which can occupy the same location at the same time. Where water and soil MPs coexist in an element, Terzaghi's effective stress principle is adopted for the soil skeleton unless the soil grains are fluidised and static liquefaction takes place. The soil MPs can thus move solely under the action of the submerged weight and drag force.

The criterion for fluidisation is defined when the local soil porosity exceeds a given limit value called the maximum porosity n_{max}. If the local porosity is less than the maximum porosity $(n < n_{max})$, then the grains are in contact and the behaviour of the mixture can be described by various constitutive models in soil mechanics. If the local porosity is greater than the maximum porosity $n > n_{max}$, then the soil grains lose their inter-granular contact and the effective stresses are set to zero. The mixture then behaves as a fluid. In the present study, the maximum porosity is taken as $n_{max} = 0.471$.

21.5.2 Free-slip boundary conditions for circular objects

The implementation of free-slip boundary conditions for circular objects consists of the following three steps.

1. Freeing the surface nodes so that they can obtain non-zero nodal properties

2. Correction of nodal values according to relative position of the circle centre in the Lagrangian phase

3. Correction of nodal values before mapping them back to the MPs element-specific in the convective phase

Freeing surface nodes and nodal correction in Lagrangian phase

In order to avoid unrealistic restrictions on the MPs within elements containing surface nodes, the nodes need to be able to move freely in the defined domain. Therefore, no horizontal or vertical fixities are placed on the surface nodes such that they can move freely in the XY plane. Hence, at the end of the Lagrangian phase, the surface nodes can obtain non-zero accelerations, for which the normal components have to be removed keeping only the tangential components.

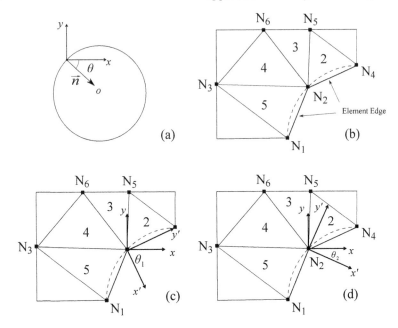

FIGURE 21.12: Sketch of a parent element angle.

Fig. 21.12 shows the definitions of a nodal angle and parent element angle, respectively. These two angles are used to correct the nodal values on the round surface. The nodal angle θ is obtained by rotating the original x-axis clockwise to the nodal normal direction $\vec{n}$ pointing towards the circle centre O as shown in Fig. 21.12 (a). The range of θ is $[0, 2\pi]$. After obtaining the nodal values in the Lagrangian phase, the circle surface nodal values are corrected according to Eq. 21.7.

$$\alpha_x = \alpha_{x0} \sin^2 \theta + \alpha_{y0} \sin \theta \cos \theta \tag{21.7a}$$

$$\alpha_y = \alpha_{x0} \sin \theta \cos \theta + \alpha_{y0} \cos^2 \theta \tag{21.7b}$$

where $\alpha = a, v$ or Δu with a, v and Δu are the acceleration, velocity and displacement components in different directions.

Nodal correction in convective phase

As shown in Fig. 21.12 (b-c), N_1, N_2, and N_4 are three nodes on a round surface and connected by element edges. For example, there are four elements containing the node N_2 – Element 2, 3, 4 and 5. MPs in Element 3 and 4 will not leave the mesh since the nodal properties have been corrected during the Lagrangian phase according to Eq. 21.7. However, MPs in Element 2 and 5 may leave the mesh once they have been updated with the nodal values obtained from the Lagrangian phase since the real element edges do not coincide with the theoretical round surface. Therefore, the nodal values have to be corrected again before being mapped to the MPs inside the surrounding elements, and the corrections have to be element-specific.

Considering Element 2, a new coordinate system is established at N_2 with the positive x' pointing inwards to the centre such that the parent element angle θ' of node N_2 is obtained by rotating the original x-axis clockwise to the positive x' direction and the range of θ_1 is $[0, 2\pi)$ (Fig. 21.12 c). The parent element angle θ_2 of node N_2 for Element 5 is obtained in the same way. Then, the nodal values are corrected before updating the MPs with the nodal values and according to the parent element angle for the element of the node using Eq. 21.8.

$$\alpha_x = \alpha_{x0} \sin^2 \theta + \alpha_{y0} \sin \theta \cos \theta \qquad (21.8a)$$

$$\alpha_y = \alpha_{x0} \sin \theta \cos \theta + \alpha_{y0} \cos^2 \theta \qquad (21.8b)$$

where $\alpha = a, v$ and Δu. The expressions are the same as Eq. 21.7 in the Lagrangian phase differing only by values of θ.

21.5.3 Point vortex

Point vortex can be viewed as a 2D case of a straight vortex filament of infinite length. For a straight vortex filament of infinite length, the magnitude of the induced velocity v_B at an arbitrary point B can be expressed as Eq. 21.9.

$$v_B = \frac{\Gamma}{2\pi h} \qquad (21.9)$$

where Γ is the vortex strength which is constant and h is the distance between point B and the straight vortex filament; Schobeiri [260] provides additional information for the derivation.

Eq. 21.9 gives the magnitude of velocity v_B as proportional to $1/h$ and the theoretical relationship should be linear as shown in Fig. 21.13 where k is a constant.

MPM model of point vortex

MPM simulations of a point vortex are carried out using the proposed free-slip boundary conditions. The dimensions of the model are given in Fig. 21.14. It has a structured mesh with 54,020 elements and 198,216 material points (4 MPs/cell). The pipe has a radius of 1 m. Water flows around the pipe under the action of the hydraulic gradient created by water head difference between both sides of the pipeline.

The boundary conditions are fixed around the box. Both no-slip and free-slip boundary conditions are applied around the pipeline. An observation zone is defined in order to obtain velocity profiles for the result analysis as indicated in Fig. 21.14.

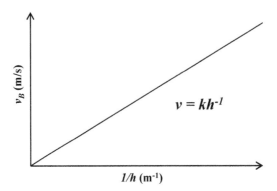

FIGURE 21.13: Theoretical induced velocity distribution profile around point vortex.

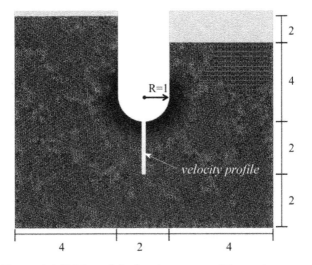

FIGURE 21.14: MPM model of point vortex. Dimensions are in [m].

The material parameters used in the point vortex calculations are the same as those used in dam breaking flood simulations, given in Table 21.1. The pressure is initiated from the water surface at the right side. The calculations are conducted with a Gauss integration, a Courant number of 0.8 together with strain smoothing and liquid pressure smoothing (see Chapter 3).

Results of point vortex

The velocity profiles are plotted in Fig. 21.15 at different instants. The data in Fig. 21.15 (a) are obtained from the simulation with no-slip boundary conditions while Fig. 21.15 (b) shows the data obtained from the new implementation with free-slip boundary conditions applied. In all plots, as mentioned before, h is the distance from the particle to the pipe centre and v is the horizontal component of the particle velocity. In theory, the velocity magnitude should grow proportionally to $h^{-1} \in (0.33, 1)$ as shown in Fig. 21.13 so linear lines with a fixed intercept of 0.0 are used to fit the data, and their equations are also shown in Fig. 21.15.

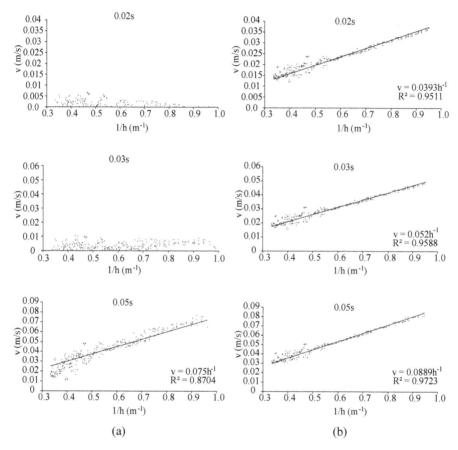

FIGURE 21.15: Velocity profiles of point vortex simulations: (a) with no-slip boundary conditions, (b) with free-slip boundary condition.

From the comparison between Fig. 21.15 (a) and (b), it can be observed that for the calculation with no-slip boundary conditions, at 0.02 s and 0.03 s, the MPs near the pipe surface $(1/h = 1)$ are locked due to the fixities applied on the pipe surface. Only as the calculation carries on, can the MPs near the pipe surface gain enough momentum to move. But in Fig. 21.15 (b), since the fixities on the pipe surface are removed entirely, the MPs near the pipe surface can acquire velocities once the simulation starts.

Additionally, it can be seen from the comparison that the linear lines can better fit the data from the simulation with free-slip boundary conditions as

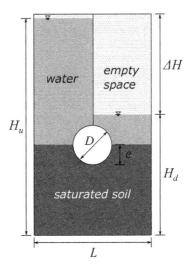

FIGURE 21.16: An illustration of MPM model of scour initiation simulations.

the values of R^2 are much closer to 1.0, which indicates better agreement with analytical relationships as shown in Fig. 21.13.

21.5.4   Pipeline erosion study

MPM with the proposed free-slip boundary conditions are further employed to study the onset of scour around embedded pipelines. According to Sumer and Fredsøe [288], the onset of scouring around pipelines initiated by the piping process is caused by the pressure gradient between the upstream and downstream sides of the pipe. The initiation of scour was investigated experimentally by Chiew [63] by artificially creating a static pressure gradient across a pipe, and it was observed that the values of the critical pressure gradient i_{crit} for soil failure are very close to the soil flotation gradient i_{f}.

$$i_{\text{f}} = (1 - n)(G_{\text{s}} - 1) \tag{21.10}$$

where n is the soil porosity and G_{s} the specific gravity.

MPM model of pipeline erosion

Numerical replications of the experiments are conducted using both the original and the proposed implementations. The computational geometry of the calculations is shown in Fig. 21.16 in which H_{u} is the upstream water depth, H_{d} is the downstream water depth, $\Delta H = H_{\text{u}} - H_{\text{d}}$ is the water head difference, L is the total length of the computational domain, e is the embedment depth and D is the diameter of the pipe. 4 liquid MPs/cell is adopted for water which 4 liquid MPs/cell and 4 solid MPs/cell are adopted for the saturated soil. The calculations are carried out with smoothing techniques.

TABLE 21.3

Material parameters for the calculation

Material	Parameter	Symbol	Unit	Value
Soil:	Specific gravity	G_s	-	2.6
	Young modulus	E	kPa	10,000
	Poisson ratio	ν	-	0.3
	Initial porosity	n	-	0.43
	Grain size	D_p	mm	1.7
	Maximum porosity	n_{max}	-	0.471
Water:	Density	ρ_w	kg/m^3	1,000
	Bulk modulus	K_w	kPa	20,000
	Dynamic viscosity	μ_w	kPa $\cdot$ s	$1 \cdot 10^{-6}$

The material parameters are listed in Table 21.3, so according to Eq. 21.10, the soil flotation gradient $i_f = (1 - 0.43)(2.6 - 1) = 0.912$. Erosion around the pipe is modelled with two different embedment ratios ($e/D = 0.25$ and 0.5) and three hydraulic gradients ($i = 1.0$, 1.3 and 1.5).

Results of pipeline erosion

Two representative simulations with embedment ratios of 0.25 and 0.5 are shown in Figs. 21.17 and 21.18, respectively, for an illustration. The hydraulic gradients i in the two figures are the same as 1.5. The contours represent the magnitude of solid displacements under the action of seepage force under static hydraulic gradients. From the comparison, it can be seen that with free-slip boundary conditions, the soil can move freely near the pipe which is physically reasonable, meaning the simulation quality is significantly improved.

It is worth mentioning that the proposed implementation does increase the computational cost because of the two-step correction of nodal values in both the Lagrangian and convective phase. However, the increase is very limited due to the fact that the number of round surface nodes is relatively small.

21.6 Closure

The chapter demonstrates the use of MPM for hydraulic engineering by simulating three different cases using both the single- and double-point formulations. These cases are namely dam break flows, Scott Russell's wave generator and free overfall. The simulation results are compared with other numerical methods, experimental data and theoretical solutions and they show good agreement. Thus, through the three case studies, the MPM has been shown to be reliable to solve hydrodynamic problems.

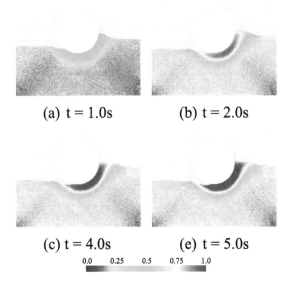

FIGURE 21.17: Erosion around embedded pipeline with $e/D = 0.25$, $i = 1.5$.

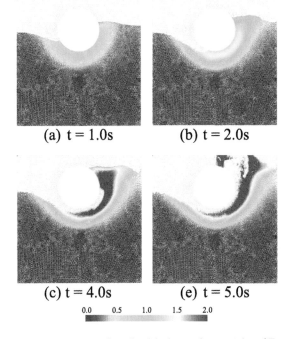

FIGURE 21.18: Erosion around embedded pipeline with $e/D = 0.5$, $i = 1.5$.

Additionally, this chapter presents a implementation of free-slip boundary conditions for circular objects, in which the round surface nodes are allowed to move freely in the tangential direction while their movements in the normal direction are forbidden. This new implementation is applied to case studies of point vortex and local scour around pipeline to evaluate its performance. By comparing with the simulation results from the original implementation, it can be seen that the proposed boundary treatment can act as a remedy for the unphysical blockage of the MPs near the circular surface and therefore produce better agreement with analytical results. This will dramatically expand the capacities of MPM to model boundaries with circular shapes. In terms of disadvantage, the proposed implementation adds computational cost but the effects are trivial for the number of nodes to be corrected is relatively small.

A

Appendix: Thermal Interaction in Shear Bands

M. Alvarado, N.M. Pinyol and E.E. Alonso

A.1   Governing equation including embedded bands

A.1.1   Momentum balance equation of the mixture

The equation of motion is formulated both fluid and solid material points (MPs) assuming the saturation of pores.

$$\rho^{M} \vec{a}_{s} = \nabla \cdot \vec{\sigma} + \rho^{M} \vec{b} \tag{A.1}$$

A.1.2   Conservation of momentum of liquid

Dynamic equilibrium equation for fluid is established in Eq. A.2.

$$\rho^{M}_{L} \vec{a}_{l} = \nabla p^{M}_{l} - \frac{n\mu_{l}}{k_{int}}(\vec{v}_{l} - \vec{v}_{s}) + \rho^{M}_{l} \vec{b} \tag{A.2}$$

where $\vec{a}_{l}$ is the acceleration of the fluid in the matrix, which depends on (1) the gradient of pore water pressure p^{M}_{l} associated with the matrix; (b) on the filtration forces $n\mu_{l}/k_{int} \cdot (\vec{v}_{l} - \vec{v}_{s})$ expressed in terms of the relative velocity of the liquid $\vec{v}_{l}$ with respect of the solid MPs velocity $\vec{v}_{s}$, the porosity of the matrix n, dynamic viscosity coefficient of the fluid μ_{l}, and the intrinsic permeability k_{int}, and (3) on the body forces $\vec{b}$.

The relative acceleration of the liquid with respect to the solid skeleton is assumed negligible ($\vec{a}_{l/s} = \vec{a}_{s} - \vec{a}_{l} = 0, \vec{a}_{s} = \vec{a}_{l}$). The formulation is then simplified to a $\vec{u} - p_{l}$ formulation commonly used in FE implementation in which the primary unknowns of the problem are the displacement of the solid skeleton and the pore water pressure. Under this assumption, Eq. A.2 becomes Eq. A.3, which corresponds to the generalised equation for the Darcy flow $\vec{q}_{l}$.

$$\vec{q}_{l} = -\frac{k_{int}}{\mu_{l}}\left(\nabla p^{M}_{l} + \rho^{M}_{l} \vec{b} + \rho^{M}_{l} \vec{a}^{M}_{s}\right) \tag{A.3}$$

A.1.3   Mass balance equations

The mass balance equation for the matrix of both solid MPs and liquid is modified as Eq. A.4.

$$n\frac{d\rho_l^M}{dt} + \frac{\rho_l^M}{\rho_s^M}(1-n)\frac{d\rho_s^M}{dt} + \rho_l^M \nabla \cdot \vec{v}_s + \nabla \cdot (\rho_l^M \vec{q_l}) = \psi_1 (p_l^B - p_l^M) \quad \text{(A.4)}$$

where the Darcy's flow $\vec{q_l}$ has been defined in Eq. A.3. A source term has been added which corresponds to the rate mass of liquid coming from the embedded band.

Constitutive equations defining the variation of solid and liquid density are introduced with the aim of expressing the governing equations in terms of the primary unknown variables. Exponential functions have been selected to define the density variations (Eqs. A.5 and A.6).

$$\rho_s^M = \rho_s^{M_0} \exp\left[\frac{1}{K_s}\left(\text{tr}(\vec{\sigma}) - \text{tr}(\vec{\sigma})^0\right) + \beta_s \left(\Theta_T^M - \Theta_T^{M_0}\right)\right] \quad \text{(A.5)}$$

$$\rho_l^M = \rho_l^{M_0} \exp\left[\alpha_1 \left(p_l^M - p_l^{M_0}\right) + \beta_1 \left(\Theta_T^M - \Theta_T^{M_0}\right)\right] \quad \text{(A.6)}$$

where $\rho_s^{M_0}$ and $\rho_l^{M_0}$ are the solid and liquid density of the matrix at reference mean stress $\text{tr}(\vec{\sigma})^0$, the reference liquid pressure in the matrix, and the reference temperature in the matrix $\Theta_T^{M_0}$, respectively. $1/K_s$ is the compressibility of solid phase against mean stress changes. The parameter α_1 defines the liquid phase compressibility and β_s and β_1 are the volumetric thermal expansion coefficients for solid and liquid phase, respectively. These coefficients have negative values to reproduce the dilation (density decrease) due to heating (temperature increase).

Since the constitutive parameter involved in Eqs. A.5 and A.6 are assumed constant, the time derivative of solid and fluid phase density for the matrix and the embedded band can be expressed as Eqs. A.7 and A.8.

$$\frac{d\rho_s^M}{dt} = \rho_s^M \beta_s \frac{d\Theta_T^M}{dt} \quad \text{(A.7)}$$

$$\frac{d\rho_l^M}{dt} = \rho_l^M \alpha_1 \frac{dp_l^M}{dt} + \rho_l^M \beta_1 \frac{d\Theta_T^M}{dt} \quad \text{(A.8)}$$

where the compressibility of the solid MPs against changes in stress has been assumed negligible.

Replacing the previous equations into the mass balance Eq. A.4 and assuming that the spatial variation of the liquid density in the matrix is small ($\nabla \rho_l^M \approx 0$), the mass balance equation becomes Eq. A.9.

$$\rho_l^M \frac{dp_l^M}{dt} + \rho_l^M Q \left(\beta \frac{d\Theta_T^M}{dt} + \nabla \cdot \vec{v}_s + \nabla \cdot \vec{q_l}\right) = Q\psi_1 (p_l^B - p_l^M) \quad \text{(A.9)}$$

where a volumetric thermal expansion coefficient for the mixture $\beta = (1 - n)\beta_s + n\beta_l$ has been defined and the notation $Q = 1/n\alpha_l$ has been included.

Similarly, the mass balance equation is applied at band level. This balance equation is applied at the local level and, for a given reference volume, the embedded band interacts exclusively with the matrix. Under such a hypothesis, the term of Darcy's flow should be neglected in the local mass balance equation associated to the band (Eq. A.10).

$$\rho_l^B \frac{dp_l^B}{dt} + \rho_l^B Q\beta \frac{d\Theta_T^B}{dt} = -Q\psi_l \left(p_l^B - p_l^M\right) \tag{A.10}$$

The effect of the volumetric strain on the pore water pressure at the band has been neglected.

A.1.4 Energy balance equations

At matrix level, the internal energy balance per unit of volume can be written as Eq. A.11.

$$\frac{d}{dt}\left[\left(n\rho_l^M c_l + (1-n)\rho_s^M c_s\right)\Theta_T^M\right] + \nabla \cdot \left[-\Gamma_\Theta \nabla\Theta_T^M\right]$$
$$+\nabla \cdot \left[\rho_l^M c_l\Theta_T^M(\vec{q}_l + n\vec{v}_s) + (1-n)\rho_s^M c_s\Theta_T^M \vec{v}_s\right] \tag{A.11}$$
$$= \psi_{\Theta_T}\left(\Theta_T^B - \Theta_T^M\right)$$

Eq. A.11 states that the external supply of heat rate ψ_{Θ_T} should be equal to the sum of the following terms: (1) internal energy in solid and liquid phase depending on their specific heats c_s and c_l, respectively, (2) the heat flow conduction driven by temperature gradients (Fourier's law) which depends on the thermal conductivity coefficient Γ_Θ, and (3) the convective heat transport due to liquid and solid flows.

Assuming that the specific heat of the phases remains constant, Eq. A.11 can be simplified to Eq. A.12.

$$\frac{d\Theta_T^M}{dt} + \frac{1}{(\rho c)_m^M}\left(\nabla \cdot \vec{q}_h + \rho_l^M c_l\Theta_T^M \nabla \cdot \vec{q}_l\right) + \Theta_T^M \nabla \cdot \vec{v}_s =$$
$$\frac{1}{(\rho c)_m^M}\psi_{\Theta_T}\left(\Theta_T^B - \Theta_T^M\right) \tag{A.12}$$

where the specific heat of the mixture at the matrix has been defined $(\rho c)_m^M = n\rho_l^M c_l + (1-n)\rho_s^M c_s$ and Eq. A.13 has been selected for the heat flow conduction.

$$\vec{q}_h = -\Gamma_\Theta \nabla\Theta_T^M \tag{A.13}$$

At band level, the energy balance is defined locally. The exchange of energy exclusively occurs with the matrix and the convective, and advective flow terms are neglected (Eq. A.14).

$$\frac{d\Theta_T{}^B}{dt} = \frac{1}{(\rho c)_m^B} \left[\dot{H}^B - \psi_{\Theta_T} \left(\Theta_T{}^B - \Theta_T{}^M \right) \right] \tag{A.14}$$

where the specific heat of the mixture of the band has been defined $(\rho c)_m^B = n\rho_l^B c_l + (1-n)\rho_s^B c_s$.

Changes of temperature at band are generated by the external supply heat rate $\dot{H}$, provided by the mechanical work given in the next sub-section.

A.1.5   First law of thermodynamics

It is assumed that the plastic work dissipates into heat. The plastic work at the embedded band can be calculated as the product between the effective stress tensor and the plastic strain rate associated with the band. The strain rate calculated for a given reference volume should be scaled taking into account the assumption that the strains are actually localised at the embedded bands and assuming that the total mechanical work generated in the volume (characterised by a reference length L_{ref} is concentrated at the shear band characterised by L_B. According to this, the heat generated by the mechanical work can be defined by Eq. A.15.

$$\dot{H}^B = \frac{L_{ref}}{L_B}(\sigma_i' \dot{\varepsilon}_i^P) \tag{A.15}$$

The relationship between effective stress and total stress is defined in Eq. A.16.

$$\vec{\sigma}' = \vec{\sigma} - p_l \vec{m} \tag{A.16}$$

A.2   Computational algorithm

In this Appendix, the computational cycle of the MPM code is presented step by step. The governing equations presented in Section A.1 (balance equation of momentum, heat and mass) are solved explicitly as boundary value problems using the weighted residual method, in particular the Galerkin method. The time integration follows an explicit Euler approach. Solid acceleration, increment of temperature, liquid pressure and stress at time $^{k+1}t =^k t + \Delta t$ are calculated as a function of the variables evaluated at the previous time step. The computational algorithm consists of the following steps.

1. For time step ^0t: definition of the computational mesh and initial material properties, initialisation of variables at MP such as the initial position

$^0\vec{x}_{\mathrm{p}}$ at the Gauss points, volume $^0V_{\mathrm{p}}$, mass m_{p}, velocity $^0\vec{v}_{\mathrm{p}}$, stress $^0\vec{\sigma}_{\mathrm{p}}$, temperature $^0\Theta_{\mathrm{Tp}}$, liquid pressure $^0p_{l_{\mathrm{p}}}$ and embedded band variables and assignation of constitutive properties and history variables.

2. For time step $^k t$: calculation of the shape functions and their gradient associated to nodes evaluated at the position of MPs.

$$^k N_i^{\mathrm{p}} = N_i \left(^k \vec{x}_{\mathrm{p}}\right) \tag{A.17}$$

$$^k \left(\nabla N_i^{\mathrm{p}}\right) = \nabla N_i \left(^k \vec{x}_{\mathrm{p}}\right) \tag{A.18}$$

3. Calculation of the lumped mass matrix and lumped volume matrix.

$$^k m_i = \sum_{p=1}^{N_{\mathrm{p}}} {}^k N_i^{\mathrm{p}\,k} m_{\mathrm{p}} \tag{A.19}$$

$$^k V_i = \sum_{p=1}^{N_{\mathrm{p}}} {}^k N_i^{\mathrm{p}\,k} V_{\mathrm{p}} \tag{A.20}$$

4. Calculation of internal forces at nodes using element-wise stress averaging [340, 341] and using selective reduced integration strategy [188].

$$^k \vec{f}_i^{\mathrm{int}} = \sum_{p=1}^{N_{\mathrm{p}}} {}^k \left(\nabla N_i^{\mathrm{p}}\right) \cdot {}^k \vec{\sigma}_{\mathrm{p}} \frac{^k m_{\mathrm{p}}}{^k m_{\mathrm{el}}} {}^k V_{\mathrm{el}} \tag{A.21}$$

5. Calculation of body forces at nodes.

$$^k \vec{b}_i = \sum_{i=1}^{N_{\mathrm{p}}} {}^k m_{\mathrm{p}} \vec{g} {}^k N_i^{\mathrm{p}} \tag{A.22}$$

6. Calculation of external forces at nodes.

$$^k \vec{f}_i^{\mathrm{ext}} = \int_{\Gamma_{\mathrm{t}}} \rho^k N_i^{\mathrm{p}} \cdot {}^k \hat{\vec{t}} \mathrm{d}\Gamma_{\mathrm{t}} + {}^k \vec{b}_i \tag{A.23}$$

7. Calculation of the increment of momentum at nodes.

$$^k \Delta \left(m_i \vec{v}_i\right) = {}^k \vec{f}_i^{\mathrm{ext}} \Delta t - {}^k \vec{f}_i^{\mathrm{int}} \Delta t \tag{A.24}$$

8. Update of the position, velocity and acceleration of MP.

$$^{k+1} \vec{x}_{\mathrm{p}} = {}^k \vec{x}_{\mathrm{p}} + \frac{\Delta t}{^k m_i} \sum_{i=1}^{N_{\mathrm{n}}} \left[{}^k \left(m_i \vec{v}_i\right) + {}^k \Delta \left(m_i \vec{v}_i\right)\right] {}^k N_i^{\mathrm{p}} \tag{A.25}$$

$$^{k+1} \vec{v}_{\mathrm{p}} = {}^k \vec{v}_{\mathrm{p}} + \frac{1}{^k m_i} \sum_{i=1}^{N_{\mathrm{n}}} \left[{}^k \Delta \left(m_i \vec{v}_i\right)\right] {}^k N_i^{\mathrm{p}} \tag{A.26}$$

$$^{k+1} \vec{a}_{\mathrm{p}} = \frac{1}{^k m_i \Delta t} \sum_{i=1}^{N_{\mathrm{n}}} \left[{}^k \Delta \left(m_i \vec{v}_i\right)\right] {}^k N_i^{\mathrm{p}} \tag{A.27}$$

9. Update of the nodal velocities following the Modified Update Stress Last (MUSL) [285].

$$^{k+1}(m_i\vec{v_i}) = \sum_{p=1}^{N_p} {}^k m_p {}^{k+1}\vec{v}_p {}^k N_i^p \tag{A.28}$$

This step is required to avoid divisions by a shape function which may involve numerical problems when a MP is near a node. For a discussion on this, the readers are referred to [343].

10. Calculation of strain increment using selective reduced integration strategy [188] and update of the strain tensor at MPs.

$$^k\Delta\vec{\varepsilon}_p = \Delta t \sum_{i=1}^{N_n} \frac{{}^k(\nabla N_i^p) \cdot {}^{k+1}(m_i\vec{v_i})}{{}^k m_i} \tag{A.29}$$

$$^{k+1}\vec{\varepsilon}_p = {}^k\vec{\varepsilon}_p + {}^k\Delta\vec{\varepsilon}_p \tag{A.30}$$

11. Map temperature of the matrix from MP to nodes.

$$^k\Theta_{Ti} = \frac{1}{{}^k m_i} \sum_{p=1}^{N_p} {}^k m_p {}^k\Theta_{Tp}^{Mk} N_i^p \tag{A.31}$$

12. Calculation to the Fourier's heat flow associated with the MP.

$$^k\vec{q}_{h_p} = -\Gamma_\Theta \sum_{n=1}^{N_n} {}^k(\nabla N_i^p) {}^k\Theta_{Ti} \tag{A.32}$$

13. Map liquid pressure of the matrix in MP to nodes.

$$^k p_{l_i} = \frac{1}{{}^k m_i} \sum_{p=1}^{N_p} {}^k m_p {}^k p_{l_p}^{Mk} N_i^p \tag{A.33}$$

14. Calculation of gradient of liquid pressure associated with the MP.

$$^k\nabla p_{l_p} = \sum_{i=1}^{N_n} \nabla N_i^{p\,k} p_{l_i} \tag{A.34}$$

15. Calculation of Darcy's flow.

$$^k\vec{q}_p = -\frac{k}{\gamma_l}\left({}^k\nabla p_{l_p} + {}^k\rho_l^{Mk}\vec{b}_p + {}^k\rho_l^{Mk}\vec{a}_p\right) \tag{A.35}$$

16. Calculation of temperature increment at nodes.

$$\Delta\Theta_{\mathrm{T}i} = \frac{1}{(\rho c)_{\mathrm{m}}^{\mathrm{M}k} V_i} \left[\sum_{p=1}^{N_p} {}^k(\nabla N_i^{\mathrm{p}}) \, {}^k \vec{q}_{\mathrm{h}_p} \Delta t \, {}^k V_p - \int_{\Gamma_{q_h}} {}^k N_i \, {}^k \hat{q}_{\mathrm{h}} \Delta t d\Gamma_{q_h} + \right.$$

$$\sum_{p=1}^{N_p} {}^k \rho_l^{\mathrm{M}} c_l \, {}^k \Theta_{\mathrm{T}_p}^{\mathrm{M}k} (\nabla N_i^{\mathrm{p}}) \, {}^k \vec{q}_{\mathrm{p}} \Delta t \, {}^k V_p - \int_{\Gamma_{qL}} {}^k \rho_l^{\mathrm{M}} c_l \, {}^k \Theta_{\mathrm{T}_p}^{\mathrm{M}k} N_i \, {}^k \hat{q}_{\mathrm{l}} \Delta t d\Gamma_{qL} -$$

$$\left. \sum_{p=1}^{N_p} (\rho c)_{\mathrm{m}}^{\mathrm{M}k} N_i^{\mathrm{p}k} \Theta_{\mathrm{T}_p}^{\mathrm{M}k} \Delta\varepsilon_{\mathrm{vol}_p} \, {}^k V_p + \sum_{p=1}^{N_p} {}^k N_i^{\mathrm{p}k} \psi_{\Theta\mathrm{T}} \left({}^k \Theta_{\mathrm{T}_p}^{\mathrm{B}} - {}^k \Theta_{\mathrm{T}_p}^{\mathrm{M}} \right) V_p \right] \tag{A.36}$$

17. Update temperature of the matrix at MP.

$${}^{k+1}\Theta_{\mathrm{T}_p}^{\mathrm{M}} = {}^k \Theta_{\mathrm{T}_p}^{\mathrm{M}} + \sum_{i=1}^{N_n} \Delta\Theta_{\mathrm{T}i} \, {}^k N_i^{\mathrm{p}} \tag{A.37}$$

18. Calculation of increment of liquid pressure of the matrix at nodes.

$${}^k \Delta p_{l_i} = \frac{1}{{}^k V_i} \left[\sum_{p=1}^{N_p} {}^k Q \, {}^k N_i^{\mathrm{p}} \left({}^k \beta \, {}^k \Delta\Theta_{\mathrm{T}_p}^{\mathrm{M}} + {}^k \Delta\varepsilon_{\mathrm{vol}_p} \right) {}^k V_p + \right.$$

$$\int_{\Gamma_q} {}^k Q \, {}^k N_i \, {}^k \hat{q}_{\mathrm{l}} \Delta t d\Gamma_q - \sum_{p=1}^{N_p} {}^k Q \, {}^k \nabla N_i^{\mathrm{p}} \cdot {}^k \vec{q}_{\mathrm{l}_p} \Delta t \, {}^k V_p +$$

$$\left. \sum_{p=1}^{N_p} {}^k N_i^{\mathrm{p}} \frac{{}^k Q}{\rho_l^{\mathrm{M}}} \psi_{\mathrm{l}} \left({}^k p_{l_p}^{\mathrm{B}} - {}^k p_{l_p}^{\mathrm{M}} \right) \right] \tag{A.38}$$

19. Update pore pressure of the matrix at MP.

$${}^{k+1}p_{l_p}^{\mathrm{M}} = {}^k p_{l_p}^{\mathrm{M}} + \sum_{i=1}^{N_n} \Delta p_{l_i} \, {}^k N_i^{\mathrm{p}} \tag{A.39}$$

20. Calculation of increment of temperature and increment of liquid pressure of embedded bands.

$${}^k \Delta\Theta_{\mathrm{T}_p}^{\mathrm{B}} = \frac{\Delta t}{(\rho c)_{\mathrm{m}}^{\mathrm{B}}} \left[\frac{L_{\mathrm{ref}}}{L_{\mathrm{B}}} ({}^k \vec{\sigma}_{\mathrm{p}}' \cdot {}^k \vec{\varepsilon}_{\mathrm{p}}^{\mathrm{p}}) - {}^k \psi_{\Theta\mathrm{T}} \left({}^k \Theta_{\mathrm{T}_p}^{\mathrm{B}} - {}^k \Theta_{\mathrm{T}_p}^{\mathrm{M}} \right) \right] \tag{A.40}$$

$${}^k \Delta p_{l_p}^{\mathrm{B}} = {}^k Q \beta \, {}^k \Delta\Theta_{\mathrm{T}_p}^{\mathrm{B}} - \Delta t \frac{{}^k Q}{{}^k \rho_l^{\mathrm{B}}} \psi_{\mathrm{l}} \left({}^k p_{l_p}^{\mathrm{B}} - {}^k p_{l_p}^{\mathrm{M}} \right) \tag{A.41}$$

21. Update temperature and liquid pressure of embedded bands.

$${}^{k+1}\Theta_{\mathrm{T}_p}^{\mathrm{B}} = {}^k \Theta_{\mathrm{T}_p}^{\mathrm{B}} + {}^k \Delta\Theta_{\mathrm{T}_p}^{\mathrm{B}} \tag{A.42}$$

$${}^{k+1}p_{l_p}^{\mathrm{B}} = {}^k p_{l_p}^{\mathrm{B}} + {}^k \Delta p_{l_p}^{\mathrm{B}} \tag{A.43}$$

22. Calculation of effective and total stress increment at MPs.

$$^k\Delta\vec{\sigma'_p} = {^k}\mathbf{D} \cdot {^k}\Delta\vec{\varepsilon_p} \tag{A.44}$$

$$^k\Delta\vec{\sigma_p} = {^k}\Delta\vec{\sigma'_p} + \sum_{i=1}^{Nn} {^k}\Delta p_{l_i} {^k}N_i^p \tag{A.45}$$

23. Update total stress.

$$^{k+1}\vec{\sigma_p} = {^k}\vec{\sigma_p} + {^k}\Delta\vec{\sigma_p} \tag{A.46}$$

24. Update properties such as porosity, liquid density and porosity dependent variables.

25. Go to step 2 for a new time step calculation.

Bibliography

[1] NEN 9997-1. Geotechnical Design - part 1: General Rules. Technical report, Nederlands Normalisatie Instituut, Delft, The Netherlands, 2011.

[2] J. Aaron, O. Hungr, T. Stark, and A.K Baghdady. Oso, Washington, Landslide of March 22, 2014: dynamic analysis. *Journal of Geotechnical and Geoenvironmental Engineering*, 143(9):1–10, 2017.

[3] K. Abe, K. Soga, and S. Bandara. Material point method for coupled hydromechanical problems. *Journal of Geotechnical and Geoenvironmental Engineering*, page 04013033, 2013.

[4] K. Abe, K. Soga, and S. Bandara. Material point method for hydromechanical problems and its application to seepage failure analyses. In *10th World Congress on Computational Mechanics*, pages 30–41, 2014.

[5] I.K.J. Al-Kafaji. *Formulation of a dynamic material point method (MPM) for geomechanical problems*. PhD thesis, University of Stuttgart, 2013.

[6] E.E. Alonso, A. Gens, and A. Lloret. The landslide of Cortes de Pallas, Spain. *Géotechnique*, 43(4):507–521, 1993.

[7] E.E. Alonso and N.M. Pinyol. Criteria for rapid sliding I. A review of Vaiont case. *Engineering Geology*, 114(3-4):198–210, 2010.

[8] E.E. Alonso and N.M. Pinyol. The Vaiont landslide revisited. In *Proceeding 2nd Ital. Work. Landslide*, pages 9–24, Napoli, 2012.

[9] E.E. Alonso, N.M. Pinyol, and A. M. Puzrin. *Geomechanics of Failures. Advanced Topics*. Springer, 2010.

[10] E.E. Alonso, A Zervos, and N.M. Pinyol. Thermo-poro-mechanical analysis of landslides: from creeping behaviour to catastrophic failure. *Géotechnique*, 66(3):202–219, 2016.

[11] K.A. Alshibli and A. Hasan. Spatial variation of void ratio and shear band thickness in sand using X-ray computed tomography. *Géotechnique*, 58(4):249–257, 2008.

[12] M. Alvarado. *Landslide motion assessment including thermal interaction. An MPM approach*. PhD thesis, Universitat Politecnica de Catalunya, 2018.

[13] A. Andreykiv, F. Keulen, D.J. Rixen, and E. Valstar. A level-set-based large sliding contact algorithm for easy analysis of implant positioning. *International Journal of Numerical Methods in Engineering*, 89(10):1317–1336, 2012.

[14] Anura3D. *Anura3D MPM software v.2017.2.* Anura3D MPM Research Community, Delft, 2017.

[15] M. Arroyo, J. Butlanska, A. Gens, F. Calvetti, and M. Jamiolkowski. Cone penetration tests in a virtual calibration chamber. *Géotechnique*, 61(6):525–531, 2011.

[16] T.C. Badger. SR 530 MP 35 to 41 Geotechnical Study. Technical report, Washington State Department of Transportation, Olympia, 2015.

[17] T.C. Badger. SR 530 Landslide: geotechnical study. Technical report, Washington State, Department of Transportation, Olympia, 2016.

[18] M.M. Baligh. Strain path method. *Journal of Geotechnical Engineering*, 111(9):1108–36, 1985.

[19] N.J. Balmforth and R.R. Kerswell. Granular collapse in two dimensions. *Journal of Fluid Mechanics*, 538:399, 2005.

[20] S. Bandara. *Material point method to simulate large deformation problems in fluid-saturated granular medium.* PhD thesis, University of Cambridge, 2013.

[21] S. Bandara and K. Soga. Coupling of soil deformation and pore fluid flow using material point method. *Computers and Geotechnics*, 63:199–214, 2015.

[22] S.G. Bardenhagen, J.U. Brackbill, and D.L. Sulsky. The material-point method for granular materials. *Computer Methods in Applied Mechanics and Engineering*, 187(3-4):529–541, 2000.

[23] S.G. Bardenhagen, J.E. Guilkey, K.M. Roessig, J.U. Brackbill, W.M. Witzel, and J.C. Foster. An improved contact algorithm for the material point method and application to stress propagation in granular material. *C. - Computational Modelling Engineering Science*, 2(4):509–522, 2001.

[24] S.G. Bardenhagen and E.M. Kober. The generalized interpolation material point method. *Computer Modeling in Engineering and Sciences*, 5(6):477–496, 2004.

[25] K.J. Bathe, E. Ramm, and E.L. Wilson. Finite element formulations for large deformation dynamics analysis. *International Journal of Numerical Methods in Engineering*, 9(2):353–386, 1975.

[26] E. Bauer. Calibration of a comprehensive hypoplastic model for granular materials. *Soils and Foundations*, 36(1):13–26, 1996.

[27] J. Bear. *Dynamics of Fluids in Porous Media*. Dynamics of Fluids in Porous Media. American Elsevier Publishing Company, 1972.

[28] K. Been and M.G. Jefferies. A state parameter for sands. *Géotechnique*, 35(2):99–112, 1985.

[29] J.E. Beget. Postglacial volcanic deposits at Glacier Peak, Washington, and potential hazards from future eruptions. Technical report, USGS, 1982.

[30] L. Beuth. *Formulation and application of a quasi-static Material Point Method*. PhD thesis, University of Stuttgart, 2012.

[31] L. Beuth, T. Benz, P.A. Vermeer, C.J. Coetzee, P. Bonnier, and P. van den Berg. Formulation and validation of a quasi-static material point method. In *10th International Symposium in Numerical Methods in GeoMechanics*, pages 189–195, Rhodes, Greece, 2007.

[32] M.A. Biot. General theory of three dimensional consolidation. *Journal of Applied Physics*, 12:155–164, 1941.

[33] M.A. Biot. Theory of propagation of elastic waves in a fluid saturated porous solid. *The Journal of the Acoustical Society of America*, 28:168–178, 1956.

[34] A.W. Bishop. Discussion on measurement of shear strength of soils. *Géotechnique*, 2(2):108–116, 1950.

[35] A.W. Bishop. The use of the slip circle in the stability analysis of slopes. *Géotechnique*, 5(1):7–17, 1955.

[36] L. Bjerrum. Embankments on soft ground. In *Perform. Earth-Supported Struct.*, pages 1–54. Purdue University, 1972.

[37] L. Bjerrum and O. Eide. Stability of strutted excavation in clay. *Géotechnique*, 6(1):32–47, 1956.

[38] K. Bock. Rock Mechanics Analyses and Synthesis: RA Experiment. Mont Terri Technical Report 2000âĽŠ02. Technical report, Q+S Consult, 2001.

[39] M. Bolognin, M. Martinelli, K.J. Bakker, and S.N. Jonkman. Validation of material point method for soil fluidisation analysis. *Journal of Hydrodynamics*, 29(3):431–437, 2017.

[40] M.D. Bolton. The strength and dilatancy of sands. *Géotechnique*, 36(1):65–78, 1986.

[41] R.W. Boulanger. Relating to relative state parameter index. *Journal of Geotechnical and Geoenvironmental Engineering*, 129(August):770–773, 2003.

[42] J.U. Brackbill, D.B. Kothe, and H.M. Ruppel. FLIP: A low-dissipation, particle-in-cell method for fluid flow. *Computer Physics Communications*, 48(1):25–38, 1988.

[43] J.D. Bray and R.B. Sancio. Assessment of the liquefaction susceptibility of fine-grained soils. *Journal of Geotechnical and Geoenvironmental Engineering*, 132(9):1165–1177, 2006.

[44] R.B.J. Brinkgreve and H.L. Bakker. Nonlinear finite-element analysis of safety factors. In *7th International Conference on Comp. Methods and Advances in Geomechanics*, pages 1117–1122, Cairns, 1991.

[45] R.B.J Brinkgreve and E. Engin. Validation of geotechnical finite element analysis. In *18th International Conference Soil Mechanics and Geotechnical Engineering*, pages 677–683, Paris, 2013. Presses des Ponts.

[46] W. Broere and A.F. Van Tol. Modelling the bearing capacity of displacement piles in sand. *Institution of Civil Engineers - Geotechnical Engineering*, 159(3):195–206, 2006.

[47] J. Burghardt, R. Brannon, and J.E. Guilkey. A nonlocal plasticity formulation for the material point method. *Computer Methods in Applied Mechanics and Engineering*, 225-228:55–64, 2012.

[48] J.B. Burland. On the compressibility and shear strength of natural clays. *Géotechnique*, 40(3):329–378, 1990.

[49] J.B. Burland, T. Chapman, H. Skinner, and M. Brown. *ICE Manual of Geotechnical Engineering*. ICE Publishing, London, 2012.

[50] M. Calvello, S. Cuomo, and P. Ghasemi. The role of observations in the inverse analysis of landslide propagation. *Computers and Geotechnics*, 92:11–21, 2017.

[51] L. Cascini, S. Cuomo, M. Pastor, and G. Sorbino. Modeling of rainfall-induced shallow landslides of the flow-type. *Journal of Geotechnical and Geoenvironmental Engineering*, 136(1):85–98, 2009.

[52] F. Ceccato. *Study of large deformation geomechanical problems with the Material Point Method*. PhD thesis, Universita Degli Studi Di Padova, 2015.

[53] F. Ceccato. Study of flow landslide impact forces on protection structures with the Material Point Method. In *Landslides Eng. Slopes. Exp. Theory Pract. 12th International Symp. Landslides*, volume 2, pages 615–620, 2016.

[54] F. Ceccato, L. Beuth, and P. Simonini. Study of the effect of drainage conditions on cone penetration with the Material Point Method. In *XV Pan-American Conference Soil Mechanics Geotechnical Engineering*, Buenos Aires, 2015.

[55] F. Ceccato, L. Beuth, and P. Simonini. Analysis of Piezocone Penetration under Different Drainage Conditions with the Two-Phase Material Point Method. *Journal of Geotechnical and Geoenvironmental Engineering*, 142(12):4016066, 2016.

[56] F. Ceccato, L. Beuth, and P. Simonini. Adhesive contact algorithm for MPM and its application to the simulation of cone penetration in clay. In *Procedia Engineering*, pages 182–188, 2017.

[57] F. Ceccato, L. Beuth, P.A. Vermeer, and P. Simonini. Two-phase Material Point Method applied to cone penetration for different drainage conditions. In K. Soga, K. Kumar, G. Biscontin, and M. Kuo, editors, *International Symp. GeoMechanics from Micro to Macro*, pages 965–970, London, 2015. Taylor & Francis.

[58] F. Ceccato, L. Beuth, P.A. Vermeer, and P. Simonini. Two-phase Material Point Method applied to the study of cone penetration. *Computers and Geotechnics*, 2016.

[59] F. Ceccato, A. Bisson, and S. Cola. Large displacement numerical study of 3D plate anchors. *European Journal of Environmental and Civil Engineering*, 2017.

[60] F. Ceccato and P. Simonini. Numerical study of partially drained penetration and pore pressure dissipation in piezocone test. *Acta Geotechnica*, 12(1):195–209, 2016.

[61] F. Ceccato and P. Simonini. Study of landslide run-out and impact on protection structures with the material point method. In *Interpraevent 2016 – Conference Proceedings*, 2016.

[62] F. Ceccato, A. Yerro, and M. Martinelli. Modelling soil-water interaction with the material point method. Evaluation of single-point and double-point formulations. In *Numerical Methods in Geotechnical Engineering IX, Volume 1: 9th European Conference on Numerical Methods in Geotechnical Engineering (NUMGE 2018)*, pages 351–258, 2018.

[63] Y.-M. Chiew. Mechanics of local scour around submarine pipelines. *Journal of Hydraulic Engineering*, 116(4):515–529, 1990.

[64] A. Chmelnizkij and J. Grabe. Untersuchungen zur Wellenausbreitung im Boden infolge impulsartiger Beanspruchungen aus Fallmassen und Explosionen. In *Tagungsband zum 12. Hans Lorenz-Symposium 2016 in Berlin, VerÃ¼ffentlichungen des Grundbauinstituts der TU Berlin*, volume 70, pages 137–148, 2016.

[65] V.T. Chow. *Open-Channel Hydraulics.* McGraw-Hill, New York, 1959.

[66] B.N. Church. Geology of Hat Creek Basin - summary of field activities. Technical report, B.C. Department of Mines, 1975.

[67] R.V. Churchill. *Operational Mathematics.* McGraw-Hill, 3rd edition, 1972.

[68] M. Ciabatta. La dinamica della frana del Vajont. *Giornale di Geologia XXXII*, I:139–154, 1964.

[69] C.R.I. Clayton, D.W. Hight, and R.J. Hopper. Progressive destructuring of Bothkennar clay: implications for sampling and reconsolidation procedures. *Géotechnique*, 42(2):219–239, 1992.

[70] B. Z. Coelho, A. Rohe, and K. Soga. Poroelastic Solid Flow with Material Point Method. *Procedia Engineering*, 175:316–323, 2017.

[71] N. Collier, D. Pardo, M. Dalcin, M. Paszynski, and V. Calo. The cost of continuity: A study on the performance of isogeometric finite elements using direct solvers. *SIAM Journal on Scientific Computing*, 2(35):767–784, 2013.

[72] M. Cortis, C. Coombs, W. Augarde, M. Brown, A. Brennan, and S. Robinson. Imposition of essential boundary conditions in the material point method. *International Journal for Numerical Methods in Engineering*, 113(1):130–152, 2018.

[73] C.A. Coulomb. *Essai sur une application des règles de maximis & minimis à quelques problèmes de statique relatif à l'architecture.* Imprimerie Royale, 1776.

[74] S. Cuomo, P. Ghasemi, M. Calvello, and V. Hoseeinnezhad. Hypoplasticity model and inverse analysis for simulation of triaxial tests. In M. D. M. Fernandes, editor, *9th European Conference on Numerical Methods in Geotechnical Engineering*, pages 307–313. Balkema, 2018.

[75] S. Cuomo, M. Pastor, C. Sacco, and L. Cascini. Analisi delle fasi di rottura e post-rottura di frane tipo flusso. In M. D. M. Fernandes, editor, *XXV CNG – AGI - Roma*. AGI, 2014.

[76] A. Daerr and S. Douady. Sensitivity of granular surface flows to preparation. *Europhys. Lett.*, 47:324–330, 1999.

[77] C. De Boor. *A Practical Guide to Splines. Applied Mathematical Sciences.* Springer, New York, 2001.

[78] T. de Gast, P.J. Vardon, and M.A. Hicks. Estimating spatial correlations under man-made structures on soft soils. In J. Huang, G.A. Fenton, L. Zhang, and D.V. Griffiths, editors, *Geo-Risk 2017*, pages 382–389, Denver, 2017. ASCE.

[79] Deltares. *Slope stability software for soft soil engineering. D-Geo Stability*, 2016. v16.1.1.

[80] R.P. Denlinger and R.M. Iverson. Flow of variably fluidized granular masses across three-dimensional terrain: 2. numerical predictions and experimental tests. *Journal of Geophysical Research: Solid Earth*, 106(B1):553–566, 2001.

[81] C. Detournay and E. Dzik. Nodal Mixed Discretization for tetrahedral elements. In *4th International FLAC Symp. Numer. Model. GeoMechanics*, pages 07–02, Madrid, 2006. Itasca Consulting Group, Inc.

[82] S. Dey. Free overall in open channels: State-of-the-art review. *Flow Measurement and Instrumentation*, 13(5-6):247–264, 2002.

[83] T.R. Dhakal and D.Z. Zhang. Material point methods applied to one-dimensional shock waves and dual domain material point method with sub-points. *Journal of Computational Physics*, 325:301–313, 2016.

[84] J. Dijkstra, W. Broere, and A.F. Van Tol. Density changes near an advancing displacement pile in sand. In *2nd BGA International Conference on Foundations (ICOF)*, pages 545–554, 2008.

[85] G.R. Dodagoudar, S. Sayed, and K. Rajagopal. Random field modeling of reinforced retaining walls. *International Journal of Geotechinal Engineering*, 9(3):229–238, 2015.

[86] D. C. Drucker and W. Prager. Soil mechanics and plastic analysis or limit design. *Quarterly Applied Mathematics*, 10(2):157–165, 1952.

[87] D.C. Drucker, R.E. Gibson, and D.J. Henkel. Soil mechanics and work-hardening theories of plasticity. *Journal of Soil Mechanics and Foundation Engineering*, 122:338–346, 1957.

[88] P.M.D. Duff and R.D. Gilchrist. Correlation of lower Cretaceous coal measures, Peace River Coalfield. Technical report, Ministry of Energy, Mines and Petrelium ressources, British Columbia, 1983.

[89] J.M. Duncan. State of the art: Limit equilibrium and finite-element analysis of slopes. *Journal of Geotechnical Engineering*, 122(7):577–596, 1996.

[90] J.M. Duncan and C.Y. Chang. Nonlinear analysis of stress and strain in soils. *Journal of Soil Mechanics and Foundation Division*, 96(5):1629–1653, 1970.

[91] A.P. Dykes and E.N. Bromhead. The Vaiont landslide: re-assessment of the evidence leads to rejection of the consensus. *Landslides*, May 2017:1–18, 2018.

[92] D. Eckersley. Instrumented laboratory flowslides. *Géotechnique*, 40(3):489–502, 1990.

[93] J.D. Eckersley. Flowslides in stockpiled coal. *Engineering Geology*, 22(1):13–22, 1985.

[94] H.K. Engin, R.B.J. Brinkgreve, and A.F. Van Tol. Simplified numerical modelling of pile penetration - the press-replace technique. *International Journal of Numerical and Analytical Methods in Geomehcanics*, 39(15):1713–1734, 2015.

[95] S. Ergun. Fluid flow through packed column. *Chemical Engineering Progress*, 48:89–94, 1952.

[96] V. Escario and F. Juca. Strength and deformation of partly saturated soils. In *12th International Conference Soil Mech and Found. Eng.*, pages 43–46, 1989.

[97] H. Faheem, F. Cai, K. Ugai, and T. Hagiwara. Two-dimensional base stability of excavations in soft soils using FEM. *Computers and Geotechnics*, 30(2):141–163, 2003.

[98] M.M. Farias and D.J. Naylor. Safety analysis using finite elements. *Computers and Geotechnics*, 22(2):165–181, 1998.

[99] G.A. Fenton, D. Griffiths, and W. Cavers. The random finite element method (rfem) in settlement analyses. In D.V. Griffiths and G.A. Fenton, editors, *Probabilistic methods in geotechnical engineering. CISM courses and lectures 491*, pages 243–270, Vienna, 2007. Springer.

[100] G.A. Fenton and E.H. Vanmarcke. Simulation of random fields via local average subdivision. *Journal of Engineering Mechanics*, 116(8):1733–1749, 1990.

[101] E.J. Fern. *Constitutive modelling of unsaturated sand and its application to large deformation modelling*. PhD thesis, University of Cambridge, 2016.

[102] E.J. Fern, D.A. de Lange, C. Zwanenburg, J.A.M. Teunissen, A. Rohe, and K. Soga. Experimental and numerical investigations of dyke failures involving soft materials. *Engineering Geology*, 2016.

[103] E.J. Fern and K. Soga. The role of constitutive models in MPM simulations of granular column collapses. *Acta Geotechnica*, 11(3):659–678, 2016.

[104] E.J. Fern and K. Soga. Granular Column Collapse of Wet Sand. *Procedia Engineering*, 175:14–20, 2017.

[105] I.M.S. Finnie and M.F. Randolph. Punch-through and liquefaction induced failure of shallow foundations on calcareous sediments. In *International Conference on Behaviour of Offshore Structures, Boston, MA*, pages 217–230, 1994.

[106] L. Fraccarollo and E.F. Toro. Experimental and numerical assessment of the shallow water model for two-dimensional dam-break type problems. *Journal of Hydraulic Research*, 33(6):843–864, 1995.

[107] D.G. Fredlund, N.R. Morgenstern, and R.A. Widger. The shear strength of unsaturated soils. *Canadian Geotechnical Journal*, 15(3):313–321, 1978.

[108] K. Gahalaut, J. Kraus, and S. Tomar. Multigrid methods for isogeometric discretizations. *Computer Methods in Applied Mechanics and Engineering*, 253:413–425, 2013.

[109] V. Galavi, L. Beuth, B. Z. Coelho, F.S. Tehrani, P. Hirscher, and F. Van Tol. Numerical simulation of pile Installation in saturated sand using Material Point Method. *Procedia Engineering*, 175:72–79, 2017.

[110] K. Gavin, D. Cadogan, A. Tolooiyan, and P. Casey. The base resistance of non-displacement piles in sand. part i: field tests. *Instition of Civil Engineers – Geotechnical Engineering*, 166(6):540–548, 2013.

[111] R. Genevois and P. Tecca. The Vajont Landslide: State-of-the-Art. *Italian Journal of Engineering Geology and Environment*, 6:15–40, 2013.

[112] A. Gens and D.M. Potts. Formulation of quasi-axisymmetric boundary value problems for finite element analysis. *Engineering with Computers*, 1(2):144–150, 1984.

[113] D. Givoli and D. Cohen. Nonreflecting boundary conditions based on Kirchhoff-type formulae, 1995.

[114] G. Golub and C. van Loan. *Matrix Computations*. The Johns Hopkins University Press, Baltimore, 1996.

[115] J.L. Gonzalez Acosta, P.J. Vardon, and M.A. Hicks. Composite material point method (cmpm) to improve stress recovery for quasi-static problems. *Procedia Engineering*, 175:324–331, 2017.

[116] J. Grabe, T. Pucker, and T. Hamann. Numerical simulation of pile installation processes in dry and saturated granular soils. In *International Conference on Numerical Methods in Geotechnical Engineering (NUMGE)*, pages 663–668, 2014.

[117] D. Griffiths. Computation of bearing capacity factors using finite elements. *Géotechnique*, 32(3):195–202, 1982.

[118] D. V. Griffiths and P. A. Lane. Slope stability analysis by finite elements. *Géotechnique*, 49(3):387–403, 1999.

[119] D.V. Griffiths and G.A. Fenton. Probabilistic slope stability analysis by finite elements. *Journal Geotechnichal and Geoenvironmental Engineering*, 130(5):507–518, 2004.

[120] G. Gudehus. A visco-hypoplastic constitutive relation for soft soils. *Soils and Foundations*, 44(4):11–25, 2004.

[121] F. Hamad, D. Stolle, and P. Vermeer. Modelling of membranes in the material point method with applications. *International Journal for Numerical and Analytical Methods in Geomechanics*, 39(8):833–853, 2015.

[122] F. Hamad, Z. Więckowski, and C. Moormann. Interaction of fluid–solid–geomembrane by the material point method. *Computers and Geotechnics*, 81:112 – 124, 2017.

[123] T. Hamann, G. Qiu, and J. Grabe. Application of a coupled Eulerian-Lagrangian approach on pile installation problems under partially drained conditions. *Computers and Geotechnics*, 63:279–290, 2015.

[124] K. Hashiguchi and Z. Chen. Elastoplastic constitutive equation of soils with the subloading surface and the rotational hardening. *International Journal for Numerical and Analytical Methods in Geomechanics*, 22(3):197–227, 1998.

[125] R.A. Haugerud. Preliminary Interpretation of Pre-2014 Landslide Deposits in the Vicinity of Oso, Washington. Technical report, USGS, Reston (VA), 2014.

[126] A.J. Hendron and F.D. Patton. The Vaiont slide, a geotechnical analysis based on new geologic observations of the failure surface. Vol. I. *US Army Corps of Engineers*, 2:187–188, 1985.

[127] B. Henn, Q. Cao, D.P. Lettenmaier, C.S. Magirl, C. Mass, J.B. Bower, M. St Laurent, Y.X. Mao, and S. Perica. Hydroclimatic Conditions Preceding the March 2014 Oso Landslide. *Journal of Hydrometerology*, 16(3):1243–1249, 2015.

[128] I. Herle and G. Gudehus. Determination of parameters of a hypoplastic constitutive model from properties of grain assemblies. *Mechanics of Cohesive-Frictional Materials*, 4(5):461–486, 1999.

[129] M.A. Hicks, J.D. Nuttall, and J. Chen. Influence of heterogeneity on 3d slope reliability and failure consequence. *Computers and Geotechnics*, 61:198–208, 2014.

[130] M.A. Hicks and K. Samy. Influence of heterogeneity on undrained clay slope stability. *Quarterly Journal of Engineering Geology and Hydrogeology*, 35(1):41–49, 2002.

[131] M.A. Hicks and W. Wong. Static liquefaction of loose slopes. In *Numerical Methods in Geomechanics*, pages 1361–1367. Balkema, 1988.

[132] R. Hill. Some basic principles in the mechanics of solids without a natural time. *Journal of the Mechanics and Physics of Solids*, 7(3):209–225, 1959.

[133] E. Hoek. Strength of jointed rock masses. *Géotechnique*, 33(3):187–223, 1983.

[134] E. Hoek. Rocscience lecture series, Lecture 2: "The Art of Tunnelling", 2015.

[135] E. Hoek, C. Carranza, and B. Corkum. Hoek-brown failure criterion – 2002 edition. In Reginald Hammah, editor, *Narms-Tac*, pages 267–273, Toronto, 2002. University of Toronto Press.

[136] C. Hofreither, S. Takacs, and W. Zulehner. A robust multigrid method for isogeometric analysis in two dimensions using boundary correction. *Computer Methods in Applied Mechanics and Engineering*, 316:22–42, 2017.

[137] P. Holscher, A.F. Van Tol, and N.Q. Huy. Rapid pile load tests in the geotechnical centrifuge. In *9th International Conference on Testing and Design Methods for Deep Foundations, Japanese Geotechnical Society*, pages 257–263, 2012.

[138] Y. Honjo. Challenges in geotechnical reliability based design. In N. Vogt, B. Schuppener, D. Straub, and G. Bräu, editors, *3rd International Symposium on Geotechnical Safety and Risk*, pages 11–27, 2011.

[139] R. Hooke. *De Potentia Restitutiva or of Spring.* J. Martyn, London, 1678.

[140] V. Hosseinnezhad, M. Rafiee, M. Ahmadian, and M.T. Ameli. Species-based quantum particle swarm optimization for economic load dispatch. *International Journal of Electrical Power & Energy Systems*, 63:311–322, 2014.

[141] W. Hu and Z. Chen. A multi-mesh MPM for simulating the meshing process of spur gears. *Computers and Structures*, 81(20):1991–2002, 2003.

[142] Y. Hu and M. F. Randolph. A practical numerical approach for large deformation problems in soil. *International Journal for Numerical and Analytical Methods in Geomechanics*, 22(5):327–350, 1998.

[143] P. Huang, X. Zhang, S. Ma, and X. Huang. Contact algorithms for the material point method in impact and penetration simulation. *International Journal of Numerical Methods in Engineering*, 85(4):498–517, 2011.

[144] M. Huber. *Soil variability and its consequences in geotechnical engineering*. PhD thesis, University of Stuttgart, 2013.

[145] O. Hungr, S. Leroueil, and L. Picarelli. The Varnes classification of landslide types, an update. *Landslides*, 11(2):167–194, 2014.

[146] R.M. Iverson and D.L. George. Modelling landslide liquefaction, mobility bifurcation and the dynamics of the 2014 Oso disaster. *Géotechnique*, 66(3):175–187, 2016.

[147] R.M. Iverson, D.L. George, K. Allstadt, M.E. Reid, B.D. Collins, J.W. Vallance, S.P. Schilling, Jonathan W. Godt, C.M. Cannon, C.S. Magirl, R.L. Baum, J.a. Coe, W.H. Schulz, and J.B. Bower. Landslide mobility and hazards: implications of the 2014 Oso disaster. *Earth and Planetary Science Letters*, 412:197–208, 2015.

[148] M.G. Jefferies. Nor-Sand: a simple critical state model for sand. *Géotechnique*, 43(1):91–103, 1993.

[149] M.G. Jefferies and K. Been. *Soil Liquefaction - a Critical State Approach*. Taylor & Francis, London, 1st edition, 2006.

[150] M.G. Jefferies and K. Been. *Soil Liquefaction - a Critical State Approach*. CRC Press, London, 2nd edition, 2016.

[151] M.G. Jefferies and D.A. Shuttle. Dilatancy in general Cambridge-type models. *Géotechnique*, 52(9):625–638, 2002.

[152] M.G. Jefferies and D.A. Shuttle. On the operating critical friction ratio in general stress states. *Géotechnique*, 61(8):709–713, 2011.

[153] J.R. Keaton, J. Wartman, S. Anderson, J. Benoît, J. de la Chappelle, Robert Gilbert, and D.R. Montgomery. The 22 March 2014 Oso Landslide, Snohomish County, Washington. Technical report, Geotechnical Extreme Events Reconnaissance, 2014.

[154] K. Kim, M. Prezzi, R. Salgado, and W. Lee. Effect of penetration rate on cone penetration resistance in saturated clayey soils. *Journal of Geotechnical and Geoenvironmental Engineering*, 134(8):1142–1153, 2008.

[155] J. Kirstein, J. Grabe, and A. Chmelnizkij. Numerische Berechnungen und messtechnische Begleitung zur Dynamischen Intensivverdichtung. In *Tagungsband zur 34. Baugrundtagung 2016 in Bielefeld*, pages 297–304, 2016.

[156] E.U. Klotz and M.R. Coop. An investigation of the effect of soil state on the capacity of driven piles in sands. *Géotechnique*, 51(9):733–751, 2001.

[157] J.J. Kolbuszewski. An experimental study of the maximum and minimum porosities of sands. In *6th International Conference Soil Mechanics Foundation Engineering*, pages 158–165, Rotterdam, 1948.

[158] D. Kolymbas. A generalized hypoelastic constitutive law. In *XI International Conference Soil Mechanics and Foundation Engineering*, volume 5, page 2626, San Francisco, 1985. Balkema.

[159] E. Lajeunesse, A. Mangeney-Castelnau, and J.P. Vilotte. Spreading of a granular mass on a horizontal plane. *Physics of Fluid*, 16(7):2371–2381, 2004.

[160] E. Lajeunesse, J.B. Monnier, and G.M. Homsy. Granular slumping on a horizontal surface. *Physics of Fluid*, 17(10):103302, 2005.

[161] T.W. Lambe. Predictions in soil engineering. *Géotechnique*, 23(2):151–202, 1973.

[162] A. Lees. *Geotechnical finite element analysis: a practical guide*. Institution of Civil Engineers, London, 2016.

[163] A.M. Legendre. *Nouvelles méthodes pour la détermination des orbites des comètes*. F. Didot, Paris, 1805.

[164] S. Leroueil. Compressibility of clays: fundamental and practical aspects. *Journal of Geotechnical Engineering*, 122(7):534–543, 1996.

[165] J.N. Levadoux and M.M. Baligh. Consolidation after undrained piezocone penetration. i: Prediction. *Journal of Geotechnical Engineering*, 112(7):707–726, 1986.

[166] D.Q. Li, T. Xiao, Z.J. Cao, C.B. Zhou, and L.M. Zhang. Enhancement of random finite element method in reliability analysis and risk assessment of soil slopes using subset simulation. *Landslides*, 13(2):293–303, 2016.

[167] L.C. Li, C.A. Tang, W.C. Zhu, and Z.Z. Liang. Numerical analysis of slope stability based on the gravity increase method. *Computers and Geotechnics*, 36(7):1246–1258, 2009.

[168] Y.L. Li, M.A. Hicks, and J.D. Nuttall. Comparative analyses of slope reliability in 3d. *Engineering Geology*, 196:12–23, 2015.

[169] Y.L. Li, M.A. Hicks, and P.J. Vardon. Uncertainty reduction and sampling efficiency in slope designs using 3d conditional random fields. *Computers and Geotechnics*, 79:159–172, 2016.

[170] D. Liang. Evaluating shallow water assumptions in dam-break flows. *ICE - Water Management*, 163(5):227–237, 2010.

[171] D. Liang, R.A. Falconer, and Binliang Lin. Comparison between TVD-MacCormack and ADI-type solvers of the shallow water equations. *Advances in Water Resources*, 29(12):1833–1845, 2006.

[172] D. Liang, X. Zhao, and M. Martinelli. MPM simulations of the interaction between water jet and soil bed. In *Procedia Engineering*, volume 175, pages 242–249, Delft, the Netherlands, 2017. Elsevier B.V.

[173] L.J. Lim, A. Andreykiv, and R.B.J. Brinkgreve. On the application of the material point method for offshore foundations. In M.A. Hicks, R.B.J. Brinkgreve, and A. Rohe, editors, *Numerical Methods in Geotechnical Engineering*, pages 253–258, Delft, 2014. Taylor & Francis.

[174] G. Lombardi. Les ouvrages souterrains du CERN. *Publication de la Société Suisse de Mécanique des Sols et des Roches*, 104:47–53, 1981.

[175] S. Lòpez-Querol, J.A. Fernández-Merodo, P. Mira, and M. Pastor. Numerical modelling of dynamic consolidation on granular soils. *International J. Numer. Anal. Meth. GeoMechanics*, 32(12):1431–1457, 2008.

[176] E. Love and D.L. Sulsky. An unconditionally stable, energy–momentum consistent implementation of the material-point method. *Computer Methods in Applied Mechanics and Engineering*, 195:3903–3925, 2006.

[177] G. Lube, H. Huppert, R.S.J. Sparks, and A. Freundt. Collapses of two-dimensional granular columns. *Phys. Rev. E - Stat. Nonlinear, Soft Matter Phys.*, 72(4):1–10, 2005.

[178] G. Lube, H. Huppert, R.S.J. Sparks, and A. Freundt. Static and flowing regions in granular collapses down channels. *Physics of Fluid*, 19(4):043301, 2007.

[179] G. Lube, H. Huppert, R.S.J. Sparks, and M.A. Hallworth. Axisymmetric collapses of granular columns. *Journal of Fluid Mechanics*, 508:175–199, 2004.

[180] T. Lunne, J. M. Powell, P. K. Robertson, et al. *Cone Penetration Testing in Geotechnical Practice*. CRC Press, 1999.

[181] H. Luo, J.D. Baum, and R. Löhner. A discontinuous Galerkin method based on a Taylor basis for the compressible flows on arbitrary grids. *Journal of Computational Physics*, 227(20):8875–8893, 2008.

[182] M. Luong and A. Touati. Sols grenus sous fortes contraintes. *Revue Française de Géotechnique*, 24:51–63, 1983.

[183] J. Lysmer and R.L. Kuhlemeyer. Finite dynamic model for infinite media. In *ASCE*, pages 859–877, 1969.

[184] J. Ma, D. Wang, and M.F. Randolph. A new contact algorithm in the material point method for geotechnical simulations. *International Journal for Numerical and Analytical Methods in Geomechanics*, pages 1197–1210, 2014.

[185] S. Ma, X. Zhang, and X.M. Qiu. Comparison study of MPM and SPH in modeling hypervelocity impact problems. *International Journal of Impact Engineering*, 36(2):272–282, 2009.

[186] H. Mahmoodzadeh and M. F Randolph. Penetrometer testing: effect of partial consolidation on subsequent dissipation response. *Journal of Geotechnical and Geoenvironmental Engineering*, 140(6):04014022, 2014.

[187] K. Mahutka, F. Konig, and J. Grabe. Numerical modelling of pile jacking, driving and vibro driving. In *International Conference on Numerical Simulation of Construction Processes in Geotechnical Engineering for Urban Environment*, pages 235–246, 2006.

[188] D.S. Malkus and T.J.R. Hughes. Mixed finite element methods - Reduced and selective integration techniques: A unification of concepts. *Computer Methods in Applied Mechanics and Engineering*, 15(1):63–81, 1978.

[189] L.E. Malvern. *Introduction to the Mechanics of a Continuous Medium*. Prentice-Hall, Englewood Cliffs, New Jersey, 1996.

[190] S. Mandal and R. Maiti. *Semi-quantitative Approaches for Landslide Assessment and Prediction*. Springer, London, 2015.

[191] J.C. Martin and W.J. Moyce. An experimental study of the collapse of liquid columns on a rigid horizontal plane. *Philosophical Transactions of Royal Society London, Series A*, 244(882):312–324, 1952.

[192] M. Martinelli. Soil-water interaction with material point method. Technical report, Deltares, Delft, the Netherlands, 2016.

[193] M. Martinelli, S.T. Faraz, and V. Galavi. Analysis of crater development around damaged pipelines using the material point method. *Procedia Engineering*, 175:204 – 211, 2017. 1st International Conference on the Material Point Method (MPM 2017).

[194] M. Martinelli and A. Rohe. Modelling fluidisation and sedimentation using material point method. In *1st Pan-American Congress on Computational Mechanics*, pages 1–12, 2015.

[195] M. Martinelli, A. Rohe, and K. Soga. Modeling dike failure using the Material Point Method. *Procedia Engineering*, 175(2016):341–348, 2017.

[196] C.M. Mast, P. Mackenzie-Helnwein, P. Arduino, G.R. Miller, and W. Shin. Mitigating kinematic locking in the material point method. *Journal of Computational Physics*, 231(16):5351–5373, 2012.

[197] H. Matsuoka and T. Nakai. Stress deformation and strength characteristics of soil uner threee different principal stresses. *Soil Found.*, 232:59–70, 1974.

[198] P.W. Mayne and J.S. Jones, Jr. Impact stresses during dynamic compaction. *Journal of Geotechnical Engineering*, 109(10):1342–1346, 1983.

[199] G. Mesri and M. Ajlouni. Engineering properties of fibrous peats. *Journal of Geotechnical and Geoenvironmental Engineering*, 133(7):850–866, 2007.

[200] V. Milligan. Geotechnical aspects of glacial tills. In R.F. Legget, editor, *Glacial Till an Inter-disciplinary Study*, pages 269–290. Royal Society of Canada, Ottawa, 1976.

[201] J.K. Mitchell and K. Soga. *Fundamentals of Soil Behavior*. Wiley & Sons, New York, 3rd edition, 2005.

[202] C.O. Mohr. Welche Umstaad Bedingen des Elastizitaetsgrenzen und den Bruch eines material? In *Abhandlingen aus dem Gebiete der Technischen Mechanik*, pages 192–240. Ernst und Sohn, Berlin, 3rd edition, 1928.

[203] J.J. Monaghan and A. Kos. Scott Russell's wave generator. *Physics of Fluids*, 12(3):622–630, 2000.

[204] L. Monforte, J.M. Carbonell, M. Arroyo, and A. Gens. Performance of mixed formulations for the particle finite element method in soil mechanics problems. *Computational Particle Mechanics*, 4(3):269–284, 2017.

[205] J.S. Montes. Potential flow solution to 2D transition from mild to steep slope. *Journal of Hydraulic Engineering*, 120(5):601–621, 1994.

[206] K.F. Mostafa and R.Y. Liang. Numerical modeling of dynamic compaction in cohesive soils. In *Geo-Frontiers 2011*, pages 738–747, Reston, VA, 2011. American Society of Civil Engineers.

[207] D. Muir Wood. *Soil Behavior and Critical State Soil Mechanics*. Cambridge University Press, Cambridge, 1990.

[208] D. Muir Wood and K. Belkheir. Strain softening and state parameter for sand modelling. *Géotechnique*, 44(2):335–339, 1994.

[209] L. Müller. The rock slide in the Vajont Valley. *Rock Mechanics and Engineering Geology*, 2:148–212, 1964.

[210] E. Murray and J.D. Geddes. Uplift of anchor plates in sand. *Journal of Geotechnical Engineering*, 113(3):202–215, 1987.

[211] F. Naghibi, G.A. Fenton, and D.V. Griffiths. Probabilistic considerations for the design of deep foundations against excessive differential settlement. *Canadian Geotechechnical Journal*, 53:1167–1175, 2016.

[212] K. Nakagawa, K. Soga, and J. K. Mitchell. Observation of biot compressional wave of the second kind in granular soils. *Géotechnique*, 47(1):133–147, 1997.

[213] P.L. Newland and B.H. Allely. Volume changes in drained tiaxial tests on granular materials. *Géotechnique*, 7(1):17–34, 1957.

[214] A. Niemunis and I. Herle. Hypoplastic model for cohesionless soils with elastic strain range. *Mechanics of Cohesive-frictional Materials: An International Journal on Experiments, Modelling and Computation of Materials and Structures*, 2(4):279–299, 1997.

[215] E. Nonveiller. The Vajont reservoir slope failure. *Engineering Geology*, 24:493–512, 1987.

[216] R. Nova. A constitutive model for soil under monotonic and cyclic loading. In G. N. Pande and C. Zienkiewicz, editors, *Soil Mechanics - Transient Cyclic Loading*, pages 343–373, Chichester, 1982. Wiley.

[217] R. Nova and D.M. Wood. A constitutive model for sand in triaxial compression. *International Journal for Numerical and Analytical Methods in Geomechanics*, 3(3):255–278, 1979.

[218] J.R.M.S. Oliveira, M.S.S. Almeida, H.P.G. Motta, and M.C.F. Almeida. Influence of penetration rate on penetrometer resistance. *Journal of Geotechnical and Geoenvironmental Engineering*, 137(7):695–703, 2011.

[219] D.J. Palladino and R.B. Peck. Slope failures in an overconsolidated clay, Seattle, Waashington. *Géotechnique*, 22(4):563–595, 1972.

[220] J.L. Pan and A.R. Selby. Simulation of dynamic compaction of loose granular soils. *Advances in Engineering Software*, 33(7-10):631–640, 2002.

[221] K.L. Pankow, J.R. Moore, J.M. Hale, K.D. Koper, T. Kubacki, K.M. Whidden, and M.K. McCarter. Massive landslide at Utah copper mine generates wealth of geophysical data. *GSA Today*, 24(1):4–9, 2014.

[222] P. Paronuzzi and A. Bolla. Gravity-induced rock mass damage related to large en masse rockslides: Evidence from Vajont. *Geomorphology*, 234:28–53, 2015.

[223] M. Pastor, T. Blanc, B. Haddad, S. Petrone, M.S. Morles, V. Drempetic, D. Issler, G.B. Crosta, L. Cascini, G. Sorbino, et al. Application of a sph depth-integrated model to landslide run-out analysis. *Landslides*, 11(5):793–812, 2014.

[224] N.T.V. Phuong, A.F. van Tol, A.S.K. Elkadi, and A. Rohe. Numerical investigation of pile installation effects in sand using material point method. *Computers and Geotechnics*, 73:58–71, 2016.

[225] N.M. Pinyol and E.E. Alonso. Fast planar slides. A closed-form thermo-hydro-mechanical solution. *International Journal for Numerical and Analytical Methods in Geomechanics*, 34(1):27–52, 2010.

[226] N.M. Pinyol, M. Alvarado, E.E. Alonso, and F. Zabala. Thermal effects in landslide mobility. *Géotechnique*, 68(6):528–545, 2018.

[227] T.J. Plona. Observation of a second bulk compressional wave in a porous medium at ultrasonic frequencies. *Applied Physics Letters*, 36(4):259–261, 1980.

[228] E.P. Poeter and M.C. Hill. *Documentation of UCODE: A computer code for universal inverse modeling*, volume 98. DIANE Publishing, 1998.

[229] D. Potts, K. Axelsson, L. Grande, H. Schweiger, and M. Long. *Guidelines For the Use of Advanced Numerical Analysis*. Thomas Telford, London, 2002.

[230] D.M. Potts. Numerical analysis: a virtual dream or practical reality? *Géotechnique*, 53(6):535–573, 2003.

[231] D.M. Potts and L Zdravković. *Finite Element Analysis in Geotechnical Engineering: Application*. Thomas Telford, London, 1999.

[232] D.M. Potts and L. Zdravković. *Finite Element Analysis in Geotechnical Engineering: Theory*. Thomas Telford Ltd, London, 2001.

[233] D.M. Potts and L. Zdravkovic. Computer analysis principles in geotechnical engineering. In J.B. Burland, T. Chapman, H. Skinner, and Brown. M., editors, *ICE Manual of Geotechnical Engineering*, chapter 6, pages 35–58. ICE Publishing, London, 2012.

[234] J.G. Potyondy. Skin friction between various soils and construction materials. *Géotechnique*, 11(4):339–353, 1961.

[235] M.R. Pyles, J.D. Rogers, J.D. Bray, A. Skaugset, R. Storesund, and G. Schlieder. 2014 Oso Landslide - Expert report. June 30, 2016. Technical report, Superior Court of Washington for King County, King County, 2016.

[236] G. Qiu, S. Henke, and J. Grabe. Application of a coupled Eulerian-Lagrangian approach on geomechanical problems involving large deformations. *Computers and Geotechnics*, 38(1):30–39, 2011.

[237] N. Rajaratnam and D. Muralidhar. Characteristics of the rectangular free overfall. *Journal of Hydraulic Research*, 6(3):233–258, 1968.

[238] M. Randolph. Science and empiricism in pile foundation design. *Géotechnique*, 53(10):847–75, 2003.

[239] M.F. Randolph, R. Dolwin, and R. Beck. Design of driven piles in sand. *Géotechnique*, 44(3):427–448, 1994.

[240] M.F. Randolph and S. Hope. Effect of cone velocity on cone resistance and excess pore pressures. In *International Symposium on Engineering Practice and Performance of Soft Deposits*, pages 147–152. Yodagawa Kogisha Co., Ltd., 2004.

[241] G. Remmerswaal, M.A. Hicks, and P.J. Vardon. Ultimate limit state assessment of dyke reliability using the random material point method. In *4th International Symposium on Computational Geomechanics*, pages 89–90, Assisi, 2018.

[242] J.A. Replogle. Discussion: "End depth at a drop in trapezoidal channels" by M.H. Diskin. *Journal of the Hydraulics Division*, 88(2):161–165, 1962.

[243] M.F. Riemer, B.D. Collins, T.C. Badger, C. Toth, and Y.C. Yu. Geotechnical soil characterization of intact quaternary deposits forming the March 22, 2014 SR-530 (Oso) Landslide, Snohomish County, Washington: U.S. Geological Survey Open-File Report 2015-1089. Technical report, USGS, 2015.

[244] P.K. Robertson. Cone penetration test (cpt)-based soil behaviour type (sbt) classification system – an update. *Canadian Geotechnical Journal*, 53(12):1910–1927, 2016.

[245] P.K. Robertson, J.P. Sully, D. J. Woeller, T. Lunne, J.J.M. Powell, and D.G. Gillespie. Estimating coefficient of consolidation from piezocone tests. *Canadian Geotechnical Journal*, 29(4):539–550, 1992.

[246] J.D. Rogers, M.R. Pyles, J.D. Bray, A. Skaugset, R. Storesund, and G. Schlieder. 2014 Oso Landslide – Expert report January 22, 2016. Technical report, Superior Court of Washington for King County, King County, 2016.

[247] K.H. Roscoe. The Influence of strains in soil mechanics. *Géotechnique*, 20(2):129–170, 1970.

[248] K.H. Roscoe and J.B. Burland. On the generalised stress-strain behaviour of 'wet' clay. In J. Heyman and F.A. Leckie, editors, *Engineering Plasticity*, pages 535–609, Cambridge, 1968. Cambridge University Press.

[249] K.H. Roscoe and A.N. Schofield. Mechanical behaviour of an idealised wet clay. In *2nd Eur. Conference Soil Mechanics Found. Eng.*, pages 47–54, Wiesbaden, 1963.

[250] K.H. Roscoe, A.N. Schofield, and C. P. Wroth. On the yielding of soil. *Géotechnique*, 8:22–52, 1958.

[251] C.S. Ross and E.V. Shannon. The minerals of bentonite and related clays and their physical properties. *J. Am. Ceram. Soc.*, 9(2):77–96, 1926.

[252] H. Rouse. Discharge characteristics of the free overfall: Use of crest section as a control provides easy means of measuring discharge. *Civil Engineering ASCE*, 6(4):257–260, 1936.

[253] P.W. Rowe. The stress-dilatancy relation for static equilibrium of an assembly of particles in contact. In *Proceedings of the Royal Society. Mathematical, Physical and Engineering Sciences*, volume 269, pages 500–527. The Royal Society, 1962.

[254] Y. Saad. *Iterative Methods for Sparse Linear Systems*. SIAM, Philadelphia, 2003.

[255] R. Salgado, J. Mitchell, and M. Jamiolkowski. Calibration chamber size effects on penetration resistance in sand. *J. Geotech. Geoenviron. Eng.*, 124(9):878–88, 1998.

[256] R. Salgado and M. Prezzi. Computation of cavity expansion pressure and penetration resistance in sands. *International Journal of Geomechanics*, 7(4):251–265, 2007.

[257] G. Sangalli and M. Tani. Isogeometric preconditioners based on fast solvers for the sylvester equation. *SIAM Journal on Scientific Computing*, 6(38):3644–3671, 2016.

[258] W.Z. Savage, M.M. Morrissey, and R.L. Baum. Geotechnical Properties for Landslide-Prone Seattle Area Glacial Deposits: Open-File Report 00-228. Technical report, USGS, Denver, 2000.

[259] J.A. Schneider, B.M. Lehane, and F. Schnaid. Velocity effects on piezocone measurements in normally and over consolidated clays. *International Journal of Physical Modelling in Geotechnics*, 7(2):23–34, 2007.

[260] M.T. Schobeiri. *Fluid Mechanics for Engineers*. Springer, Berlin, 2010.

[261] A.N. Schofield. Cambridge geotechnical centrifuge operations. *Géotechnique*, 30(3):227–268, 1980.

[262] A.N. Schofield and P. Wroth. *Critical State Soil Mechanics*. McGraw-Hill, London, 2nd edition, 1968.

[263] H.F. Schweiger. An introduction to numerical modelling in geotechnical engineering. Technical report, ISSMGE, 2017.

[264] R.A. Scott and R.W. Pearce. Soil compaction by impact. *Geotechnique*, 25(1):19–30, 1975.

[265] E. Semenza. *La Storia del Vaiont Raccontata del Geologo che ha Scoperto la Frana*. Tecomproject Editore Multimediale, 2001.

[266] E. Semenza and M. Ghirotti. History of the 1963 Vaiont slide: the importance of geological factors. *Bulletin of Engineering Geology and the Environment*, 59:87–97, 2000.

[267] W.D. Shannon. Slide on North Fork Stillaguanish River near Hazel. Technical report, William D. Shannon & Associates, Washington, 1952.

[268] S. Shao and E.Y.M. Lo. Incompressible SPH method for simulating Newtonian and non-Newtonian flows with a free surface. *Advances in Water Resources*, 26(7):787–800, 2003.

[269] M.F. Silva, D.J. White, and M.D. Bolton. An analytical study of the effect of penetration rate on piezocone tests in clay. *International journal for numerical and analytical methods in geomechanics*, 30(6):501–527, 2006.

[270] A.W. Skempton. The bearing capacity of clays. In *Building Research Congress, London*, pages 180–189, 1951.

[271] A.W. Skempton. The colloidal 'activity' of clays. In *3rd International Conference Soil Mechanics Found. Eng.*, pages 57–61, Zurich, 1953. ICOSOMEF.

[272] A.W. Skempton. Residual strength of clays in landslides, folded strata and the laboratory. *Géotechnique*, 35(1):3–18, 1985.

[273] A.W. Skempton and A.W. Bishop. The measurement of the shear strength of soils. *Géotechnique*, 2(2):90–108, 1950.

[274] S. Sloan and M.F. Randolph. Numerical prediction of collapse loads using finite element methods. *International Journal of Numerical and Analytical Methods in Geomechanics*, 6(1):47–76, 1982.

[275] S.W. Sloan. Substepping schemes for the numerical integration of elasto-plastic stress-strain relations. *International Journal of Numerical Methods in Engineering*, 24(5):893–911, 1987.

[276] S.W. Sloan. Geotechnical stability analysis. *Géotechnique*, 63(7):531–571, 2013.

[277] J.I. Sneddon. *A study of two soils derived from volcanic ash in southwestern British Columbia and a review and determination of ash distribution in western Canada*. PhD thesis, University of British Columbia, 1970.

[278] K. Soga, E.E. Alonso, A. Yerro, K. Kumar, and S. Bandara. Trends in large-deformation analysis of landslide mass movements with particular emphasis on the material point method. *Géotechnique*, 66(3):248–273, 2016.

[279] A. Srivastava, G.L.S. Babu, and S. Haldar. Influence of spatial variability of permeability property on steady state seepage flow and slope stability analysis. *Engineering Geology*, 110:93–101, 2010.

[280] T. Stark, Ahmed K. Baghdady, O. Hungr, and J. Aaron. Case study: Oso, Washington, Landslide of March 22, 2014 - material properties and failure mechanism. *Journal of Geotechnical and Geoenvironmental Engineering*, 143(5):05017001, 2017.

[281] T. Stark and H. Eid. Drained residual strength of cohesive soils. *Journal of Geotechnical Engineering*, 120(5):856–871, 1994.

[282] M. Steffen, R.M. Kirby, and M. Berzins. Analysis and reduction of quadrature errors in the material point method (MPM). *International Journal for Numerical Methods in Engineering*, 76(6):922–948, 2008.

[283] M. Steffen, R.M. Kirby, and M. Berzins. Decoupling and balancing of space and time errors in the material point method (MPM). *International Journal for Numerical Methods in Engineering*, 82(10):1207–1243, 2010.

[284] D. Sulsky and H.L. Schreyer. Axisymmetric form of the material point method with applications to upsetting and Taylor impact problems. *Computer Methods in Applied Mechanics and Engineering*, 139:409–429, 1996.

[285] D.L. Sulsky, Z. Chen, and H.L. Schreyer. A particle method for history-dependent materials. *Computer Methods in Applied Mechanics and Engineering*, 118:179–196, 1994.

[286] D.L. Sulsky and M. Gong. Improving the material-point method. In K. Weinberg and A. Pandolfi, editors, *Innovative Numerical Approaches for Multi-Fields and Multi-Scale Problems*, pages 217–240, Berlin, 2016. Springer.

[287] D.L. Sulsky, S.-J. Zhou, and H.L. Schreyer. Application of a particle-in-cell method to solid mechanics. *Journal of Computational Physics Community*, 87:236–252, 1995.

[288] B.M. Sumer and J. Fredsøe. *The Mechanics of Scour in the Marine Environment*. World Scientific, 2002.

[289] C.C. Swan and Y.-K. Seo. Limit state analysis of earthen slopes using dual continuum/FEM approaches. *International Journal for Numerical and Analytical Methods in Geomechanics*, 23(12):1359–1371, 1999.

[290] R.G. Tart. Why the Oso Landslide Caused So Much Death and Destruction. In *Geotechnical and Structural Engineering Congress 2016*, volume 1, pages 1545–1554, Reston, VA, 2016. American Society of Civil Engineers.

[291] F. Tatsuoka. Discussion: The strength and dilatancy of sands. *Géotechnique*, 37(2):219–226, 1987.

[292] D.W. Taylor. *Fundamentals of Soil Mechanics*. Wiley, New York, 1948.

[293] C.I. Teh and G.T. Houlsby. Analytical study of the cone penetration test in clay. *Géotechnique*, 41(1):17–34, 1991.

[294] F.S. Tehrani and V. Galavi. Comparison of Cavity Expansion and Material Point Method for Simulation of Cone Penetration in Sand. In *4th International Symposium on Cone Penetration Testing (CPT'18)*, pages 611–615, Delft, The Netherlands, 2018.

[295] K. Terzaghi. *Theorie der Setzung von Tonschichten : eine Einfuhrung in die analytische Tonmechanik (in German)*. Franz Deuticke, Leipzig, 1936.

[296] K. Terzaghi. *Theoretical Soil Mechanics*. John Wiley & Sons, Inc., New York, first edition, 1943.

[297] H.S. Thilakasiri, M. Gunaratne, G. Mullins, P. Stinnette, and B. Jory. Investigation of impact stresses induced in laboratory dynamic compaction of soft soils. *International Journal for numerical and analytical methods in geomechanics*, 20(10):753–767, 1996.

[298] E.L. Thompson and H. Huppert. Granular column collapses: further experimental results. *Journal of Fluid Mechanics*, 575:177–186, 2007.

[299] M. Thury and P. Bossart. Mont Terri rock laboratory: a new international research project in a Mesozoic shale formation in Switzerland. *Engineering Geology*, 52:347–359, 1999.

[300] R. Tielen, M. Möller, and C. Vuik. Efficient multigrid based solvers for isogeometric analysis. In *6th European Conference on Computational Mechanics and 7th European Conference on Computational Fluid Dynamics (ECCM-ECFD) Conference 2018*, 2018.

[301] R. Tielen, E. Wobbes, M. Möller, and L. Beuth. A high order material point method. *Procedia Engineering*, 175:265–272, 2017.

[302] T.E. Tika and J.N. Hutchinson. Ring shear tests on soil from the Vaiont landslide slip surface. *Géotechnique*, 49(1):59–74, 1999.

[303] A. Tolooiyan and K. Gavin. Modelling the cone penetration test in sand using cavity expansion and arbitrary Lagrangian-Eulerian finite element methods. *Computers and Geotechnics*, 38(4):482–490, 2011.

[304] C. Truesdell and R.A. Toupin. *The Classical Field Theories of Mechanics*. Springer, New York, 1960.

[305] B. Ukritchon, A.J. Whittle, and S.W. Sloan. Undrained stability of braced excavations in clay. *Journal of Geotechnical and Geoenvironmental Engineering*, 129(8):738–755, 2003.

[306] S. Uriel and R. Molinia. Kinematic aspects of Vaiont slide. In National Academy of Sciences, editor, *3rd International Conference of the ISRMR*, volume 2B, pages 865–870, Denver, USA, 1977.

[307] A.P. van den Eijnden and M.A. Hicks. Efficient subset simulation for evaluating the modes of improbable slope failure. *Computers and Geotechnics*, 88:267–280, 2017.

[308] J.G.M. van der Grinten, M.E.H. van Dongen, and H. van der Kogel. Strain and pore pressure propagation in a water-saturated porous medium. *Journal of Applied Physics*, 62(12):4682–4687, 1987.

[309] J.M. van Esch, D. Stolle, and I. Jassim. Finite element method for coupled dynamic flow-deformation simulation. In *2nd International Symposium in Computational Geomechanics (COMGEO II)*, pages 415–524, Cavtat-Dubrovnik, Croatia, 2011.

[310] P.J. Vardanega and Stuart K. Haigh. The undrained strength-liquidity index relationship. *Canadian Geotechnical Journal*, 200(1):40, 2014.

[311] I. Vardoulakis. Shear band inclination and shear modulus of sand in biaxial tests. *International Journal of Numerical and Analytical Methods in Geomechanics*, 4(2):113–119, 1980.

[312] I. Vardoulakis. Catastrophic landslides due to frictional heating of the failure plane. *Mechanics of Cohesive-Frictional Materials*, 5(6):443–467, 2000.

[313] I. Vardoulakis. Dynamic thermo-poro-mechanical analysis of catastrophic landslides. Géotechnique. *Géotechnique*, 52(3):157–171, 2002.

[314] J. Vaunat, E. Romero, C. Marchi, and C Jommi. Modelling of the shear strength of unsaturated soils. In JFT Jucá, TMP de Campos, and FAM Marinho, editors, *3rd International Conference of Unsaturated Soils, UNSAT 2002*, pages 245–251. Swets & Zeitinger, 2002.

[315] R. Verdugo and K. Ishihara. The steady state of sandy soils. *Soils and Foundations*, 36(2):81–91, 1996.

[316] P.A. Vermeer, Y. Yuan, L. Beuth, and P. Bonnier. Application of interface elements with the Material Point Method. In *18th International Conference Computer Methods Mechanics*, pages 477–478, 2009.

[317] A. Veronese. Rilievi sperimentali sugli sbocchi liberi. *L'Energia Elettrica*, pages 441–638, 1948.

[318] A. Verruijt. *An Introduction to Soil Dynamics*. Springer, Dordrecht, 2010.

[319] E. Veveakis, I. Vardoulakis, and G. Di Toro. Thermoporomechanics of creeping landslides: The 1963 slide, northern Italy. *Journal of Geophysical Research*, 112(F3):2156–2202, 2007.

[320] B. Voight and C. Faust. Frictional heat and strength loss in some rapid slides. *Géotechnique*, 32(1):43–54, 1982.

[321] J. Von Neumann and R.D. Richtmyer. A method for the numerical calculation of hydrodynamic shocks. *Journal of Applied Physics*, 21:232–237, 1950.

[322] P.A. Von Wolffersdorff. A hypoplastic relation for granular materials with a predefined limit state surface. *Mechanics of Cohesive-Frictional Materials*, 1(3):251–71, 1996.

[323] P.C. Wallstedt and J.E. Guilkey. An evaluation of explicit time integration schemes for use with the generalized interpolation material point method. *Journal of Computational Physics*, 227(22):9628–9642, 2008.

[324] B. Wang, M.A. Hicks, and P.J. Vardon. Slope failure analysis using the random material point method. *Geotechnics Letters*, 6(2):113–118, 2016.

[325] L. Wang, W.M. Coombs, C.E. Augarde, M. Brown, J. Knappett, A. Brennan, D. Richards, and A. Blake. Modelling screwpile installation using the MPM. *Procedia Engineering*, 175:124–132, 2017.

[326] J. Wartman, D.R. Montgomery, S. Anderson, J.R. Keaton, J. Benoît, J. de la Chappelle, and R. Gilbert. The 22 March 2014 Oso landslide, Washington, USA. *Geomorphology*, 253(March):275–288, 2016.

[327] D. Wegener and I. Herle. Prediction of permanent soil deformations due to cyclic shearing with a hypoplastic constitutive model. *Geotechnik*, 37(2):113–122, 2014.

[328] Z. Wieckowski. Enhancement of the Material Point Method for Fluid-Structure Interaction and Erosion. Technical report, Deltares Ltd, Delft, 2013.

[329] E. Wobbes, M. Möller, V. Galavi, and C. Vuik. Taylor least squares reconstruction for material point methods. In *6th European Conference on Computational Mechanics and 7th European Conference on Computational Fluid Dynamics (ECCM-ECFD) Conference 2018*, 2018.

[330] A. Wolter, D. Stead, and J.J. Clague. A morphologic characterisation of the 1963 Vajont Slide, Italy, using long-range terrestrial photogrammetry. *Geomorphology*, 206:147–164, 2014.

[331] Y. Wu and Y.-M. Chiew. Three-dimensional scour at submarine pipelines. *Journal of Hydraulic Engineering*, 138(9):788–795, 2012.

[332] A. Yerro. *MPM modelling of landslides in brittle and unsaturated soils.* PhD thesis, Universitat Politecnica de Catalunya, 2015.

[333] A. Yerro, E. Alonso, and N. Pinyol. Modelling large deformation problems in unsaturated soils. In *E-UNSAT 2016*, pages 1–6, 2016.

[334] A. Yerro, E.E. Alonso, and N.M. Pinyol. The material point method for unsaturated soils. *Géotechnique*, 65(3):201–217, 2015.

[335] A. Yerro, E.E. Alonso, and N.M. Pinyol. Run-out of landslides in brittle soils. *Computers and Geotechnics*, 80:427–439, 2016.

[336] A. Yerro, N.M. Pinyol, and E.E. Alonso. Internal progressive failure in deep-seated landslides. *Rock Mechanics Rock Eng.*, 49(6):2317–2332, 2016.

[337] A. Yerro, K. Soga, and J.D. Bray. Runout evaluation of the Oso landslide with the Material Point Method. *Canadian Geotechnical Journal*, accepted, 2018.

[338] J.T. Yi, S.H. Goh, F.H. Lee, and M.F. Randolph. A numerical study of cone penetration in fine-grained soils allowing for consolidation effects. *Géotechnique*, 62(8):707, 2012.

[339] A.R.II York, D.L. Sulsky, and H.L. Schreyer. The material point method for simulation of thin membranes. *International Journal of Numerical Methods in Engineering*, 44(10):1429–1456, 1999.

[340] F. Zabala. *Modelación de problemas geoternicos hidromecanicos utilizando el metodo del punto material.* PhD thesis, Polytechnic University of Catalonia, 2010.

[341] F. Zabala and E.E. Alonso. Progressive failure of Aznalcóllar dam using the material point method. *Géotechnique*, 61(9):795–808, 2011.

[342] D.Z. Zhang, X. Ma, and P.T. Giguere. Material point method enhanced by modified gradient of shape function. *Journal of Computational Physics*, 230(16):6379–6398, 2011.

[343] X. Zhang, Z. Chen, and Y. Liu. *The Material Point Method: A Continuum-Based Particle Method*. Springer, 2017.

[344] X. Zhao and D. Liang. MPM modelling of seepage flow through embankments. In *Twenty-Sixth (2016) International Ocean and Polar Engineering Conference*, pages 1161–1165, Rodos, Greece, 2016.

[345] X. Zhao, D. Liang, and M. Martinelli. MPM simulations of dam-break floods. *Journal of Hydrodynamics, Ser. B*, 29(3):397–404, 2017.

[346] H. Zheng, L.G. Tham, and D. Liu. On two definitions of the factor of safety commonly used in the finite element slope stability analysis. *Computers and Geotechnics*, 33(3):188–195, 2006.

[347] D. Zhu, D.V. Griffiths, J. Huang, and G.A. Fenton. Probabilistic stability analyses of undrained slopes with linearly increasing mean strength. *Géotechnique*, 67(8):733–746, 2017.

[348] O.C. Zienkiewicz, A.H.C. Chan, M. Pastor, B.A. Schrefler, and T. Shiomi. *Computational Geomechanics with Special Reference to Earthquake Engineering*. Wiley, Chichester, UK, 1999.

[349] O.C. Zienkiewicz and T. Shiomi. Dynamic behaviour of saturated porous media; the generalized Biot formulation and its numerical solution. *International Journal for Numerical and Analytical Methods in Geomechanics*, 8(1):78–96, 1984.

[350] O.C. Zienkiewicz, Y.M. Xie, B.A. Schrefler, A. Ledesma, and N. Bicanic. Static and Dynamic Behaviour of Soils: A Rational Approach to Quantitative Solutions. II. Semi-Saturated Problems. *Royal Society A: Mathematical, Physical and Engineering Sciences*, 429(1877):311–321, 1990.

[351] B. Zuada Coelho, A. Rohe, A. Aboufirass, J. Nuttall, and M. Bolognin. Assessment of dike safety within the framework of large deformation analysis with the material point method. In A.S. Cardoso, J.L. Borges, P.A. Costa, A.T. Gomes, J.C. Marques, and C.S. Vieira, editors, *Numerical Methods in Geotechnical Engineering IX*, pages 657–663. CRC Press, 2018.

[352] B. Zuada Coelho, A. Rohe, and K. Soga. Poroelastic solid flow with double point material point method. *Journal of Hydrodynamics*, 29(3):423–430, 2017.

Index

Printed and bound by CPI Group (UK) Ltd, Croydon, CR0 4YY

26/10/2024

01779796-0001